AF848983

ENVIRONMENTAL SCIENCE, ENGINEERING AND TECHNOLOGY

CYCLONES

FORMATION, TRIGGERS AND CONTROL

Environmental Science, Engineering and Technology

Additional books in this series can be found on Nova's website under the Series tab.

Additional e-books in this series can be found on Nova's website under the e-book tab.

Natural Disaster Research, Prediction and Mitigation

Additional books in this series can be found on Nova's website under the Series tab.

Additional e-books in this series can be found on Nova's website under the e-book tab.

ENVIRONMENTAL SCIENCE, ENGINEERING AND TECHNOLOGY

CYCLONES

FORMATION, TRIGGERS AND CONTROL

KAZUYOSHI OOUCHI
AND
HIRONORI FUDEYASU
EDITORS

nova publishers
New York

Copyright © 2012 by Nova Science Publishers, Inc.

All rights reserved. No part of this book may be reproduced, stored in a retrieval system or transmitted in any form or by any means: electronic, electrostatic, magnetic, tape, mechanical photocopying, recording or otherwise without the written permission of the Publisher.

For permission to use material from this book please contact us:
Telephone 631-231-7269; Fax 631-231-8175
Web Site: http://www.novapublishers.com

NOTICE TO THE READER

The Publisher has taken reasonable care in the preparation of this book, but makes no expressed or implied warranty of any kind and assumes no responsibility for any errors or omissions. No liability is assumed for incidental or consequential damages in connection with or arising out of information contained in this book. The Publisher shall not be liable for any special, consequential, or exemplary damages resulting, in whole or in part, from the readers' use of, or reliance upon, this material. Any parts of this book based on government reports are so indicated and copyright is claimed for those parts to the extent applicable to compilations of such works.

Independent verification should be sought for any data, advice or recommendations contained in this book. In addition, no responsibility is assumed by the publisher for any injury and/or damage to persons or property arising from any methods, products, instructions, ideas or otherwise contained in this publication.

This publication is designed to provide accurate and authoritative information with regard to the subject matter covered herein. It is sold with the clear understanding that the Publisher is not engaged in rendering legal or any other professional services. If legal or any other expert assistance is required, the services of a competent person should be sought. FROM A DECLARATION OF PARTICIPANTS JOINTLY ADOPTED BY A COMMITTEE OF THE AMERICAN BAR ASSOCIATION AND A COMMITTEE OF PUBLISHERS.

Additional color graphics may be available in the e-book version of this book.

Library of Congress Cataloging-in-Publication Data

Cyclones : formation, triggers, and control / [edited by] Kazuyoshi Oouchi and Hironori Fudeyasu.
p. cm.
Includes bibliographical references and index.
ISBN 978-1-61942-976-5 (hardcover)
1. Cyclones. I. Oouchi, Kazuyoshi. II. Fudeyasu, Hironori.
QC941.C935 2011
551.55'13--dc23
2012001049

Published by Nova Science Publishers, Inc. † New York

CONTENTS

PREFACE

A tropical cyclone is a key extreme phenomenon of the climate system. Since it threatens safety of our lives and infrastructure on various time scales, precise forecasting is strongly required. High-performance supercomputers are indispensable technological tools for satisfying the demands, and give additional flavor to the research direction. Recent development of supercomputers has prompted a rapid advancement in the research of our climate and its future change, and tropical cyclone researches have evolved rapidly and deeply as well, being in concert with this trend.

How has been and will be the tropical cyclone research directed in the high-performance supercomputer era? We tend to stay alert on the specific new findings related to our own interests. However, we would be equally benefited from learning other findings and issues of tropical cyclones in the whole picture of research progress, given the enhanced interconnections of tropical cyclone research communities and international guidance of research projects. Tropical cyclone researches have evolved on a variety of topics including, genesis, development structure, from modeling, and observational standpoints. To be involved in such research progress would be immensely stimulating, yet reviewing all of them is not easy as there are wide-ranging research aspects and long-years accumulation of outcomes.

The aim of this book is to provide an available overview of recent tropical cyclone research related to formation, triggers and control of tropical cyclone. It has been difficult to define the scope of this publication in terms of the above motivation in a sensible way. Chapters are devoted to either an up-to-date review or report of new research on the tropical cyclone, encompassing the problems of the genesis, development, climate control of tropical cyclone based on every types of method -- theoretical, modeling, and observational analysis. Not only comprehensive climate model but stand-alone theoretical and conceptual modeling researches will continue to play important role in illustrating the essential physics of tropical cyclone, and observational studies will support their specific features. The topics that have increasingly attracted our vital attention include assessment and projection of the tropical cyclone characteristics under future climate, chapters are devoted to a theory and an application and use of probabilistic model to this issue. Editors hope that the book will help overview a current state of some aspects of tropical cyclone research, and would increase breadth and depth of research topics and interests on tropical cyclone, and serve as seeds for future research development.

Kazuyoshi Oouchi and Hironori Fudeyasu
November 2011

In: Cyclones: Formation, Triggers and Control
Editors: K. Oouchi and H. Fudeyasu
ISBN: 978-1-61942-976-5
©2012 Nova Science Publishers, Inc.

Chapter 1

TROPICAL CYCLOGENESIS IN WIND SHEAR: CLIMATOLOGICAL RELATIONSHIPS AND PHYSICAL PROCESSES

David S. Nolan* and Michael G. McGauley
Rosenstiel School of Marine and Atmospheric Science,
University of Miami, Miami, FL, US

ABSTRACT

The formation of tropical cyclones remains a topic of great interest in the field of tropical meteorology. A number of influential studies have considered the process of tropical cyclone formation (also known as TC genesis) from a pre-existing, weak tropical disturbance in a quiescent atmosphere from theoretical perspectives and using numerical simulations. However, it is shown that the large majority of TC genesis events occur under the influence of significant vertical wind shear. The effects of wind shear on TC genesis is explored from both a climatological perspective and from the statistics of wind shear in environments around individual TC genesis events. While earlier studies suggested that moderate wind shear values, in the range of 5 to 10 ms^{-1}, were the most favorable states for genesis, it is shown that small values of wind shear in the range of 1.25 to 5 ms^{-1} are the most favorable, and very little shear (less than 1.25 ms^{-1}) is not unfavorable. Statistically, easterly shear appears to be more favorable than westerly shear.

The physical process of TC genesis in wind shear is explored with high-resolution numerical simulations using a mesoscale model in an idealized framework. The transformation of a weak, mid-level vortex into a warm-cored tropical cyclone is simulated in environments with no flow, with mean flow and no wind shear, and with mean flow and wind shear. The simulations show that in terms of the formation of a closed, low-level circulation, moderate wind shear is indeed more conducive to genesis, but is also prohibitive to further development. However, in contrast to the statistical findings and some previous results, westerly shear is found to be significantly more favorable for TC genesis than easterly shear. The reasons for the greater favorableness of

---

*E-mail address: dnolan@rsmas.miami.edu

wind shear versus no wind shear, and of westerly shear versus easterly shear, are discussed within the context of the numerical simulations. Further statistical analysis suggests that the greater favorableness for easterly shear in the real atmosphere may be due to a correlation between easterly shear and more favorable thermodynamic conditions.

1. Introduction

The formation of a tropical cyclone (TC), a phenomenon that is often referred to as tropical cyclogenesis (hereafter, TC genesis), is a process by which some pre-existing, synoptic-scale or mesoscale weather feature in the tropics evolves so as to take on the characteristics of a tropical cyclone. There is no universally accepted set of criteria for what defines a tropical cyclone, but fairly similar sets of criteria have been adopted by operational forecasting centers [1, 2] and also for research purposes [3]. Common criteria are the requirement of a quasi-circular, closed circulation at or near the surface; the existence of a "warm core," which is another way of saying that the strongest circulating winds must be within or near the top of the atmospheric boundary layer; and the presence of deep moist convection near the center of the circulation.

Many disturbances travel across the tropical ocean basins during their respective tropical cyclone seasons, and yet relatively few become TCs. Since the pioneering work by Gray [4], it has been known that a number of environmental conditions are favorable for genesis: a sufficiently warm sea surface temperature (SST), a sufficiently moist environment in the lower and middle troposphere, and a relatively small value of vertical wind shear, which is the change with height of the horizontal wind (hereafter, "wind shear," unless otherwise specified). Many studies have emphasized the need for sufficient instability of the atmospheric environment to moist convection, measured either in terms of convective available potential energy (CAPE) [5, 6] or in terms of the decrease of the equivalent potential temperature θ_e between the surface and mid-troposphere [7]. In more recent studies, measures of instability have been replaced with measures of the thermodynamic disequilibrium between the ocean surface and the atmosphere, such as the maximum potential intensity (MPI) of Emanuel [8, 9], or the "relative" SST, which is the difference between local SST and the global tropical mean SST [10, 11].

The thermodynamic conditions (by which we mean the combined state of the SST and the overlying vertical profiles of temperature and humidity) tend to vary slowly over time and space, and saying that these parameters are favorable for TC genesis is usually synonymous with saying that it is tropical cyclone season. Wind shear, on the other hand, is much more volatile: it is the parameter that is most likely to vary substantially from day to day. On a practical level, when considering the possibility of TC genesis for a particular disturbance, wind shear is the first, and often the last, environmental parameter that is considered. We say "last" because if the wind shear is sufficiently large, the possibility of genesis is usually dismissed outright.

As we will show below, there is little debate about the relationship between TC genesis and wind shear in a climatological sense. All other things being equal, a region with lower values of mean monthly or seasonal wind shear will be much more favorable for TC genesis. However, when considering the likely evolution of an individual disturbance, there remains uncertainty as to whether less shear is always more favorable for genesis. Both conceptual and quantitative theories of TC genesis suggest that, indeed, low wind shear and even zero wind shear is the most favorable condition for TC genesis. However, a significant body of work has appeared in the literature that argues that some wind shear - but not too much - is actually more favorable for TC genesis than no wind shear or even very little wind shear.

Despite the importance of wind shear, much of the fundamental research on TC genesis and early intensification considers these processes in a primarily axisymmetric framework, either by using an axisymmetric theoretical model [12, 13], an axisymmetric numerical model [14-16], or by using three-dimensional numerical simulations in an environment with no motion and no wind shear [17, 18]. These studies and some of the resulting theories of TC genesis focus on the dynamical and thermodynamical processes that cause a weak, broad, synoptic or mesoscale vortex with a peak circulation above the boundary layer to transform into a smaller-scale vortex with peak winds within the boundary layer. An excellent discussion of the findings of these and other studies is presented in the review by Tory and Frank [19].

While the axisymmetric mechanics of TC genesis will undoubtedly remain an important component of the process, it is worthwhile to further investigate the role of wind shear and the structural changes that it causes to pre-genesis disturbances. In this chapter we will attempt to clarify current understanding about the relationship between large-scale wind shear and the favorableness for TC genesis, both on climatological scales and for individual disturbances. Numerical simulations will be presented which illustrate some of the processes by which, depending on the circumstances, wind shear can either accelerate or suppress genesis.

2. Prior Research and Current Understanding

Early observational studies

Riehl [20] noted that TC genesis from low-level disturbances in the Pacific seemed to be timed with overhead passage of upper-level anticyclones. Genesis was seen to occur as the low-level trough was overtaken from the east by the upper level anticyclone. While the presence of the upper-level anticyclone itself was (and still is) considered to be favorable to upper-level outflow and thus deep convection [21], the passage of the center of the anticyclone overhead would likely have been associated with a minimum in wind shear over the disturbance. Palmen and Newton [22,23] argued that low vertical wind shear was necessary for the development of a vertically upright warm core, and this argument was supported by contemporary climatological analyses [4] that found a very close correlation between areas of frequent TC genesis and near-zero monthly values of zonal wind shear.

Ramage [24] and later Sadler [25,26] argued that TC genesis events could be triggered by the extension of a climatological upper-level trough in the middle of the Pacific toward the southwest, bringing it partially over the low-level easterlies. The southwesterly flow on the southeast side of the trough was believed to create an “outflow channel” favorable to persistent deep convection. Perhaps more importantly, the impingement of the upper-level flow associated with the base of the trough into the climatological mean easterly and north-easterly upper level flow caused a region of local anticyclonic flow and reduced wind shear.

McBride and Zehr [27,28] analyzed wind fields derived from rawinsonde observations around 80 tropical disturbances in the Pacific and Atlantic oceans. After compositing the wind fields according to developing and non-developing disturbances, they found the composite-mean flow over the developing systems had an axis of near-zero wind shear over the low-level disturbances. They also found westerly wind shear to the north and easterly wind shear to the south, consistent with the presence of an anticyclone overhead. Using an expanded data set and similar analysis methods, Lee [29] also found decreased wind shear in developing system, but more specifically, the developing systems were associated with a deep profile of very light easterly shear (less than 5 ms^{-1} between 850 and 200 hPa), while the non-developing systems had strong westerly shear overlying low-level easterly shear[1].

Numerical modeling

Many papers have been published based on numerical simulations of TCs embedded in wind shear. Very few of these studies focused on TC genesis, but rather, have investigated the effects of wind shear on intensifying and mature TCs. A common feature of such simulations is the use of a fairly strong initial vortex for the purpose of "jump-starting" the cyclone to strong tropical storm or hurricane intensity. As examples, Wang and Holland [30] used vortices that were already mature TCs; Frank and Ritchie [31] used a vortex with maximum winds at the surface of 15 ms^{-1}; Kimball and Evans [32] used 20 ms^{-1}; Wong and Chan [33] used 30 ms^{-1}; and Riemer et al. [34] used 18 ms^{-1}. Furthermore, in some cases, the wind shear was introduced after 48 hours of simulation, rather than at the initial time [31,34].

In recent years, there have been quite a few simulations of TC genesis based on real-case storms, but many of these involved systems in light wind shear, as genesis in a favorable environment is more reliably reproduced in a mesoscale model [35-37]. A few examples of case studies for genesis in stronger wind shear will be discussed later in this chapter.

Tuleya and Kurihara [38], hereafter referred to as TK81, is a well-known idealized modeling study of TC genesis in wind shear. They simulated the evolution of a weak, warm-core vortex with maximum cyclonic winds of 5 ms^{-1} at the surface, nearly constant up to about 700 hPa, and then decaying to zero above 400 hPa. The radius of maximum winds (RMW) of this vortex was approximately 300 km. The vortex was embedded in an easterly flow of 5 ms^{-1}, centered at 18° N in a domain with a realistically varying Coriolis parameter. In addition, the large-scale flow in their simulations contained both meridional and vertical wind shear. The meridional shear, when present, caused the zonal flow to vary from -2.5 to -7.5 ms^{-1} from south to north, such that the initial vortex was embedded in cyclonic flow; for some simulations, the meridional shear reversed direction with height, so that the vortex was in cyclonic flow at low levels and anticyclonic flow aloft (similar to what had been observed in [28]). In the vertical direction, the flow was made to increase or decrease smoothly between about 850 and 200 hPa, thus creating westerly or easterly shear, respectively.

TK81 has been frequently cited due to their result that mean easterly wind shear is more favorable for TC genesis than westerly shear (for a westward moving storm in the Northern

[1] The vertical wind profiles were shown in terms of the "storm-relative" winds, rather than earth-relative, making this result difficult to compare against some results shown later in this chapter.

hemisphere). In fact, they found that the easterly shear value most favorable for the early development of the TC was 15 ms^{-1} (about 30 knots). However, their model, while highly advanced for its time, used approximately 70 km horizontal resolution, 11 vertical levels, and a fairly simple convective parameterization. The rather large value found for the most favorable shear, the resolution of the model, and the primitive structure of the initial condition, all suggest that their results deserve further investigation. Remarkably, this result has never been systematically verified with a broader set of simulations. In section 4, we will present some attempts to reproduce their results using a modern mesoscale model.

Roles for wind shear in TC genesis

The TK81 results notwithstanding, the general view until the 1990s was that lower values of environmental wind shear were always the more favorable condition for TC genesis. However, this view changed with the publication of a series of papers documenting cases of TC genesis within regions of moderate wind shear, beginning with Bosart and Bartlo [39]. In this study, the authors explored the features of the synoptic environment around the formation of Hurricane Diana (1984), which grew out of a residual frontal boundary that had stalled off the east coast of Florida. The approach of an upper-level potential vorticity feature, advected by westerly upper-level winds associated with the residual baroclinicity of the synoptic-scale environment, generated large-scale ascent, leading to large-scale cooling and moistening. Subsequently, deep convection triggered along the frontal boundary (and enhanced by the mid-level cooling and moistening) led to the transformation of the titled, partly cold-cored system into a more vertically upright, warm-cored cyclone.

Briegel and Frank [40] studied the synoptic-scale environments around TC genesis events in 1988 and 1989 in the Western North Pacific. In about a third of the genesis cases, they found that an upper-level trough was present to the west or northwest of the disturbance before genesis, a configuration similar to that identified by Bosart and Bartlo. Bracken and Bosart [41] studied TC genesis over a 10 year period in the Atlantic, from which they developed composite wind fields around genesis events in two regions: the northwest tropical Atlantic (labelled as "Bahamas cases") and the eastern tropical Atlantic ("Cape Verde cases"). For the Bahamas cases composite, they found a distinct trough-ridge pattern in the upper atmosphere, with the PV of the trough lying to the west of the surface disturbance, advected toward it by southwesterly wind shear. For the Cape Verde composite, they found upper level anticyclonic relative vorticity (resulting in a negative vertical gradient of potential vorticity) being advected westward by easterly shear away from the corresponding surface disturbance. In both cases, quasi-geostrophic theory indicates that there would be large-scale ascent forced by PV advection over the disturbance, favoring or even causing TC genesis.

Continuing work by Davis and Bosart [42-44] established the approach of an upper-level potential vorticity anomaly over some pre-existing tropical or sub-tropical disturbance to be a robust mechanism for TC genesis, one that occurs much more frequently in the subtropics (20 to 30° from the equator) or occasionally on the edges of the "deep tropics" (latitudes less than 20° from the equator). A comprehensive and analytical survey of such TC genesis events, and those forming through more "tropical" mechanisms, was presented by McTaggart-Cowan et al. [45]. Using NCEP/NCAR Reanalysis fields [46], they analyzed the extent to which upper-level features and quasi-geostrophic forcing played a role in TC genesis events in the Atlantic from 1948 to 2004. Their analysis method distinguished among genesis events through both the magnitude and the time-evolution of the upper-level forcing. While a majority of the genesis events were found to be primarily tropical in nature, with little upper-level forcing, about 40% were found to have occurred through some degree of upper-level forcing.

In the deep tropics there is less influence from upper-level troughs and ridges, and the wind shear is often in a much more zonal configuration. Yet, there are reasons to believe that shearing of a weak, low or mid-level circulation associated with a pre-existing disturbance could also be favorable for genesis. Research on the evolution of mesoscale convective vortices (MCVs) generated by summer season convection over continents has identified a mechanism by which such disturbances can become self-sustaining [47,48]. The MCV is a mid-level circulation, usually with no cyclonic surface circulation (or even anticyclonic flow at low levels) that appears in the wake of a mesoscale convective system (MCS), which is a cluster of multi-celled deep convection that is often embedded in a widespread area of stratiform precipitation. The mid-level circulation is generated by the vertical dipole of heating above and cooling below the melting level, respectively, in the stratiform regions [49].

When MCVs are subjected to light or moderate wind shear, almost invariably present in the mid-latitudes, quasi-geostrophic reasoning indicates there should be ascent on the downshear side of the vortex and descent on the upshear side. Due to adiabatic cooling, the downshear side is destabilized, and furthermore, parcels of air advected around the MCV are then forced to rise up the upward sloping potential temperature surfaces, leading to the development of new convection. This convection can, in turn, sustain or even enhance the MCV, leading to a longer-lived system and further convective outbreaks. Rogers and Fritsch [50] discussed the circumstances under which an MCV, with anticyclonic flow underneath, could develop a cyclonic circulation at the surface. This was found to be possible when repeated events of deep moist convection occurred, each successively closer to the center of the MCV itself. Increasing humidity and near-saturation of the upper levels aided in expanding and intensifying the convective bursts. However, for the convective events to repeat, the system must be embedded in some wind shear: either at the mid-levels, to drive tilting of the vortex and downshear ascent [47]; or at low-levels, to bring fresh, low-level, unstable air into the system that will rise up the sloped isentropes associated with the low-level cold core [48].

A number of more recent studies on TC genesis and wind shear will be discussed later in this chapter. In the next section, we will explore the extent to which wind shear in the deep tropics, in configurations that would not be associated with significant quasi-geostrophic forcing, is favorable or unfavorable for TC genesis. The relative favorableness of easterly versus westerly shear will also be considered.

3. The Climatological Relationship between TC Genesis and Wind Shear

The relationship inferred from genesis parameters

We desire to assess the impact of wind shear on TC genesis in a way that isolates the effects of shear from other environmental parameters such as SST, MPI, or humidity. This is not trivial, as these parameters co-vary: for example, all four parameters vary in a favorable direction as one moves from the mid-latitudes to the tropics. Fortunately, there have already been a number of attempts to systematically identify mathematical expressions that describe the independent contributions of environmental parameters to TC genesis, and then to put these expression together into a single formula: such a formula is often called a "genesis parameter." Components of these expressions, such as constants or exponents, may be found through matching the formula to observed genesis frequencies through trial and error, least squares, or some other error minimization technique.

Several different genesis parameters have been developed. All indicate a steadily increasing likelihood (or frequency) for genesis with decreasing wind shear. It is interesting to compare the functions used in these genesis parameters to express the increasing favorableness. The shear parameter used by Gray [7] is:

$$S = \frac{1}{S_z + 3} \tag{1}$$

where here we use S for the "suppression parameter," and Gray defined S_z as the wind shear in ms^{-1} per 750 hPa of vertical depth. Normalizing this shear to the nearly universal practice today of expressing wind shear as the absolute vector difference of the mean winds between the 850 and 200 hPa pressure levels (hereafter V_{shear}), and subsequently normalizing the parameter to have a maximum value of one, (3.1) becomes

$$S_G = \frac{1}{1.0 + 0.385 \times V_{shear}} \tag{2}$$

Emanuel and Nolan [51] used:

$$S_{EN} = \frac{1}{(1.0 + 0.1 \times V_{shear})^2} \tag{3}$$

which is already normalized to a maximum value of one. Emanuel [52] later revised the genesis parameter, with the shear suppression changed to:

$$S = \frac{1}{(25.0 + V_{shear})^4} \tag{3.4}$$

which normalized to a maximum value of one becomes

$$S_{E10} = \frac{1}{(1.0 + 0.04 \times V_{shear})^4} \quad (3.5)$$

Tippett et al. [53] used a Poisson regression technique to derive coefficients for a log-linear predictor of TC genesis. Their factor by which shear suppresses genesis has the form:

$$S_T = \exp(b_V V_{shear}) \quad (3.6)$$

where, for their optimal set of predictors, they find $b_V = -0.13$.

Figure 1.1 shows these four suppression parameters as a function of the mean wind shear. After normalization, all four suppression parameters are surprisingly similar, and all four curves peak sharply to their maximum value as the mean wind shear approaches zero. Thus, the exercise of developing these genesis parameters, and the extent to which they are successful, indicates that zero wind shear is by far the most favorable state for TC genesis.

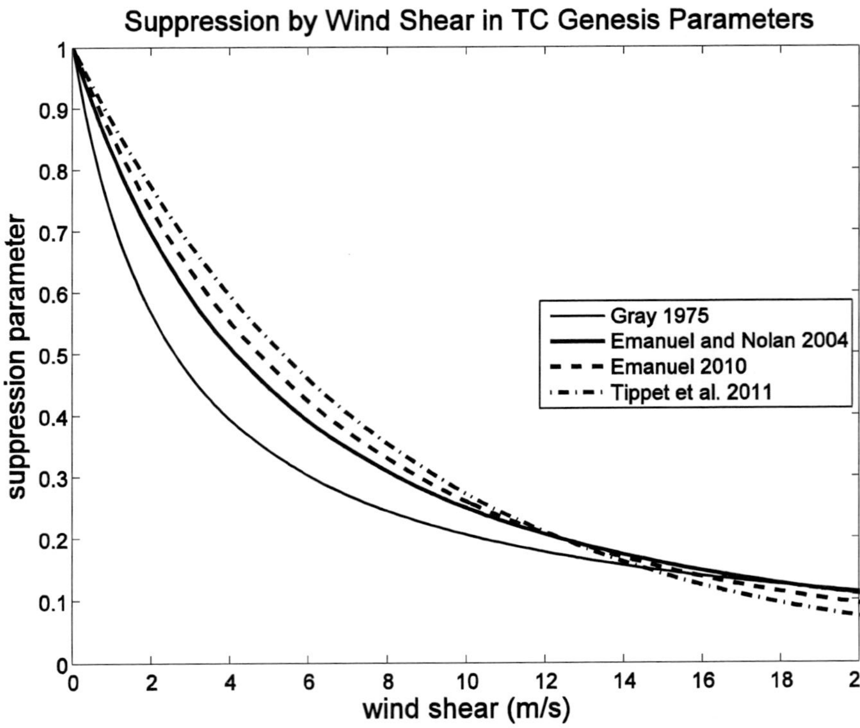

Figure 1.1. Normalized factors by which some of the genesis parameters in the literature are reduced by mean wind shear.

However, this is not a correct conclusion, for two reasons. First, the chosen forms of the "suppression" parameters reflect an *assumption* that favorableness will increase unconditionally for decreasing shear. Second, it is important to remember the context in which these curves were derived and applied. They are used to predict monthly or seasonal TC activity over large regions. As will be shown in the next section, near-zero wind shear is extremely rare, even in the tropics, and even for monthly means over small regions the smallest monthly mean wind shear values are greater than 6 ms^{-1}. This is illustrated in Fig. 1.2, which shows monthly mean wind shears in each of the 6 basins during the month that the minimum shear occurs. The "relevant ranges" of the suppression curves lie in their values for shear between 6 ms^{-1}, where their individual values range from 0.3 to 0.45, and 12.5 ms^{-1}, where they converge to a value of 0.2. In other words, to the extent which these genesis parameters are accurate, they indicate that a region with mean wind shear of 6 ms^{-1} is 1.5 to 2.25 times more favorable for TC genesis than one with mean wind shear of 12.5 ms^{-1}.

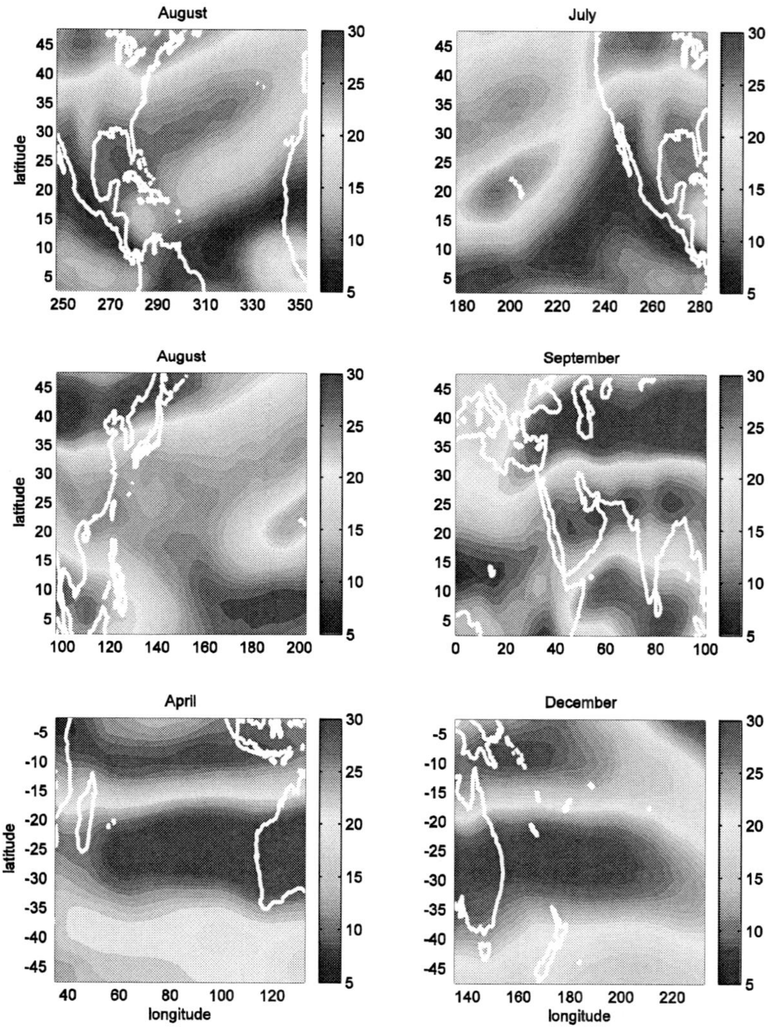

Figure 1.2. Maps of monthly mean wind shear computed from NCAR/NCEP Reanalyses for each basin in the month that its minimum point value of monthly mean shear occurs: a) August in the North Atlantic, minimum value 6.9 ms^{-1}; b) July in the East Pacific, minimum value 7.5 ms^{-1}; c) August in the Northwest Pacific, minimum value 8.5 ms^{-1}; d) September in the Northern Indian Ocean, minimum value 7.1 ms^{-1}; e) April in the Southern Indian Ocean, minimum value 7.6 ms^{-1}; f) December in the South Pacific, minimum value 8.7 ms^{-1}.

But why would a region with 12.5 ms^{-1} (i.e., 25 knots) of mean wind shear be at all favorable? Because the wind shear will vary considerably around the mean. In fact, distributions of absolute wind shear are one-sided and often (but not always) skewed to larger values (see, for example, Figure 26 of [54]). The most frequent shear values are lower than the mean. Thus, what makes a region with mean wind shear of 10 ms^{-1} quite favorable for TC genesis is the fact that it will frequently have wind shear of 5 ms^{-1}.

Knowing then the real meaning of the shear suppression factors derived from empirical studies of climatological TC genesis, we must conclude that these functions are not suitable for evaluating the favorableness of a particular shear value for a potential, individual case of TC genesis[2]. To more directly address this issue, we will look at the statistics of wind shear around actual TC genesis events.

The climatological record

Figure 1.3 shows a histogram of the mean wind shear surrounding TC genesis events as identified by times and locations of TC genesis using the International Best Track Archive for Climate Stewardship (IBTrACS; [55]) from 1969 to 2008. In this case, only genesis events within 20° of the equator have been counted, so as to eliminate almost all of the subtropical or strongly baroclinic cases. The wind shear values are computed from the NCAR/NCEP Reanalyses [46], computing the mean wind at 850 and 200 hPa from the grid point nearest to the TC genesis event and its eight neighbors, representing an area of 7.5° x 7.5°. The results show that the most frequent values of wind shear for genesis events are between 6.25 and 7.5 ms^{-1}; values of 5-6.25 ms^{-1} and 7.5-8.75 ms^{-1} are next two most common. Also, the distribution decreases slowly for higher values of shear, and there are numerous genesis events with shear between 10 and 20 ms^{-1}. A similar plot for shear over genesis events in the Atlantic was shown by Bracken and Bosart [41], with the most frequent shear values for genesis between 8 and 9 ms^{-1} (and with no events with shear less than 2 ms^{-1}).

Both the Bracken and Bosart figure and Fig. 1.3a certainly give the impression that shear values between 5 and 10 ms^{-1} are the most favorable range for genesis. They also give the impression that very low values, 2.5 ms^{-1} or less, are not favorable. That impression is deceiving, because in fact very low values of wind shear are extremely rare.

In comparison, Fig. 1.3b shows a distribution of wind shear in the tropics. For each TC genesis event, 40 randomly chosen locations of other TC genesis events in the data base (from the same hemisphere, that occurred on *different* dates) were sampled for their wind shear values (on the date of the *reference* event), using the same averaging area as above. This is not a perfect depiction of the distribution of wind shear values in the tropics, but it weighted to the distribution of locations of actual TC genesis events. Wind shear values of less than 2.5 ms^{-1} consist of 4.3% of the events, and less than 1.25 ms^{-1} are only 1.1% of the events.

[2] This finding was anticipated in the last paragraph of McBride (1981).

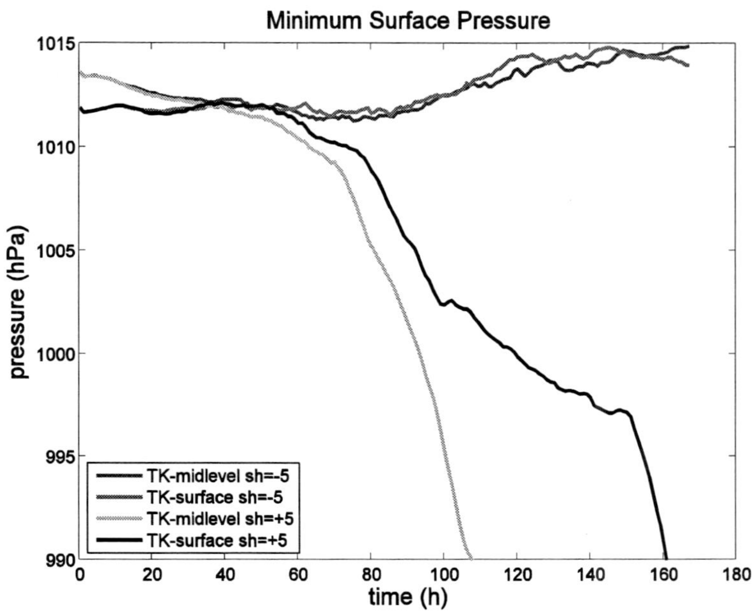

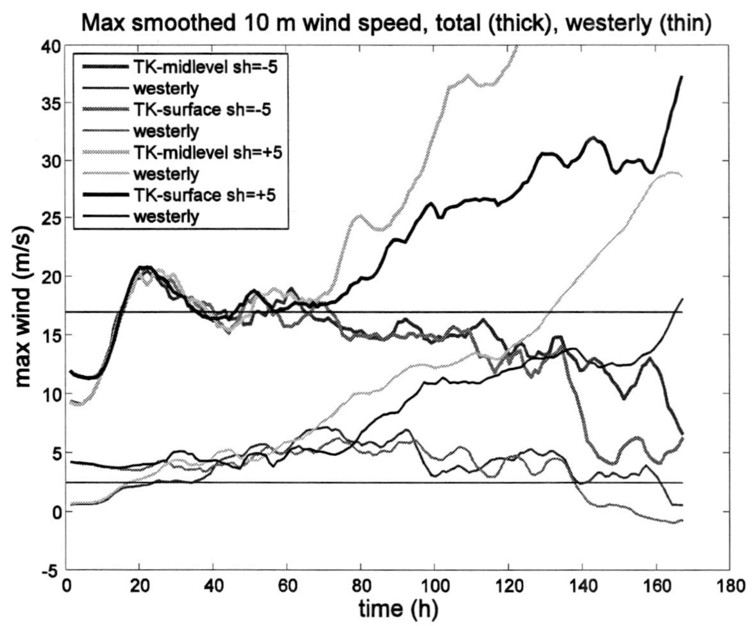

Figure 1.14. As in Fig. 1.8, but for results from simulations modeled after those of Tuleya and Kurihara (1981), with easterly (negative) and westerly (positive) wind shear, and with the initial vortex maximized at mid-levels and at the surface.

TK81 argued that the reasons for more rapid TC genesis and development were related to the extent to which the warm core of the system remained vertically aligned. They tracked the westward propagation of the low-level circulation and the upper-level warm core. They found that the low-level disturbances propagated westward rather quickly, between 4 and 6 ms^{-1} faster than the low-level flow for easterly shear, but also about 2 ms^{-1} faster *for cases with*

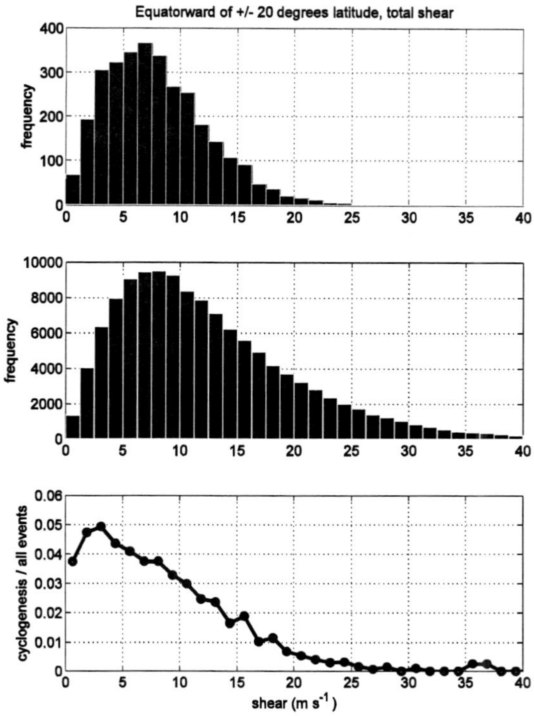

Figure 1.3. Histograms of frequencies of wind shear, in bins of 1.25 ms^{-1} (about 2.5 knots), for all TC genesis events in the IBTrACS data set from 1969 to 2008, that were equatorward of 20° (a); from randomly sampled locations in the tropics - see text for details (b); the result of dividing the two histograms (c).

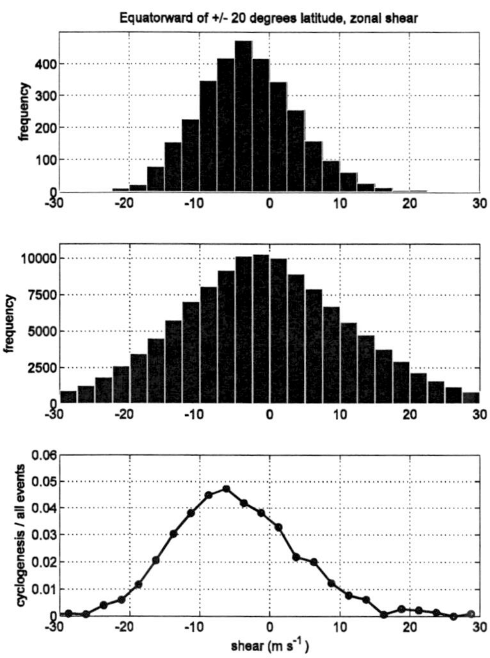

Figure 1.4. As in Fig. 1.3, but for zonal (east-west) wind shear, with bin sizes of 2.5 ms^{-1}.

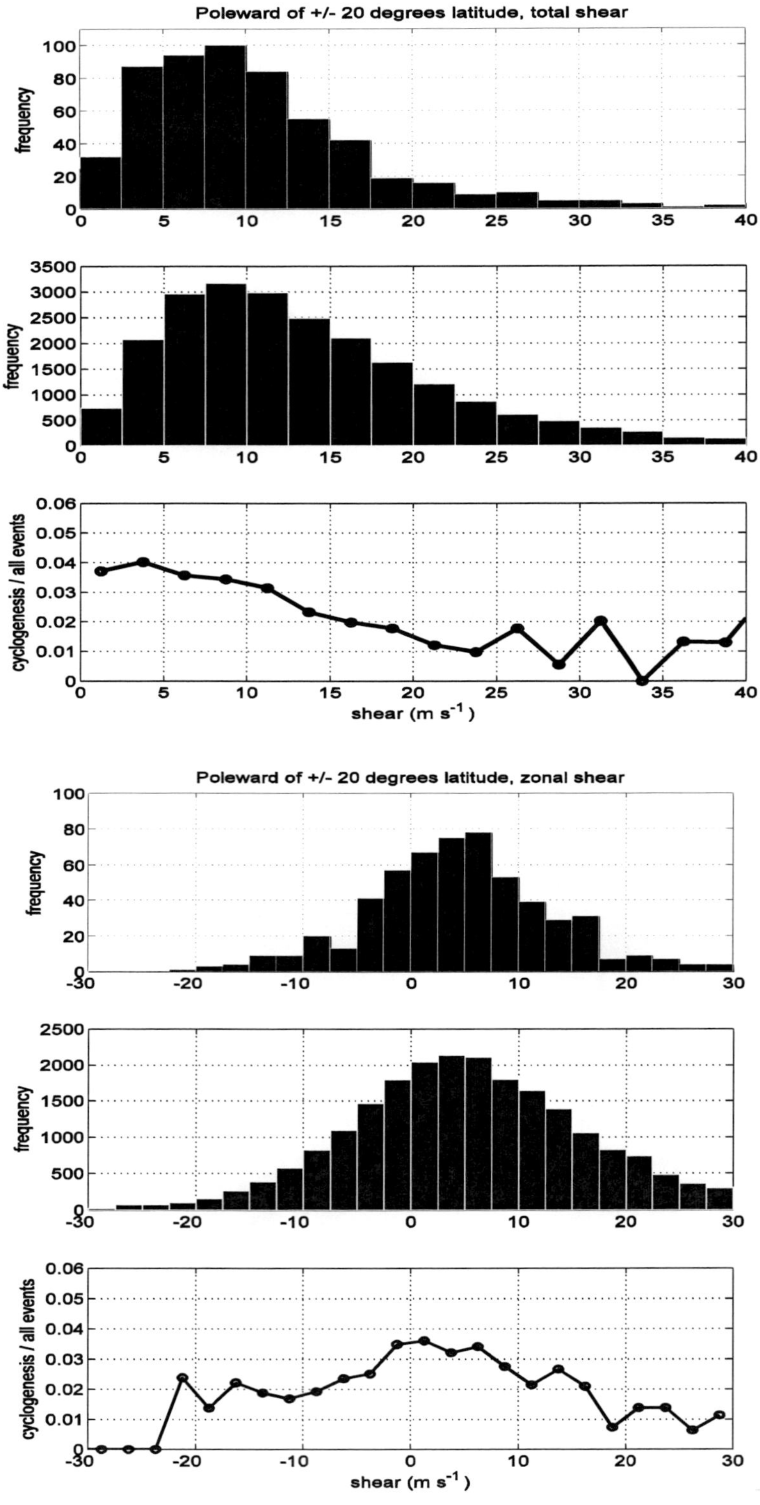

Figure 1.5. As in Fig. 1.3 and Fig. 1.4, but for TC genesis poleward of 20°.

If one divides these frequencies as an indicator of the *relative* favorableness of shear, the curve shown in Fig. 1.3c is produced. This curve indicates that 2.5-3.75 ms^{-1} of shear is the most favorable, with 1.25-2.5 ms^{-1} only slight less so. 0-1.25 ms^{-1} is also favorable for genesis, but not more so than shear values of 5.0 to 8.75 ms^{-1}. For shear values greater than 10 ms^{-1}, the favorableness decreases steadily. The local peaks in the values for wind shears greater than 15 ms^{-1} are due to the decreasing sample sizes. Figure 1.3 is quite similar to Fig. 11d of [53], which also discussed the difficulties in interpreting such figures: without more sophisticated treatment, it is difficult to distinguish the favorableness associated with decreasing shear from other factors, such as increasing SST and relative humidity. This caveat aside, Fig. 1.3 indicates that light wind shear in the range of 1.25 to 3.75 ms^{-1} (about 2.5 to 7.5 knots) is the most favorable condition for TC genesis, and that nearly zero shear and more moderate wind shear (3.75 to 6.25 ms^{-1}) are also quite favorable.

As noted above, it is also unknown whether easterly or westerly shear is more favorable for genesis. Applying the same analysis as above, but for zonal (east-west) shear rather than total shear, produces the plots shown in Fig. 1.4. After dividing by the histogram of the random shear distribution, it does appear that easterly shear is significantly more favorable for genesis than westerly shear, with the peak favorableness at 5-7.5 ms^{-1} of easterly shear (note the bins are larger in these figures). Cheung [6] presented a figure similar to Fig. 1.4a, using data and TC genesis events from the West Pacific, but did not assess the relative occurrence of easterly shear. This result appears to give strong support to the aforementioned results of TK81. But again, we must consider the possibility that this result is due at least in part to co-variation with other factors. Tory and Frank [19] speculated that easterly shear is simply more frequent in environments that are more favorable for genesis. For example, easterly shear is more common in the Pacific than the Atlantic, and at lower latitudes, where SSTs and relative humidities are greater. For completeness, we show in Fig. 1.5 similar histogram analyses for total shear and zonal shear for TCs that formed poleward of 20° N/S. In these regions, again we see that near-zero shear is quite favorable, but the maximum favorableness occurs for 2.5-5 ms^{-1}. Westerly zonal shear is the most common condition, both for the randomly sampled shear and for TC genesis. Westerly zonal shear does not appear to be any more favorable than zero zonal shear, but it is more favorable than easterly shear.

4. Sensitivity to Wind Shear as seen in Numerical Simulations

Recent simulations of TC genesis in wind shear

In recent years some idealized simulations have been designed with more realistic initial conditions that would be more representative of a pre-cursor disturbance that precedes TC genesis. Examples of such disturbances are easterly waves, monsoon depressions, mesoscale convective systems, or mesoscale circulations that have developed along a decaying frontal boundary. All of these features are associated with anomalous vorticity and a weak circulation that is maximized above the boundary layer, with the peak circulation somewhere between 900 and 600 hPa [28,56-58].

With these features in mind, Montgomery et al. [17], Nolan [18], and Nolan et al. [59] designed initial conditions with weak vortices (peak tangential wind speeds of 5 to 10 ms^{-1}) that were maximized near 700 hPa, with the wind decreasing above and below this maximum, but with non-zero circulation reaching to the ocean surface. All three of these studies simulated TC genesis in environments with no flow and no wind shear (the so-called "hurricane in a box"). As a result, the initial disturbances would always eventually achieve TC genesis, shortly followed by rapid intensification, even if the environment had been modified to be less favorable (such as by making the middle atmosphere substantially drier).

Nolan and Rappin [60] and Rappin et al. [61] extended the idealized simulations of [59] to include mean wind shear, and found much more realistic behavior. With moderate wind shear values, TC genesis from the pre-cursor, mid-level vortex was delayed, and as the wind shear was increased, genesis was eventually suppressed entirely. For a particular value of wind shear that did suppress TC genesis, genesis could still be achieved if the strength of the initial disturbance was increased, or if the thermodynamic environment was modified so as to make the environment more favorable. Some of the distinct differences between genesis in simulations with and without wind shear will be shown in the next section.

Exploring TC genesis in moderate shear with some new simulations

To further illustrate some of the processes described in the previous sections, we present simulations of TC genesis, modeled after those described above, but extended to include mean flow and wind shear, balanced by pressure and temperature gradients as required by the thermal wind equation. These simulations are performed using the Weather Research and Forecast model (WRF) version 3.1.1 [62]. The model domain is a rectangular area using 320x200 grid points with 18 km grid spacing. Within this domain, two nested grids with 6km and 2km resolution follow the vortex circulation; their sizes are 120x120 and 216x216 points, respectively. 40 vertical levels are used, equally spaced in pressure coordinates. The boundary conditions are periodic in the zonal (x) direction and there are free-slip walls on the north and south boundaries. The model physics and the numerical method for computing the balanced initial conditions are otherwise identical to what is described in [63]. The initial temperature and moisture profiles at the center of the domain, undisturbed by the initial vortex, are equal to the "moist tropical" sounding of Dunion [64]. The SST is set to 29° C throughout the domain.

For the initial condition, we use a weak, mid-level vortex, with a maximum tangential wind speed of 9 ms^{-1} at a radius of maximum winds (RMW) of 126 km; this vortex is weaker and broader than what has been have used previously [59,61,63], so as to make the initial wind fields more consistent with weak tropical cyclones [65]. The radial profile of the tangential wind is a modified Rankine vortex with decay parameter a = 0.4, along with a sharper exponential cutoff that smoothly forces the circulation to zero beyond 600 km from the initial center (cf. [18], Eq. 3). The peak tangential wind occurs at z = 3.72 km, with Gaussian decay above and below, with the decay parameter chosen so that the peak tangential surface wind is exactly half the peak mid-level wind (cf. [18], Eq. 4).

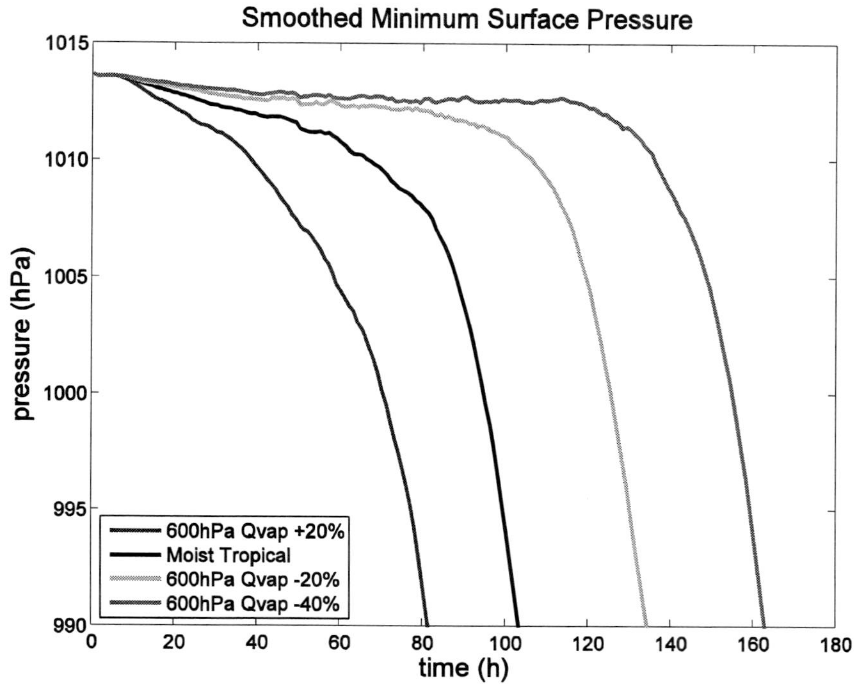

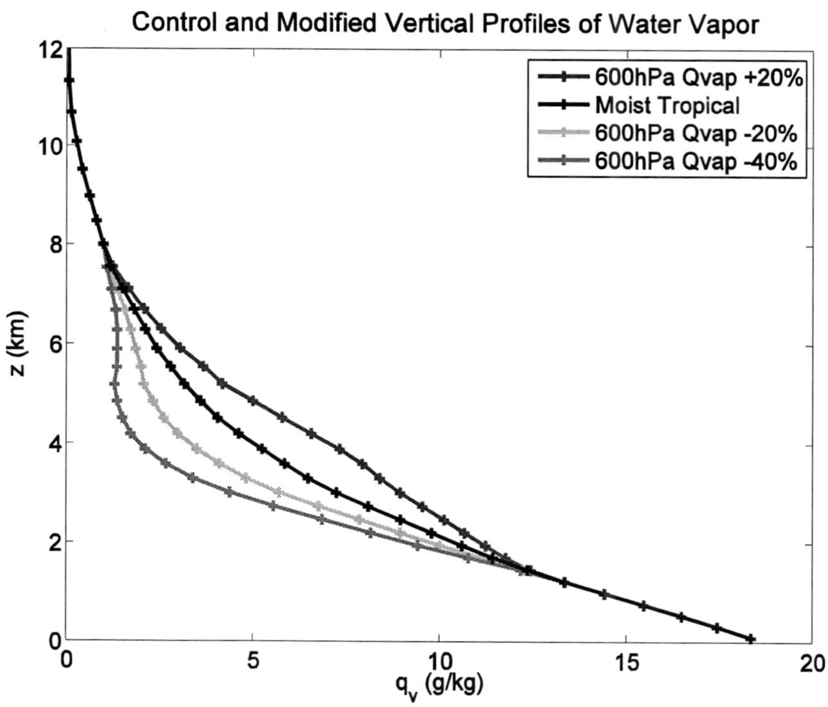

Figure 1.6. Results from simulations of TC genesis with no mean flow or wind shear: a) smoothed minimum surface pressure with time for simulations with varying humidity profiles; b) specific humidity profiles of the simulations, varying around the Dunion (2011) "moist tropical" sounding.

Before showing the effects of wind shear on genesis, we briefly present the results of "hurricane in a box" simulations with no mean flow, on an *f*-plane corresponding to 15° N. The minimum surface pressure, smoothed first in space and then in time[3], is shown for these simulations in Fig. 1.6. Also shown is the vertical profile of water vapor for the moist-tropical sounding, and 3 similar soundings where the mid-level humidity has been modified. For the modified soundings, the specific humidity at 600 hPa is changed by -40%, -20%, or +20%, with a Gaussian decay of the change with height above and below 600 hPa that goes essentially to zero above 400 hPa and below 800 hPa. Figure 1.6a illustrates the point already stated above: without mean flow or shear, even a very weak vortex in a rather unfavorable environment will eventually achieve TC genesis, followed by rapid intensification.

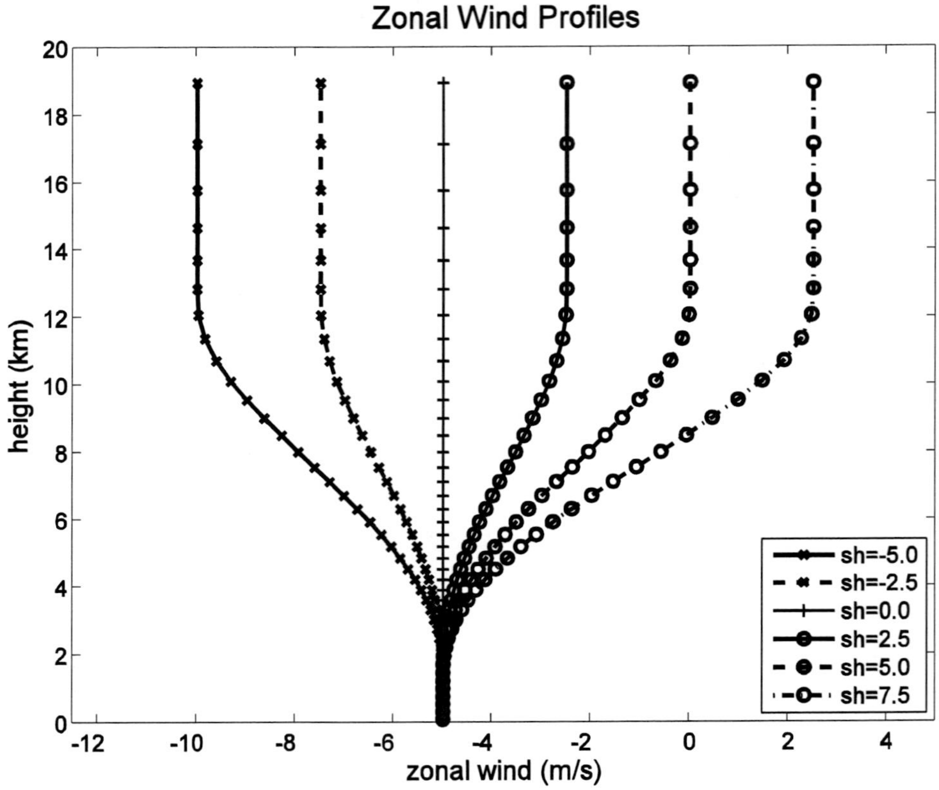

Figure 1.7. The initial zonal wind profiles used for many of the simulations, with 5 ms^{-1} of low-level easterly flow and varying values of wind shear in ms^{-1}. The markers show the heights of the model levels.

[3] Specifically, these plots are generated from hourly model output from the intermediate (6km) grid, each cell of which represents the mean of 9 grid cells from the 2km grid. The wind and pressure fields are smoothed 10 times with a 1-2-1 filter in the east and west directions. The pressures and wind speeds as a function of time are generated from a running average of the 6 most recent values. The smoothing is performed to remove the noisiness of the pressure and wind fields that is associated with simulated convection and its associated gusts, thus allowing more clear differentiation of trends among the simulations, especially at early times.

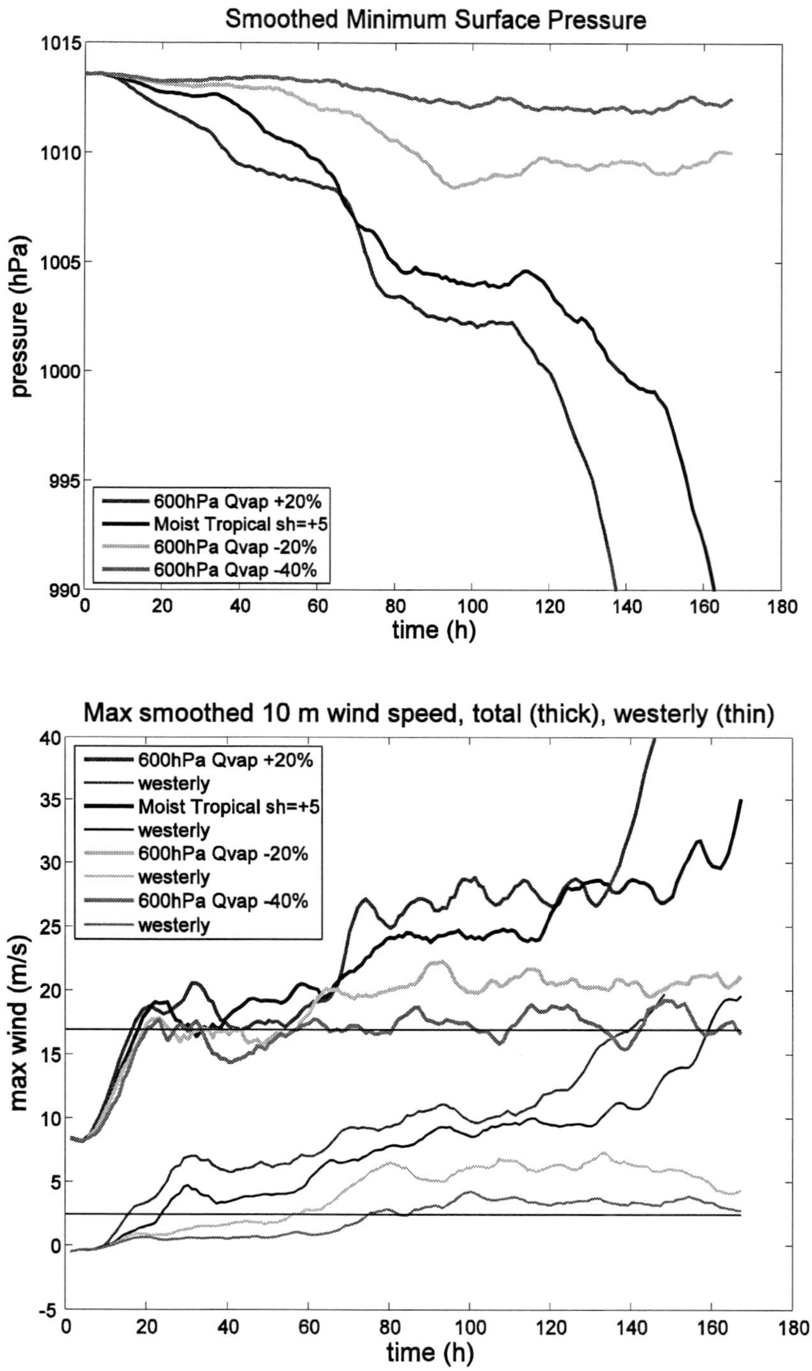

Figure 1.8. Results from simulations using 5 ms^{-1} of westerly wind shear and varying humidity profiles as shown in Fig. 1.6b: a) smoothed minimum surface pressure; b) maximum smoothed total wind (thick lines), and maximum smoothed westerly wind (thin lines). The thin black horizontal lines indicate whether the westerly wind is strong enough to identify the low-level circulation as a tropical depression, or whether the total wind exceeds tropical storm force.

Now we consider the results of similar simulations, but with mean flow and wind shear on the beta-plane centered at 15N. All the simulations presented hereafter have one of the zonal wind profiles shown in Fig. 1.7: the low-level wind is easterly at 5 ms^{-1}, and then varies according to a sinusoidal function of log-pressure height between 850 and 200 hPa. The mean flow and shear are balanced using the iterative scheme described in [63], and the specific humidity profiles are adjusted so that the vertical profile of relative humidity is constant across the domain. Figure 1.8 shows the results of simulations using this wind profile. The results show a strong dependence on humidity, but in a rather different way than what is seen with the no-flow simulations. Rather than simply delaying genesis and rapid intensification, the combination of dry air and wind shear can lead to very different outcomes. With a 20% increase in mid-level humidity, the cyclone begins to develop almost immediately, and goes through a series of stair-step intensifications. The evolution is similar for the control profile, but the intensifications are delayed by 12-24 hours. However, for a 20% reduction in mid-level humidity, the outcome of the simulation is quite different: some development occurs, but the surface pressure falls only to about 1008 hPa, after which it rises to 1010 hPa. For a 40% reduction, there appears to be almost no development, with only 1 hPa of surface pressure fall over the 7 day simulation.

To better identify the simulated stages of development, Fig. 1.8b shows the maximum surface wind (smoothed in time and space as well). All simulations show a rapid increase of peak winds from the initial, 9.5 ms^{-1} surface wind on the north side of the cyclone, to 15-20 ms^{-1} within the first 18 hours. But the peak wind remains on the north side, where the vortex circulation adds to the 5 ms^{-1} low-level flow, and is locally enhanced by vertical mixing due to convection and downdrafts. The peak wind does not indicate whether the system has achieved a closed low-level circulation. To identify the presence of a robust, closed, surface circulation, the thinner lines in Fig. 1.8b show the peak, smoothed[4], westerly component of the surface winds. Where these become sufficiently positive (so that they would be persistent and observable), we conclude that the model equivalent of a tropical depression has formed. We choose 2.5 ms^{-1} to be the "sufficiently positive" value for the smoothed westerly winds (i.e., 5 knots). In our modeling framework, this is our definition of TC genesis. This value is indicated by the lower of the two thin, horizontal, black lines on the figure. The upper horizontal line corresponds to 17 ms^{-1}, the threshold for the naming of a tropical storm.

With these metrics we can more specifically interpret the results of the shear and dry-air simulations. For mid-level humidity +20%, a depression forms in less than 20 hours, and almost immediately becomes a tropical storm. As humidity is decreased, TC genesis and the transitions to further stages of development are delayed. For mid-level humidity -40%, it takes over 3 days for a depression to form, and the peak winds do not stay consistently above tropical storm strength.

There is a more subtle effect of wind shear that can be seen by comparing Fig. 1.6a and Fig. 1.8a. The surface pressure actually falls *more* during the first 48-72 hours for all the simulations with wind shear than for the no-flow simulations. This suggests that, indeed, the presence of wind shear is favorable for genesis, but not necessarily favorable for further intensification. To further explore this possibility, we show in Fig. 1.9 the results of

[4] The westerly winds are smoothed 50 times with the 1-2-1 filter. The greater smoothing is used here to more accurately indicate a closed circulation on scales of 10 to 100 km, one that would be considered as representative by an operational center, rather than what might be indicated by a single point observation.

simulations using the unmodified moist-tropical sounding with increasing westerly wind shear. Indeed, the simulations with greater wind shear (5 and 7.5 ms^{-1}) clearly achieve the genesis threshold at an earlier time than the simulations with less shear (0 and 2.5 ms^{-1}). In fact, the low-shear simulations behave much like the no-flow simulations: the surface pressure remains nearly flat for the first 60 hours, after which genesis occurs and intensification follows. It is also interesting that the simulation with 2.5 ms^{-1} of westerly shear intensifies more quickly than the 0 ms^{-1} shear simulation.

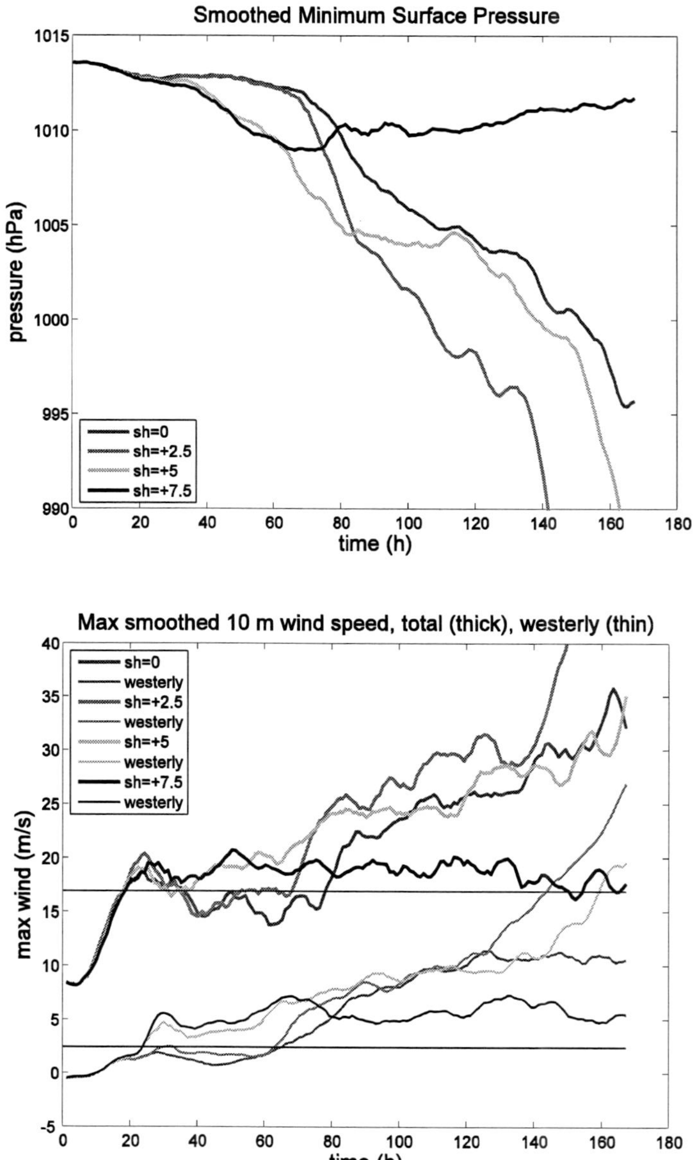

Figure 1.9. As in Fig. 8, but for simulations using the Dunion (2011) "moist tropical" sounding and with increasing westerly wind shear.

How does the presence of wind shear hasten the formation of a low-level circulation in these simulations? For zero shear and 5 ms^{-1} shear, we show in Fig. 1.10 and Fig. 1.11, respectively, the evolution of the surface circulation, mid-level circulation, and the moist convection in terms of the surface pressure and surface wind vectors, the geopotential height of the 600 hPa pressure surface[5], and the simulated reflectivity at that level.

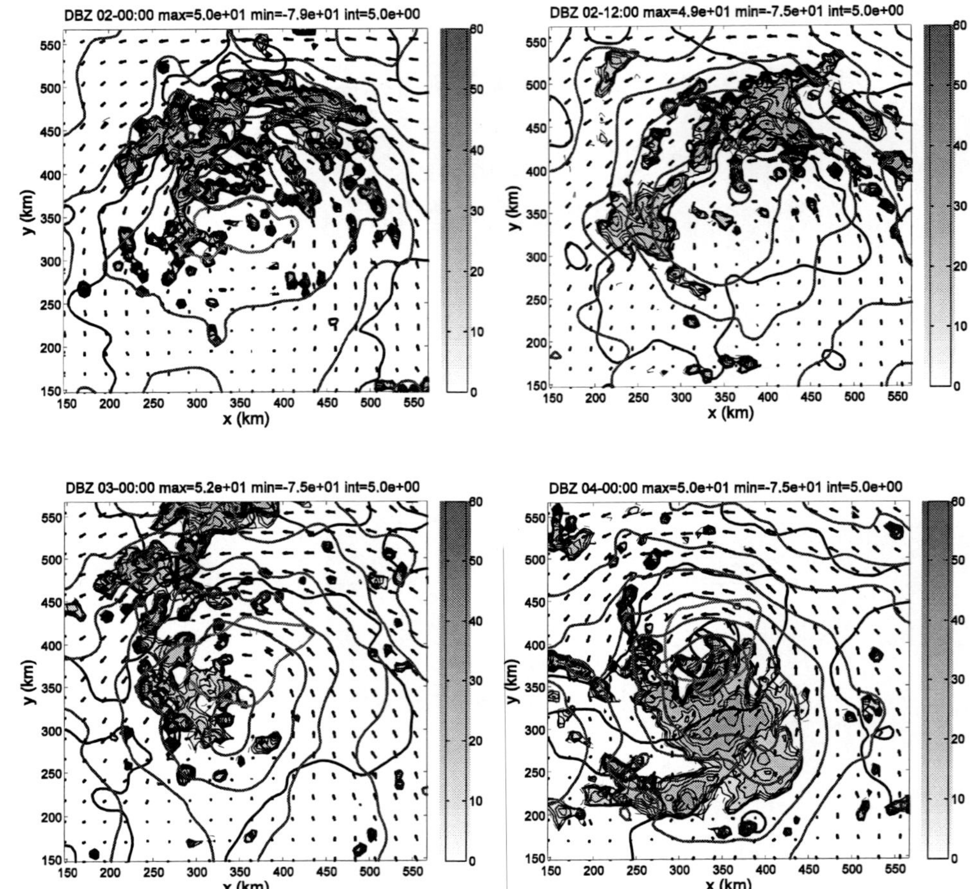

Figure 1.10. For the simulation with 5 ms^{-1} low-level flow and zero wind shear at a) $t = 24$ h; b) 36 h; c) 48 h; d) 72 h. These plots show simulated reflectivity near 600 hPa (grayscale filled contours), surface wind vectors, smoothed surface pressure (black contours), and smoothed, approximate heights of the 600 hPa pressure surface (blue contours, see text for details). The contour interval is 1 hPa for surface pressure, with the 1013 hPa contour in red, and 5 m for the 600 hPa height contours. Wind vectors are scaled so that a vector which reaches from one grid point to the next indicates 30 ms^{-1}. In these and subsequent figures, the date shown at the top of each plot refers to the time in days, hours, and minutes of a simulation that began at 00Z on day 1.

[5] Since the WRF model uses the dry hydrostatic pressure normalized by its surface value as the vertical coordinate, the heights of constant pressure surfaces are not easily derived. Instead, we compute the mean pressure value on a model level, and then use the hypsometric equation to find the vertical distance between the model level (the altitude of which can be easily computed) and that pressure value. In the figures shown, model level 18 is used, which for all the plots has a mean pressure value around 592 hPa.

Looking first at zero wind shear (Fig. 1.10), we see that at 24 hours, there is deep convection widely dispersed across the north side of the circulation. The 600 hPa height contours show a weak, mid-level circulation that is displaced slightly north of the surface circulation. At t = 36 and 48 hours, the mid-level circulation rotates around to the south side, and at these times the convection is still scattered along a cyclonically curved band around the west and north sides of the vortex.

Finally, at t = 72 hours, a more coherent area of mixed deep and stratiform convection has developed (an MCS), co-located with a stronger mid-level circulation (now an MCV). A coherent surface vortex has developed on the north side of the MCS.

With 5 ms^{-1} of westerly shear, the appearance of the MCV and then a coherent surface vortex both occur much more quickly, at t = 36 and 48 h, respectively. Even by t = 24 h the developing MCV is evident, located downshear and downshear-left of the low-level circulation. This convective structure is what would be expected to develop in association with a vortex that is being tilted by wind shear: positive vorticity advection by the thermal wind generates mesoscale ascent and low-level cooling on the downshear side [31,47,66,67]. In turn, low-level air circulating around to the downshear side of the vortex moves upward along the upward sloped potential temperature surfaces leading into the cooler air, triggering both convection and more widespread precipitation.

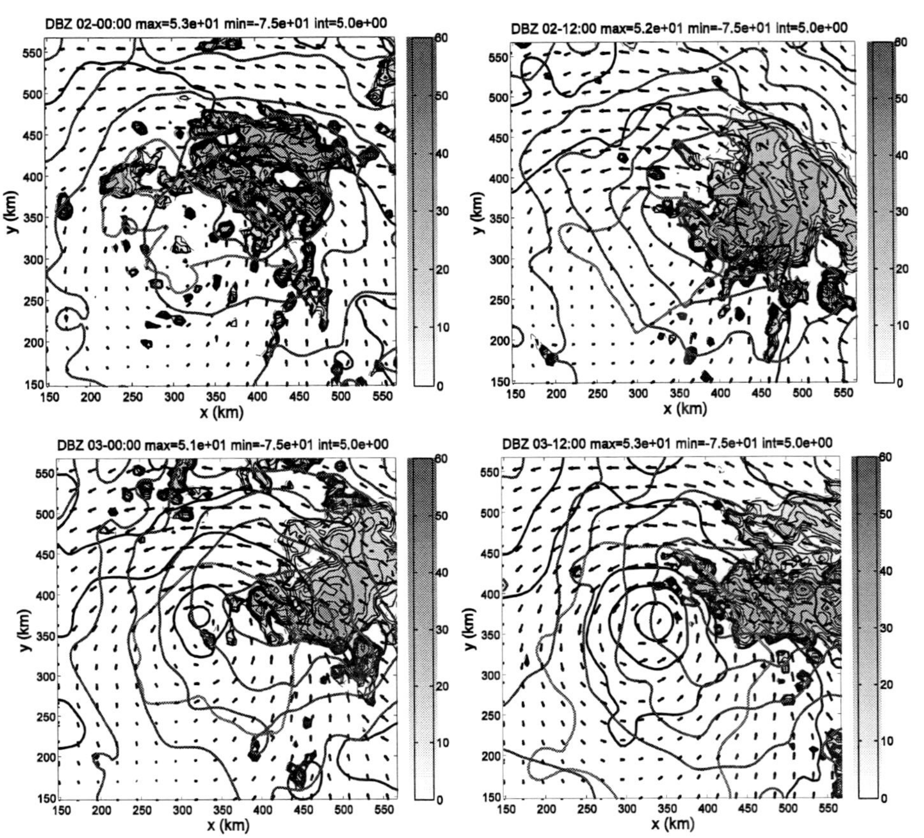

Figure 1.11. As in Fig. 1.10, but for the simulation with 5 ms^{-1} of westerly shear, at a) t = 24 h; b) 36 h; c) 48 h; d) 60 h.

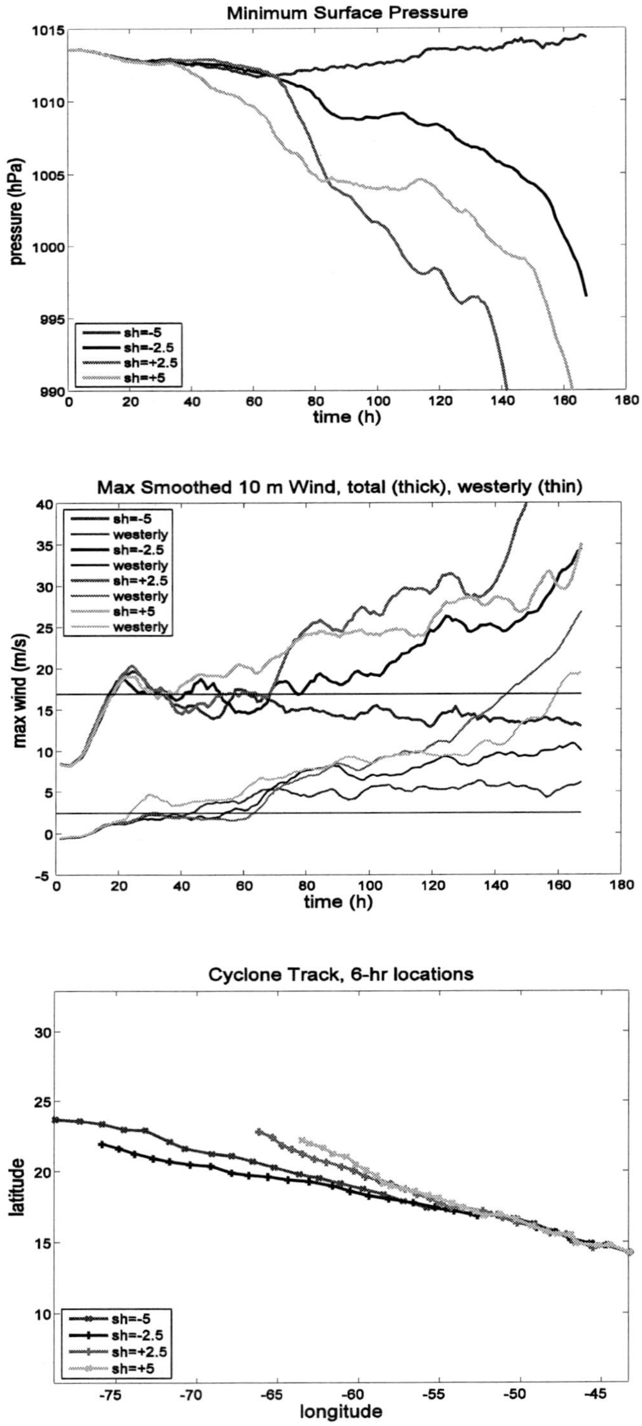

Figure 1.12. As in Fig. 1.8, but for simulations with both easterly (negative) and westerly (positive) wind shear; as above, "westerly" in the legend refers to the smoothed westerly surface wind field. Also shown in (c) are the 6-hourly locations of the surface pressure minima associated with the disturbances. The longitude values are only meant to be representative .

Without further analysis beyond the scope of what may be presented here, it is not possible to say exactly why the seemingly more coherent MCS/MCV that develops in the sheared case leads to the more rapid development of a surface circulation. We can speculate that for zero wind shear the more scattered distribution of deep convective cells, as shown in Fig. 1.10, is less effective in producing large net positive diabatic heating, perhaps due to increased entrainment of mid-level dry air into the updrafts and evaporation of precipitation between the updrafts. As shown in [13], the intensification of a pre-existing vortex by convection is mostly determined by the integrated diabatic heating and the proximity of that heating to the circulation center. For the sheared disturbance, the more robust MCS may feed back onto the intensification of the mid-level circulation, which has been hypothesized to be a necessary pre-cursor for the development of a surface vortex [16,18,68].

We proceed with the next, obvious comparison: easterly shear versus westerly shear. Results from simulations with 2.5 and 5.0 ms^{-1} of easterly shear are shown in Fig. 1.12, along with the previous westerly shear results. This brings us to our first unexpected result: in these simulations, easterly shear is not more favorable. In fact, it is considerably less favorable than westerly shear. Interestingly, genesis still comes sooner for 5 ms^{-1} of easterly shear than for 2.5 ms^{-1} of either easterly or westerly shear. Nonetheless, for 5 ms^{-1} of easterly shear, the smoothed surface pressure does not fall below 1012 hPa, and after 60 hours the system weakens steadily. Figure 1.13 shows the structure of the vortex for this case at t = 48 and 60 h, i.e., around its "peak intensity." The convection is well downshear of the surface circulation, and a coherent MCS or MCV is not present at either time.

This result stands in contrast to the climatological results shown above, and more directly in disagreement to the modeling results of TK81. However, there are some significant differences between these two sets of simulations. The TK81 simulations with varying vertical shear also contained meridional shear with the initial disturbance embedded in a zonal band of cyclonic vorticity. Also, in TK81 the initial vortices were strongest at the surface. With our idealized modeling framework, it is straightforward to reproduce these conditions. The exact same meridional variation of the easterly flow as described in TK81 is used, and additional simulations are performed with the maximum circulation at the surface (we use the same vertical structure as the "surface vortex" case of [18]). The wind shear profiles are similar, but not identical. The mathematical form of the shear is not stated in TK81, but one may see from their Fig. 3 that the variation of the zonal wind is roughly linear in their vertical (σ) coordinates, between what appears to be approximately 850 and 200 hPa.

Figure 1.14 shows the results of our attempt to reproduce the findings of TK81 using WRF 3.1.1, our vertical shear, and their meridional shear. Both the mid-level and surface vortex simulations develop quickly in westerly shear (even faster than what is shown above, perhaps due to the positive environmental vorticity of the meridional shear), while the vortices in easterly shear show marginal development followed by decay. Despite our modifications, the simulations continue to show the opposite result found by TK81. It is also interesting to note that the westerly shear simulation with the initial mid-level vortex develops a little more quickly than the simulation with a surface vortex.

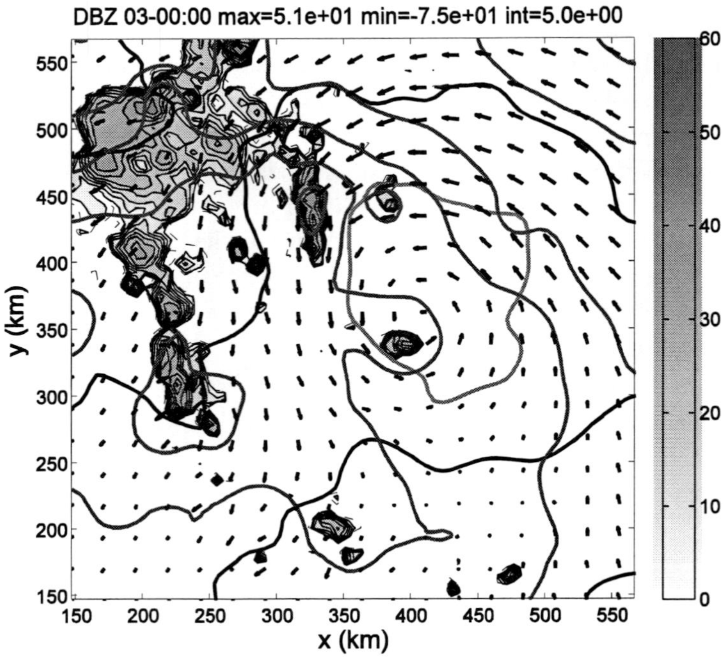

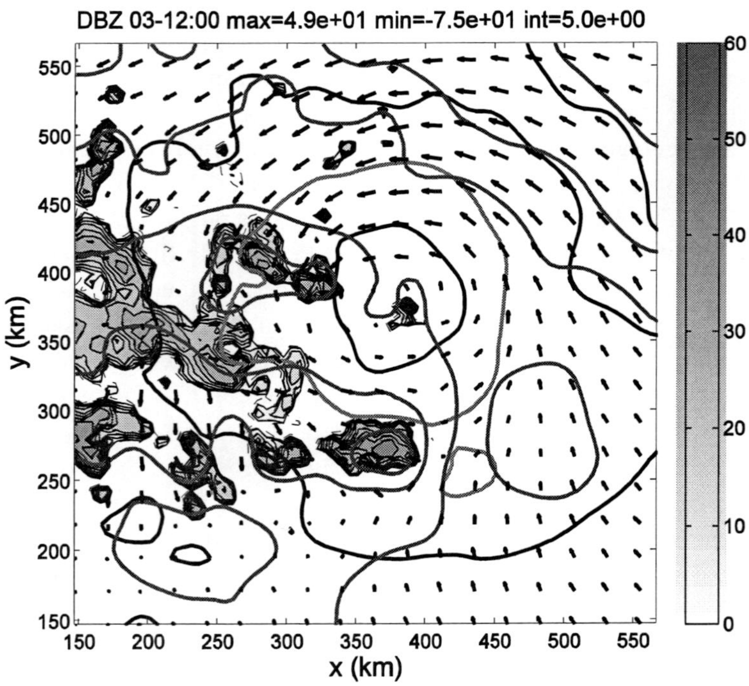

Figure 1.13. As in Fig. 1.10, for simulations with 5 ms^{-1} of easterly wind shear, at a) 48 h; b) 60 h.

westerly shear (see Table 1.2 in TK81). With moderate easterly shear, there was a favorable vertical alignment of the system, while for westerly shear, the systems were quickly sheared apart.

However, westward propagation faster than the low-level easterly flow, with *westerly* shear, seems very unlikely. Returning to our first comparison of easterly versus westerly shear, Fig. 1.12c shows the tracks of the disturbances. For 5 ms^{-1} of easterly shear the mean westward speed is 6.4 ms^{-1}, while it is 3.7 $ms^{-1}$for 5 ms^{-1} of westerly shear. The disturbance speed is close to the low-level flow speed and it is only modestly changed by the overlying wind shear. Such behavior is far more consistent with real disturbances in the tropics. Thus, it may be possible that the reason for the findings in TK81 is due to an unphysical westward propagation of the low-level disturbances.

The idealized simulations presented here, along with numerous additional simulations not shown (such as using the *f*-plane and with different sizes and strengths of the initial vortices), indicate that - with all other environmental factors identical - westerly wind shear is more favorable for genesis than easterly shear. Within the context of these idealized simulations, there are reasons that can explain this preference. By the thermal wind equation, westerly wind shear with the profile shown in Fig. 1.7 requires that there be cooler temperatures in the middle and upper troposphere to the north, and warmer temperatures to the south. On the beta-plane, the vortices drift northward due to the "beta effect" [69,70] while they are carried westward by the low-level flow. All the storms presented here drift northward 5 to 7 degrees in latitude over the 7 day periods of the simulations (Fig. 1.12c). Thus, westward moving storms in westerly shear are drifting into an environment that is more favorable for TC development, due to the cooler temperatures in the middle and upper troposphere overlying the same SST and temperatures in the lower atmosphere.

For 5 ms^{-1} of westerly shear, Fig. 1.15 shows meridional profiles of the temperature anomaly at $z = 8$ km (the level where it is largest), MPI, convective available potential energy (CAPE), and saturation deficit at 600 hPa, across the parts of the model domain that are not perturbed by the initial vortex. From south to north, MPI decreases by a small fraction due to small increases in the temperature in the lower stratosphere on the north side of the domain; this results in a warmer "outflow temperature" for the MPI calculation [8,9]. The $z = 8$ km temperature anomaly is small, remaining less than 1 C in magnitude within 1000 km of the domain center, but the anomaly extends between 4 and 12 km height, causing CAPE to increase by 10% within 500 km north of the center and to larger values beyond. Rappin et al. [61] and Nolan [63] have argued that mid-level saturation deficit is a more important parameter for genesis, as it is indicative of the amount of water that must be transported by convection from the ocean to the middle levels of the atmosphere in order to achieve near-saturation and enable TC genesis. Saturation deficit does decrease to the north, but only by a very small amount in this example. The thermodynamic variations shown here are considerably smaller than those shown in similar calculations in [63]. This is because the example in that paper considered greater shear (10 ms^{-1}) at a higher latitude (20° N). With the fairly small variations shown here, the large differences between the simulations using only 5 ms^{-1} of westerly and easterly shear suggests that other mechanisms must be at work.

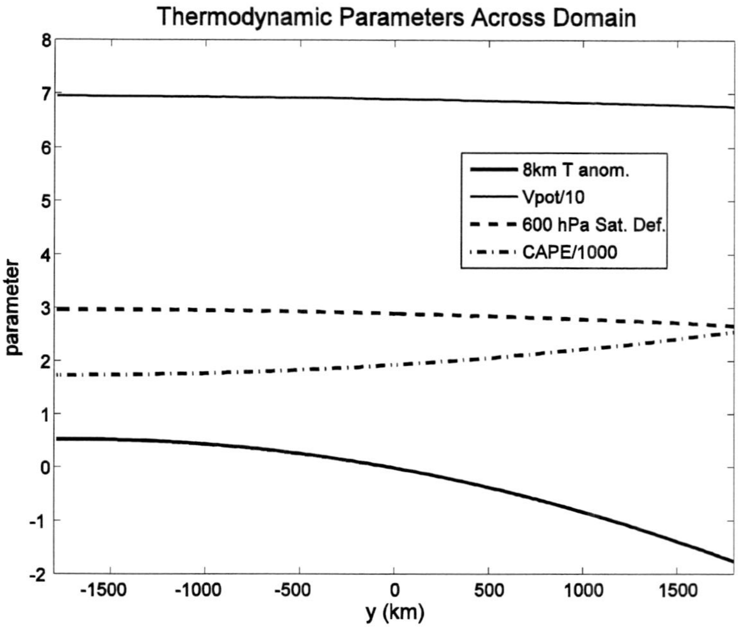

Figure 1.15. Meridional profiles of z = 8 km temperature anomaly (K), MPI (ms^{-1}), 600 hPa saturation deficit ($g\ kg^{-1}$), and convective available potential energy (CAPE, $J\ kg^{-1}$) for the simulations with 5 ms^{-1} of westerly shear.

In fact, similar results for easterly and westerly shear have been found with simulations using no meridional temperature gradient whatsoever. Using the "point-downscaling" technique [60,63], simulations of TC genesis in wind shear can be performed without the temperature gradients required by the thermal wind equations. This method uses forcing terms in the meridional momentum equation to maintain the desired wind shear. As described in detail in Rappin and Nolan [71], wind shear in the same direction as the low-level flow (equivalent to easterly shear here) was found to be much less favorable for genesis than when the shear is in the opposite direction. This result was attributed to an asymmetry in the surface fluxes on the right and left sides of the vortex (relative to the low-level flow). For easterly low-level flow, the superposition of the low-level flow and the initial vortex enhances surface fluxes on the north side and suppresses fluxes on the south side. When the shear is from the west, these enhanced fluxes favor the development of new convection upshear (in this case in the counter-clockwise direction from the MCS), bringing deep convection closer to the circulation center, and reducing the tilt of the system.

This favorable configuration between downshear convection and the surface fluxes can be seen in Fig. 1.16a, which shows the surface moisture flux at t = 36 h for the 5 ms^{-1} westerly shear case. In addition to the effect of the increased wind speed, fluxes on the north side of the vortex are further enhanced by cool and dry downdrafts that emanate westward out of the MCS. The vastly larger surface winds speeds greatly decrease the time required for surface fluxes to modify the boundary layer. Figure 1.16b shows the moisture fluxes at t = 36 h for the 5 ms^{-1} easterly shear case (note that the surface pressures and surface winds are still very similar for the two cases at this time).

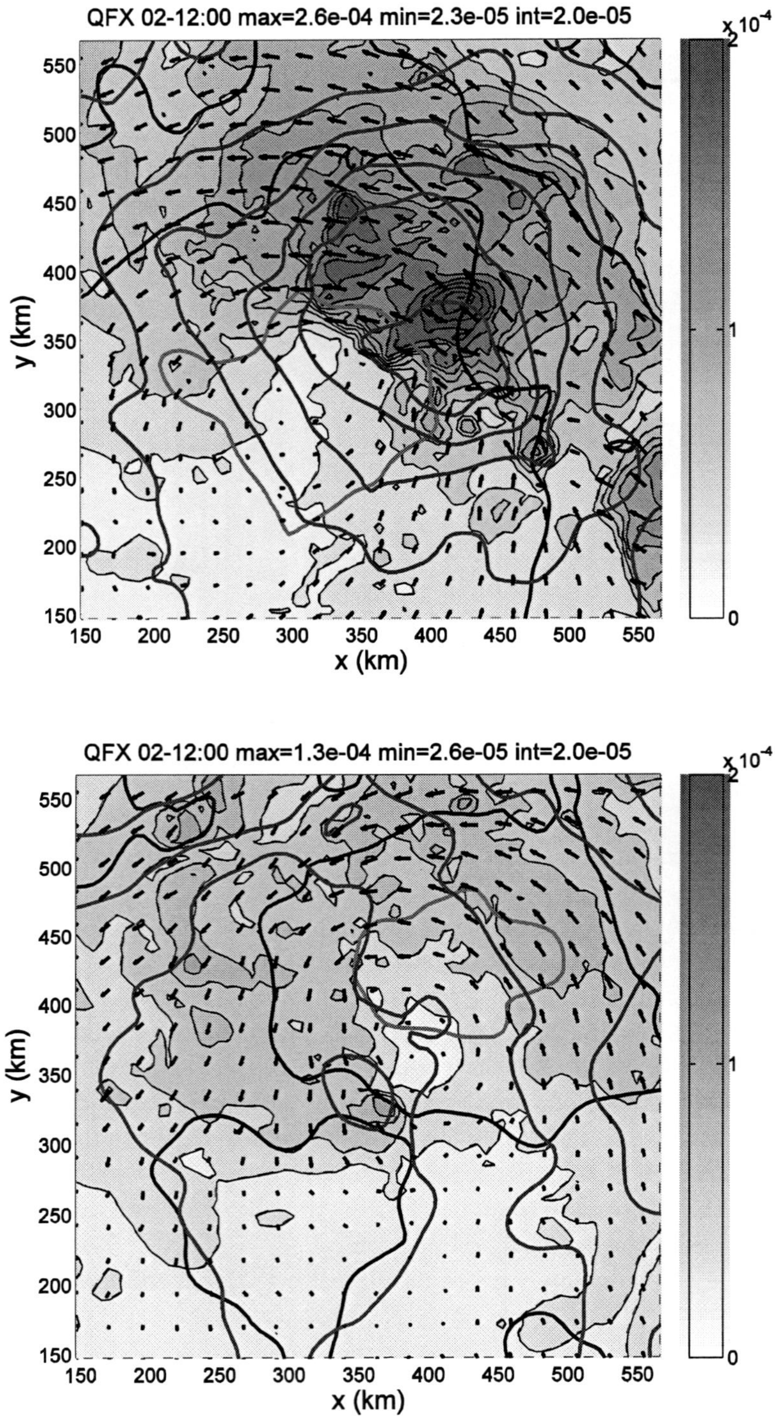

Figure1.16. As in Fig. 1.10, but with the grayscaled contours showing surface moisture flux, for a) 5 ms^{-1} of westerly shear at t = 36 h; b) 5 ms^{-1} of easterly shear at t = 36 h.

When the shear and low-level flow are aligned, the surface wind speeds downstream of the MCV are nearly zero. Any potential for larger surface fluxes associated with downdrafts and cold pools emanating from the downshear MCS is defeated by the small values of the surface wind.

Comparison to some recent studies of genesis in shear

Molinari et al. [72] investigated the genesis of Hurricane Danny (1997), which apparently formed from an MCS that drifted southward from the southern United States into the north-central Gulf of Mexico. During the genesis period, the system was under the influence of 5 to 11 ms^{-1} of wind shear. Repeated events of deep convection on the downshear side of the MCV (presumably generated by same processes discussed above in section 2c) led to the intensification of the low-level circulation and ultimately the formation of a new, stronger circulation under the MCV itself. Based on their findings, [72] proposed that Danny (1997) was an example of wind shear accelerating or perhaps even causing TC genesis.

Musgrave et al. [73] presented mesoscale model simulations of the genesis of Hurricane Gabrielle (2001), also in the Gulf of Mexico, which formed under the influence of wind shear that varied between 4 and 8 ms^{-1}. Their "control" simulation, using analyzed wind and temperature fields as initial conditions, reproduced the genesis of a tropical cyclone similar to that which was observed. An additional simulation in which the synoptic-scale flow was modified to have less wind shear over the system generated a significantly weaker system, with TC genesis (defined as the development of a low-level, warm core) occurring 36 h later than in the control case.

However, both the Danny (1997) and Gabrielle (2001) cases occurred outside of the "deep tropics" and under the influence of wind shear associated with mid-latitude systems. Analyses of the genesis of a more purely "tropical" system are presented by Montgomery et al. [74] and Raymond and Lopez-Carrillo [75], providing detailed observational analyses of the genesis of Typhoon Nuri (2008), made available by numerous flights of research aircraft into the system as part of the Tropical Cyclone Structure 2008 (TCS-08) field experiment [76]. During the first set of flights into Nuri, when it had not yet been identified as a tropical depression, areas of deep convection were to the south and south-east of the low-level circulation center (as seen in a reference frame moving with the disturbance; see Fig. 12 of [74]), thus being downshear and downshear left of the north-northeasterly wind shear around the system on that day. The low-level circulation was separated from the mid-level circulation center by about 300 km (see Fig. 5 of [75]). Analyses of data obtained on the next day, shortly after the system had been declared as a tropical depression by the Joint Typhoon Warning Center, show that the distance between the low-level and mid-level centers had decreased to less than 200 km (see Fig. 6 of [75]) while the nearby deep convection was now to the southwest and south (see Fig. 11 of [74]), again downshear and downshear-left of the wind shear that had rotated clockwise to east-northeasterly.

Also using data from TCS-08, Bell and Montgomery [77] documented the meso-scale flow and the structure of deep convection in a Pacific disturbance that later became Typhoon Hagiput (2008). Satellite images and airborne radar reflectivity fields in their paper show deep convection downshear and downshear-left of moderately strong northeasterly and then later northwesterly wind shear, with a low-level circulation exposed on the north side of the MCS. These structures are also quite similar to those that appear in our simulations.

Finally, Hogsett and Zhang [78,79] present numerical simulations of the formation of Typhoon Chanchu (2011). This system developed on the north side of a westerly wind burst moving eastward along the equator in the West Pacific. The strong cyclonic shear on the north side of the westerly flow generated the development of a cyclonic circulation, one that was also strongly tilted with height due to the low-level easterly shear of the wind burst itself. A series of simulated MCSs that develop along the system moisten the environment and generate vorticity which enhances the mid-level circulation of the tilted system. Ultimately, the convection causes the low-level and mid-level circulations to re-unite, leading to TC genesis.

Discussion

These idealized simulations, in conjunction with a series of previous studies using similar modeling frameworks [60,61,63,71], but with different versions of the WRF model (2.1.2, 2.2.1, and 3.1.1), differing horizontal resolutions (ranging from 4km to 2km), different environmental soundings (radiative-convective equilibrium versus the moist-tropical sounding), different sizes and structures of the initial vortex, and different shear profiles (linear shear or sinusoidal shear with height) appear to produce a self-consistent body of work regarding the effects of wind shear on TC genesis. These results support some ideas that have arisen in the literature: that moderate wind shear can accelerate TC genesis, but will suppress further development; that the combination of dry air and wind shear is particularly lethal; and that genesis in shear involves the interactions between a frictionally-enhanced surface vortex and a convectively enhanced mid-level vortex.

Unfortunately, we are left in a quandary regarding the relative favorableness of easterly versus westerly shear. These up-to-date simulations clearly support westerly shear as the more favorable condition but remain at odds with the results of TK81 and with the climatological statistics shown in section 3b. As noted above, a possible explanation may lie in the fact that environments with easterly shear are more favorable in other ways. To explore this possibility, Fig. 17 shows a scatter-plot of zonal wind shear and MPI computed from [9] (both computed from NCEP/NCAR Reanalysis fields, as above). While MPI predicts the peak wind speed of a hypothetical, mature cyclone in an unsheared environment, it is also a measure of the relative thermodynamic disequilibrium between the ocean and atmosphere, and as a result, it is also a useful measure of thermodynamic favorableness for TC genesis [51,52,59,61]. The black dots in Fig. 1.17 show zonal shear and MPI around TC genesis events, and they illustrate the strong preference for genesis to occur in weak to moderate easterly shear with high MPI. The smaller red dots show zonal shear and MPI from points randomly sampled in the same manner as described in section 3b. Noting that all of these points are within 20° of the equator, it is clear that there is a correlation between easterly shear and higher values of MPI. Similar plots (not shown) also show that easterly shear is correlated with lower latitudes and with some locations known to be highly favorable for genesis, such as the East Atlantic and the East Pacific.

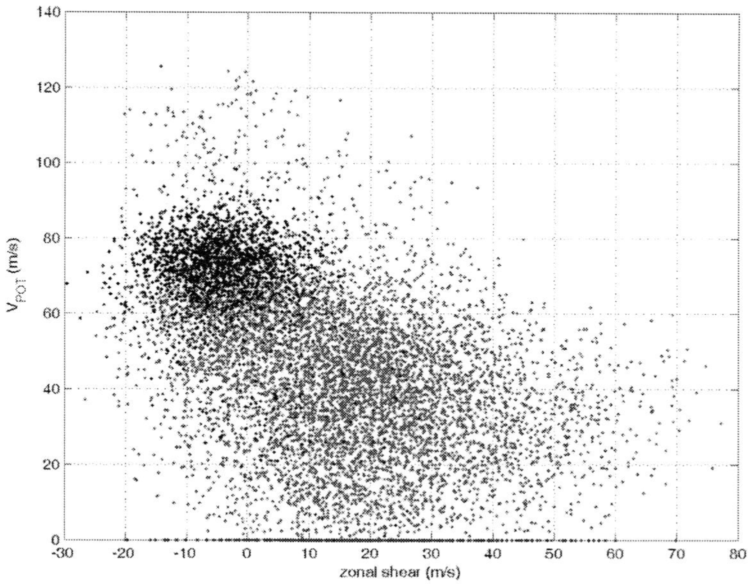

Figure 1.17. Scatterplot of values of zonal wind shear and maximum potential intensity from the theory of Emanuel (1995), for TC genesis events (black dots) and for randomly sampled locations (red dots, see text for details) equatorward of 20° latitude.

Conclusion

As noted in the Introduction, much of our conceptual understanding of TC genesis has been developed around a primarily axisymmetric framework, with no specific role for wind shear. However, a straightforward accounting of observed TC genesis events, coupled with an analysis of their environmental flow fields computed from reanalyses, shows that the large majority of TC genesis events occur in at least 5.0 ms^{-1} (about 10 knots) of 850 to 200 hPa wind shear. The numerical simulations presented in this study indicate that 5.0 ms^{-1} of shear is enough to cause significant tilt of a weak, pre-genesis circulation, and to cause significant mesoscale organization of its associated convection.

From a climatological point of view, studies of the development of genesis parameters that predict TC genesis frequency based on environmental parameters have found that less climatological wind shear corresponds to more TC genesis events (when all other climatological factors are equal). On the other hand, computing the number of TC genesis events as a function of environmental wind shear suggests that wind shear values of 7.5 to 10 ms^{-1} produce the most TC genesis events. In fact, both of these "results" are incorrect. The genesis parameter results only apply to *mean* wind shear values on monthly time scales or longer, while the basic "histogram" approach (as in [41]) does not account for the fact that very low wind shear is very uncommon. When the distribution of wind shear values not associated with TC genesis is into taken into account, we find that very low wind shear (0 to 1.25 ms^{-1}) is indeed quite favorable, as are moderate values (5.0 to 7.5 ms^{-1}), and there is a maximum of "favorableness" for light wind shear, in the range of 1.25 to 5.0 ms^{-1}. The same statistical analysis shows that, in the "deep tropics" (latitudes equatorward of 20°) easterly wind shear is significantly more favorable than westerly shear.

Three-dimensional, full-physics, numerical simulations in the idealized framework of uni-directional flow on the beta-plane were performed to explore the sensitivity of TC genesis to wind shear. These simulations found that, in terms of TC genesis as defined by the formation of a closed, low-level circulation, moderate shear values are more favorable than very low shear or no wind shear at all. However, the larger wind shear values delayed or suppressed further development, consistent with a substantial body of previous work regarding the effects of wind shear on developing and mature tropical cyclones [31,33,34,80,81].

The idealized simulations were also used to explore the effects of easterly versus westerly shear. In contrast with the results of Tuleya and Kurihara [38], and with the aforementioned statistical result, westerly shear was found to be substantially more favorable for genesis than easterly shear. Modifications to our simulations to make them more like those of [38] by adding horizontal wind shear and using an initial disturbance with maximum winds at the surface produced the same result. This result is consistent with similar findings derived from a slightly different idealized modeling framework [71]. In that study, an increased favorableness for TC genesis when the direction of the wind shear was opposite to the low-level flow (equivalent to easterly low-level flow and westerly shear) was explained by a favorable superposition of enhanced surface fluxes on one side of the circulation with the location of the down-shear left mesoscale convective complex. As discussed in [63], westerly shear should also be more favorable because it requires cooler upper-air temperatures to the north of the cyclone, into which a cyclone on the beta-plane will inevitably drift.

The reasons for the apparent disagreement between these idealized simulations and the statistical favorableness of easterly wind shear for genesis in the deep tropics remains undetermined. We have proposed that the reason for the discrepancy lies in a strong correlation between easterly shear and other favorable factors, such as increased thermodynamic favorableness (as measured by the combination of SST and the overlying temperature and humidity profiles), or with geographical locations that are associated with stronger initiating disturbances. However, we do not consider our results to be final. Rather, we hope that substantial further study on TC genesis in wind shear, both observational and numerical, will continue. As noted above, all previous theoretical frameworks and many numerical studies do not explicitly account for the effects of wind shear in either suppressing or favoring TC genesis. Until we can incorporate the role of wind shear in modulating the vertical structure of the circulation and the organization of convection, our understanding of TC genesis will be incomplete.

ACKNOWLEDGMENTS

The authors would like to thank Eric Rappin, Daniel Stern, and Yumin Moon for their comments on earlier drafts of this chapter. The research presented here was supported by the National Science Foundation under grants ATM-0851021 and ATM-0756308, and by the University of Miami.

REFERENCES

[1] National Hurricane Center, Glossary of Terms, http://www.nhc.noaa.gov /aboutgloss.shtml (2011)
[2] World Meteorological Organization, http://cawcr.gov.au/publications/BMRC_archive /tcguide. (1992)
[3] S. J. Camargo, and S. E. Zebiak, *Wea. Forecasting*, 17, 1152 (2002)
[4] W. M. Gray, *Mon. Wea. Rev.*, 96, 669 (1968)
[5] M. DeMaria, J. A. Knaff, and B. H. Connell, *Wea. Forecasting*, 16, 219 (2001)
[6] K. K. W. Cheung, *J. Climate*, 17, 466 (2004)
[7] W. M. Gray, Dept. of Atmos. Sci. Paper No. 234, Colo. State. Univ., Fort Collins, CO, 121 pp. (1975)
[8] K. A. Emanuel, *J. Atmos. Sci.*, 43, 585 (1986)
[9] K. A. Emanuel, *J. Atmos. Sci.*, 52, 3969 (1995)
[10] K. L. Swanson, *Geochem. Geophys. Geosys.*, doi: 10.1029/2007GC001844. (2008)
[11] G. A. Vecchi, K. L. Swanson, and B. J. Soden, *Science*, 322, 687 (2008)
[12] D. J. Raymond, S. L. Sessions, and Z. Fuchs, *Q. J. Roy. Meteorol. Soc.*, 133, 1743 (2007)
[13] D. S. Nolan, Y. Moon, and D. P. Stern, *J. Atmos. Sci.*, 64, 3377 (2007)
[14] R. Rotunno, and K. A. Emanuel, *J. Atmos. Sci.*, 44, 542 (1987)
[15] K. A. Emanuel, *J. Atmos. Sci.*, 46, 3431 (1989)
[16] M. Bister, and K. A. Emanuel, *Mon. Wea. Rev.*, 125, 2662 (1997)
[17] M. T. Montgomery, M. E. Nicholls, T. A. Cram, and A. B. Saunders, *J. Atmos. Sci.*, 63, 355 (2006)
[18] D. S. Nolan, *Aust. Meteorol. Mag.,* 56, 241 (2007)
[19] K. J. Tory, and W. M. Frank, *World Scientific*, 436 pp. (2010)
[20] H. Riehl, *J. Meteor.*, 5, 247 (1948)
[21] E. D. Rappin, M. C. Morgan, and G. J. Tripoli, *J. Atmos. Sci.*, 68, 177 (2011)
[22] E. Palmen, Australian Bur. Meteorol., Melbourne, 213 pp. (1956)
[23] E. Palmen, and C. W. Newton, *Academic Press*, New York, 603 pp. (1969)
[24] C. S. Ramage, *J. Meteorology*, 16, 227 (1959)
[25] J. C. Sadler, *Mon. Wea. Rev.*, 104, 1266 (1976)
[26] J. C. Sadler, *Mon. Wea. Rev.*, 106, 1137 (1978)
[27] J. L. McBride, *J. Atmos. Sci.*, 38, 1117 (1981)
[28] J. L. McBride, and R. Zehr, *J. Atmos. Sci.*, 38, 1132 (1981)
[29] C. S. Lee, *J. Atmos. Sci.*, 46, 2580 (1989)
[30] Y. Wang, and G. J. Holland, *J. Atmos. Sci.*, 53, 3313 (1996)
[31] W. M. Frank, and E. A. Ritchie, *Mon. Wea. Rev.*, 129, 2249 (2001)
[32] S. K. Kimball, and J. L. Evans, *Mon. Wea. Rev.*, 130, 2210 (2002)
[33] M. L. M. Wong, and J. C. L. Chan, *J. Atmos. Sci.*, 61, 1859 (2004)
[34] M. Riemer, M. T. Montgomery, and M. E. Nicholls, *Atmos. Chem. Phys.*, 10, 3163 (2010)
[35] E. A. Hendricks, M. T. Montgomery, and C. A. Davis, *J. Atmos. Sci.*, 61, 1209 (2004)
[36] S. A. Braun, M. T. Montgomery, K. J. Mallen, and P. D. Reasor, *J. Atmos. Sci.*, 67, 999 (2010)
[37] J. Fang, and F. Zhang, *J. Atmos. Sci.*, 67, 655 (2010)

[38] R. E. Tuleya, and Y. Kurihara, *Mon. Wea. Rev.*, 109, 2487 (1981)
[39] L. F. Bosart, and J. A. Bartlo, *Mon. Wea. Rev.*, 119, 1979 (1991)
[40] L. M. Briegel, and W. M. Frank, *Mon. Wea. Rev.*,125, 1397 (1997)
[41] W. E. Bracken, and L. F. Bosart, *Mon. Wea. Rev.*, 128, 353 (2000)
[42] C. A. Davis, and L. F. Bosart, *Mon. Wea. Rev.*, 131, 2730 (2003)
[43] C. A. Davis, and L. F. Bosart: *Bull. Amer. Meteorol. Soc.*, 85, 1657 (2004)
[44] C. A. Davis, and L. F. Bosart, *Q. J. R. Meteorol. Soc.*, 132, 2055 (2006)
[45] R. McTaggart-Cowan, G. D. Deane, L. F. Bosart, C. A. Davis, and T. J. Galarneau, *Mon. Wea. Rev.*, 136, 1284 (2008)
[46] E. Kalnay, and Coauthors, *Bull. Amer. Meteorol. Soc.*, 77, 437 (1996)
[47] D. J. Raymond, and H. Jiang, *J. Atmos. Sci.*, 47, 3067 (1990)
[48] J. M. Fritsch, J. D. Murphy, and J. S. Kain, *J. Atmos. Sci.*, 51, 1780 (1994)
[49] S. S. Chen, and W. M. Frank, *J. Atmos. Sci.*, 50, 2401 (1993)
[50] R. F. Rogers, and J. M. Fritsch, *Mon. Wea. Rev.*, 129, 605 (2001)
[51] K. A. Emanuel, and D. S. Nolan, "Tropical Cyclone activity and the global climate system", *26th Conference on Hurricanes and Tropical Meteorology, (Miami, Florida)*, 10A.2 (2004)
[52] K. A. Emanuel, *J. Adv. Model. Earth Syst.*, 2, Art. #1, 12 pp. (2010)
[53] M. K. Tippett, S. J. Camargo, and A. H. Sobel, *J. Climate*, 24, 2335 (2011)
[54] M. G. McGauley, and D. S. Nolan, *J. Climate*, 24, 5968-5997. (2011)
[55] K. R. Knapp, M. C. Kruk, D.H. Levinson, H. J. Diamon, and C. J. Neumann, *Bull. Amer. Meteorol. Soc.*, 3, 363 (2010)
[56] D. J. Raymond, C. Lopez-Carrillo, and L. L. Cavazos, *Q. J. R. Meteorol. Soc.*, 124, 2005 (1998)
[57] G. N. Kiladis, C. D. Thorncroft, N. M. J. Hall, *J. Atmos. Sci.*, 63, 2212 (2006)
[58] S. B. Hopsch, C. D. Thorncroft, and K. R. Tyle, *Mon. Wea. Rev.*, 138, 1399 (2010)
[59] D. S. Nolan, E. D. Rappin, and K. A. Emanuel, *Q. J. R. Meteorol. Soc.*, 133, 2085 (2007)
[60] D. S. Nolan, and E. D. Rappin, *Geophys. Res. Lett.*, 35, L14805, doi:10.1029/2008GL034147. (2008)
[61] E. D. Rappin, D. S. Nolan, and K. A. Emanuel, *Q. J. R. Meteorol. Soc.*, 124, 2005 (2010)
[62] W. C. Skamarock, J. B. Klemp, J. Dudhia, D. O. Gill, D. M. Barker, M. G. Duda, X.-Y. Huang, W. Wang, and J. G. Powers, NCAR technical note 475+STR, 113 pp. (2008)
[63] D. S. Nolan, *J. Adv. Model. Earth Syst.*, 3, Art. M08001, doi: 10.1029/2011MS000063. (2010)
[64] J. P. Dunion, *J. Climate*, 24, 893 (2011)
[65] K. J. Mallen, M. T. Montgomery, and B. Wang, *J. Atmos. Sci.*, 62, 408 (2005)
[66] M. DeMaria, *J. Atmos. Sci.*, 53, 2076 (1996)
[67] W. M. Frank, and E. A. Ritchie, *Mon. Wea. Rev.*, 127, 2044 (1999)
[68] D. J. Raymond, S. L. Sessions, and C. Lopez-Carrillo, *J. Geophys. Res.*, 116, D18101, doi:10.1029/2011JD015624. (2011)
[69] J. C.-L. Chan, and R. T. Williams, *J. Atmos. Sci.*, 44, 1257 (1987)
[70] M. Fiorino, and R. L. Elsberry, *J. Atmos. Sci.*, 46, 975 (1989)
[71] E. D. Rappin, and D. S. Nolan, *Q. J. Roy. Meteorol. Soc.*, in press. (2012)
[72] J. Molinari, D. Vollaro, and K. L. Corbosiero, *J. Atmos. Sci.*, 61, 2493 (2004)

[73] K. D. Musgrave, C. A. Davis, and M. T. Montgomery, *Mon. Wea. Rev.*, 136, 3151 (2008)
[74] M. T. Montgomery, L. L. Lussuer, R. W. Moore, and Z. Wang, *Atmos. Chem. Phys.*, 10, 9879 (2010)
[75] D. J. Raymond, and C. Lopez-Carrillo, *Atmos. Chem. Phys.*, 11, 147 (2010)
[76] R. L. Elsberry, and P. A. Harr, *Asia-Pac. J. Atmos. Sci.*, 44, 209 (2008)
[77] M. M. Bell, and M. T. Montgomery, *Geophys. Res. Lett.*, 37, L06802, doi:10.1029/2009GL042313. (2010)
[78] W. Hogsett, and D. L. Zhang, *J. Atmos. Sci.*, 67, 3774 (2010)
[79] W. Hogsett, and D. L. Zhang, *J. Atmos. Sci.*, 68, 1377 (2011)
[80] M. DeMaria, and J. Kaplan, *Wea. Forecast.*, 9, 209 (1999)
[81] B. Tang, and K. Emanuel, *J. Atmos. Sci.*, 67, 1817 (2010)

In: Cyclones: Formation, Triggers and Control
Editors: K. Oouchi and H. Fudeyasu

ISBN: 978-1-61942-976-5
©2012 Nova Science Publishers, Inc.

Chapter 2

ANALYSIS OF TROPICAL CYCLONE ACTIVITY IN THE SOUTHERN HEMISPHERE USING OBSERVATION AND CGCM SIMULATION

Satoshi Iizuka[1]* and Tomonori Matsuura[2]

[1]National Research Institute for Earth Science and Disaster Prevention, Japan
[2]Graduate School of Science and Engineering, University of Toyama, Japan

ABSTRACT

This chapter presents the relation between El Niño / Southern Oscillation (ENSO) and tropical cyclone (TC) activity over the Southern Hemisphere, using both observations and outputs of a high-resolution (T213) coupled ocean-atmosphere general circulation model. Both observations and simulations show that, in El Niño (La Niña) years, the TC genesis frequency is enhanced in the northeastern (southwestern) South Pacific Ocean. However, no remarkable change in the ratio of annual number of intense TC is found in either the observations or simulations. Therefore, changes in environmental conditions associated with ENSO contribute to TC frequency over the South Pacific, but they do not seem to affect TCs in terms of developing and sustaining their intensity.

The interannual variability of TC activity over the southern Indian Ocean and its relation with Indian Ocean Dipole (IOD) are also examined using partial regression technique. Both observations and simulations show the nearly zonal uniformly reduction (enhancement) in the annual TC frequency in El Niño (La Niña) years after taking away the effect of IOD, whereas the east-west dipole structure characterized by an increase (decrease) in the annual TC frequency in the western (eastern) part of the southern Indian Ocean during positive IOD events after excluding influences arising from ENSO. The results suggest that the interannual variability of TC activity over the southern Indian Ocean can be affected by both ENSO and IOD. Furthermore, observations show that the annual TC frequency formed over both the eastern and western parts of the southern Indian Ocean is related more closely to IOD events than ENSO. During positive (negative) IOD events, the TC frequency generated in the early TC season is reduced (enhanced) over the eastern part of the southern Indian Ocean. The TC genesis frequency and associated TC days in the western part of the southern Indian Ocean tend to increase

*E-mail address: iizuka@bosai.go.jp

(decrease) in positive (negative) IOD events. However, there are not significant relations for the ratio of the annual number of intense TCs to total TCs over both the eastern and western parts of the southern Indian Ocean.

INTRODUCTION

Theory and numerical results of several studies suggest that tropical cyclones (TCs) would increase in intensity in response to global warming [1–4]. Consequently, the change in TC activity under global warming has become a question of great concern in recent decades because TCs are often accompanied by extreme winds, torrential rainfall, and storm surges, causing economic damage and loss of human life. Emanuel [5] and Webster *et al.* [6] reported an increase in the frequency of intense TC occurrence over the past a few decades and related this increase to a concurrent increase in sea surface temperature (SST). However, Chan [7] claimed that the change in the number of intense TCs over the western North Pacific is unrelated to local SST but rather to similar decadal fluctuations in the atmospheric environment. Indeed, various large-scale environment factors can affect TCs by developing and sustaining their intensity, although warm SSTs are the most fundamental factor in TC formation in climatological aspect [8]. Consequently, an understanding of natural variability of TC activity on interannual and decadal time scales is important to detect the possible effects of anthropogenic climate change on TC activity.

Many reports of earlier studies have documented that El Niño / Southern Oscillation (ENSO) strongly affect the interannual variability of TC activity in most basins [9]. Increased vertical wind shear during El Niño years helps to prevent tropical disturbances from developing into TCs, resulting in fewer TC formations in the North Atlantic. La Niña years have more TCs than average because the vertical wind shear is reduced [10, 11]. In the western North Pacific, TCs have longer lifetimes and are more intense in El Niño years than La Niña years, accompanied by the southeastward shift of mean TC genesis locations [12, 13]. Moreover, TCs in the South Pacific tend to form northeastward more often in El Niño years than in La Niña years [14]. It is also observed that more (fewer) TCs exist around Australia in La Niña (El Niño) years [15].

The historical record of best track data is inhomogeneous because of advances in observation system technologies and operational procedures [16]. Consequently, numerical simulations may become a valuable tool to elucidate the interannual variability of TC activity [17]. In this chapter, we present the interannual variations of TC activity over the Southern Hemisphere using both observations and outputs of a high-resolution coupled GCM (CGCM) with nearly 60 km horizontal resolution. The model can simulate the relation between ENSO and TC intensity in the western North Pacific and the North Atlantic [18, 19], as observed. This chapter is organized as follows. A description of the data is given in the next section. The model climatology is compared with observations in the third section. The fourth section presents the relation between ENSO and TC activity over the Southern Hemisphere simulated in the model along with observations. We then present the relation between ENSO and TC intensity over the South Pacific in both the model and observation. We also examine the relation between Indian Ocean Dipole (IOD) and TC activity over the southern Indian Ocean. A summary is presented in the final section.

DATA AND MODEL

For the analysis described in this chapter, we use the best track data for the period 1971/72–2009/10, as compiled by the Joint Typhoon Warning Center (JTWC) (http://www.npmoc.navy.mil/jtwc.html). The JTWC data include locations and maximum wind speeds of TCs for successive six-hour intervals. The starting time is chosen as 1971/72 because the data in the pre-satellite era may be less reliable. For analysis of the TC intensity, however, the starting time is chosen as 1984/85 because of incomplete information of intensity in the data. For the Southern Hemisphere, the TC season spans calendar years and is generally regarded as the 12 months of July–June. In this chapter, the first year identifies the TC season (e.g., 2009/10 is designated as the 2009 season). To examine the relation between TC activity and environmental conditions, wind fields and SST, taken from the National Center for Environmental Prediction–National Center for Atmospheric Research (NCEP–NCAR) global atmospheric reanalysis data [20] and the extended reconstructed sea surface temperature (ERSST) ver.3 [21], are used as observations.

To confirm the relation between ENSO and TC intensity, we also use the outputs of the CGCM simulation during a period from year 20 to year 100. The atmospheric component of the CGCM has a horizontal resolution of T213 corresponding to a grid distance of about 60 km. In this chapter, we used the following definitions to identify the locations of TCs simulated in the CGCM:

1. The grid point with minimum sea level pressure in a 9 x 9 grid-point box is defined as the center of a TC,
2. The difference in sea level pressure between the center and each grid point on the boundaries at 4 grid points away from the center is greater than 5 hPa.

   Then, we used the following definitions to exclude tracks of extratropical cyclones:
3. Maximum surface wind speeds at 10 m in a 9 x 9 grid-point box around the center are greater than 17m/s,
4. The difference in the vorticity of surface winds between the center and each grid point on the boundaries at 4 grid points away from the center is greater than $1.5 \times 10^{-4}\ s^{-1}$,
5. The lifetime is more than 2 days,
6. The genesis location is located to the north of 40°S,
7. The vorticity of surface winds around the center has a value greater than $5.0 \times 10^{-4}\ s^{-1}$ at least 4 times throughout its lifetime.

It is noted that there are no TCs with maximum surface winds more than 43 m/s, belonging to the category 2 on the Saffir-Simpson scale, because the horizontal resolution of the CGCM has no ability to simulate such strong TCs. The model simulated quasi-periodic (about five years) ENSO-like events. The time and spatial structure in the model anomalies show good agreement with those in the observed data. However the amplitudes of Niño34 SST anomaly in the CGCM are twice as large as observed ones. A more detailed description of the CGCM is presented in Iizuka and Matsuura [18, 19].

CLIMATOLOGY

The spatial distribution of TC genesis locations over the Southern Hemisphere observed during all TC seasons of 1971–2009 is presented in Fig. 1a. The climatological annual TC genesis density is obtained by counting of TC generation in each 4° × 5° latitude–longitude grid box. The TC genesis locations are observed over the range of longitudes of 40°E–160°W

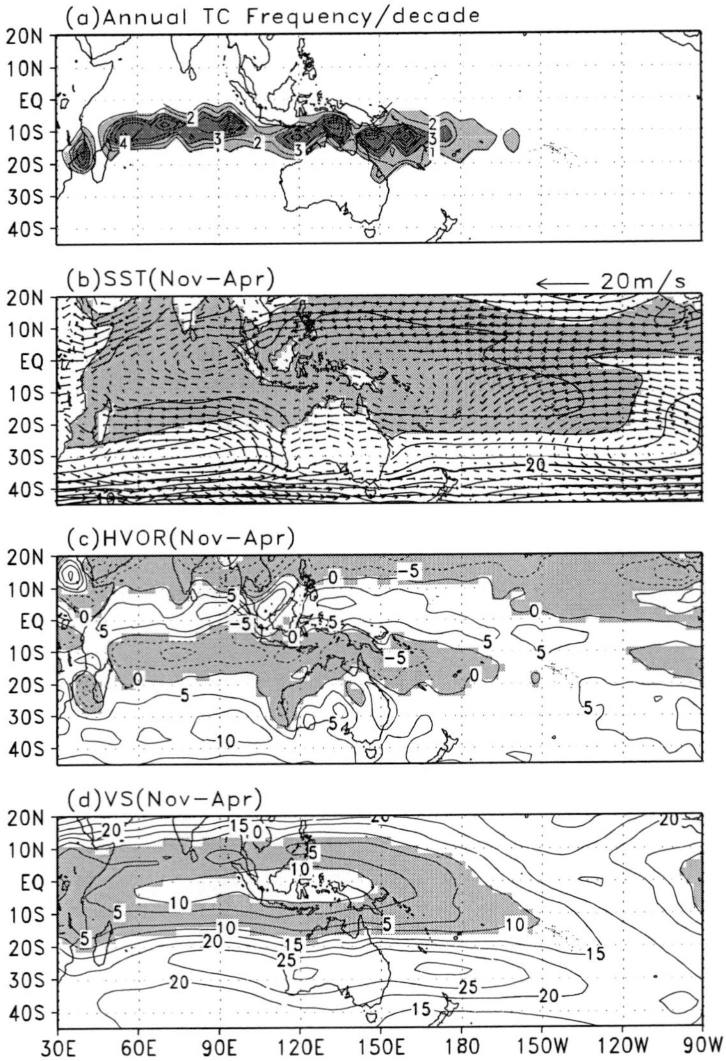

Figure 1. (a) Climatology of the density of observed annual TC genesis number per decade. The contour interval is 1, and light (dark) shaded regions show the values greater than 1 (3). Zero line is omitted. (b) Climatology of the observed SST averaged for November−April. The contour interval is 2°C; shaded regions indicate values greater than 26°C. The vector of wind at 850 hPa is also shown. (c) Climatology of the observed horizontal vorticity at 850 hPa averaged for November−April. The contour interval is 5×10^{-6} s^{-1}; regions with negative values are shaded. (d) Climatology of the observed vertical wind shear defined as the magnitude of difference of the 200 hPa and 850 hPa horizontal winds averaged for November−April. The contour intervals are 5 m s^{-1}; regions with values less than 10 m s^{-1} are shaded.

within a range of latitudes and equatorward from around 25°S. A noticeable southwestward displacement of TC genesis locations is apparent in the area between the African coast and 60°E. The TC genesis number is generally higher in the southern Indian Ocean than in the South Pacific Ocean. Climatological TC genesis active regions are observed in the central tropical southern Indian Ocean, off the northwestern coast of Australia, and in the Coral Sea off the northeastern coast of Australia. The TC genesis locations are spread over the warm SST region exceeding 26°C (Fig. 1b), which suggests that warm ocean thermal energy plays a crucial role in TC formation in climatological aspect [8]. The locations are also observed over the region with large negative low-level horizontal vorticity and weak vertical shear (Figs. 1c and 1d), which help tropical disturbances to develop into TC [8, 22].

Figure 2 presents the spatial distribution of simulated TC genesis locations over the Southern Hemisphere during all TC seasons from years 20/21 to 99/100, along with the climatology of the model SST, low-level horizontal vorticity, and vertical shear averaged for

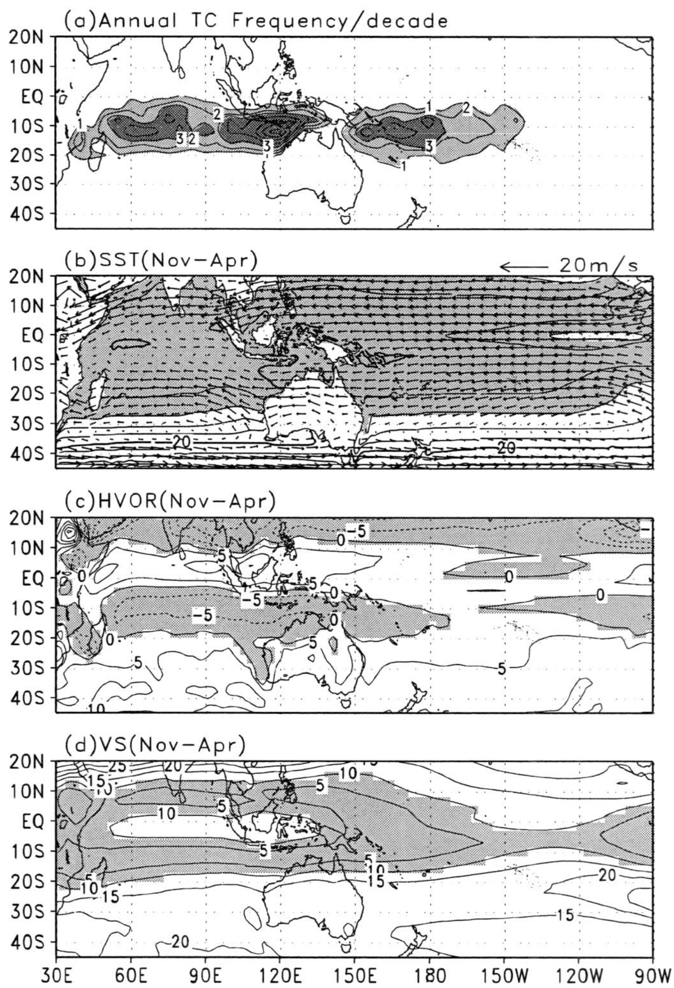

Figure 2. Same as in Figure 1, but for the model.

November–April. The overall patterns and magnitudes of model warm SST regions exceeding 26°C show agreement with the observations, although the model SSTs over the eastern part of the South Pacific are warmer than observations (cf. Figs. 1b and 2b), partially because of a lack of low-level marine stratus in the model [18]. This model bias is accompanied by the larger negative low-level horizontal vorticity and weaker vertical shear over the region of 160°W–100°W along 5°S (Figs 2c and 2d). However, the overall patterns and magnitudes of the model low-level horizontal vorticity and vertical wind shear fields over the region of 30°E–160°W are in general agreement with the observations (cf. Figs. 1c and 1d).

The climatological model TC genesis locations are observed over the range of longitudes of 40°E–140°W within a range of latitudes and equatorward from around 25°S (Fig. 2a). The locations in the model extend more eastward than the observed TC genesis locations (cf. Figs 1a). This bias is apparently related to the deficiency in the model SST, low-level horizontal vorticity, and vertical wind shear over the eastern part of the South Pacific. We also note here that the model TCs are formed even over the South Atlantic, as is often found in other GCMs [23], but we exclude the South Atlantic from analysis in this chapter because actual TCs are merely observed over the South Atlantic. Despite these discrepancies in the model, however, the active TC genesis locations in the model over the southern Indian Ocean and South Pacific Ocean show generally agreement with the observations.

VARIABILITY

Genesis Location and Frequency

We first focus on the relation between ENSO and TC genesis frequency in the Southern Hemisphere. Although various indices have been proposed to define ENSO years in previous studies, here we define El Niño (La Niña) years when SST anomalies in the Niño34 region (5°S–5°N, 170°W–90°W) averaged during October–December are more (less) than 0.9 times the standard deviation. The three months of October–December is chosen because of the large variance of the Niño34 index in the months. Consequently, the following TC seasons are selected as El Niño years: 1972, 1982, 1986, 1991, 1994, 1997, 2002, 2006, and 2009. The La Niña years are the following: 1971, 1973, 1975, 1988, 1998, 1999, and 2007.

Figure 3 shows the TC genesis locations that respectively observed in the El Niño and La Niña years, in addition to the composite seasonal evolution of TC frequency over the several domains in both El Niño and La Niña years in Fig. 4. The definition of the domain is the following: eastern South Pacific Ocean (ESP) (90°W–165°E), Australia (AUS) (165°E–110°E), southeastern Indian Ocean (SEI) (110°E–80°E), and southwestern Indian Ocean (SWI) (80°E–30°E). These domains are selected based on results of previous studies [14, 15, and 24]. It is noteworthy that the three-month averaged TC frequency is used for the composite because of the small size of monthly TC data sample.

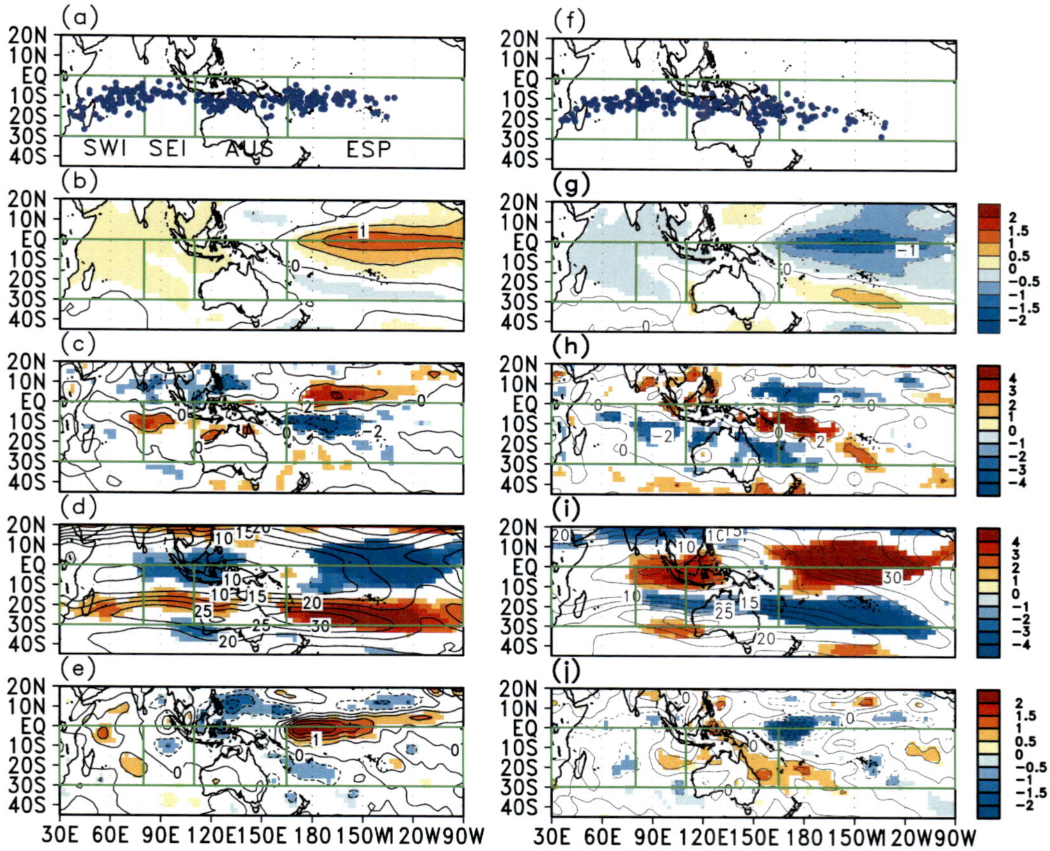

Figure 3. (a) Composite map of observed TC genesis locations in El Niño years. (b) Composite map of observed SST anomaly averaged for **November–April**. The contour interval is 0.5°C. (c) Same as in (b) but for observed horizontal vorticity anomaly at 850 hPa. The contour interval is 2×10^{-6} s^{-1}. (d) Same as in (b) but for observed vertical wind shear. The contour interval is 5 m s^{-1}. Anomaly is shown in color. (e) Same as in (b) but for observed divergence anomaly at 200 hPa. The contour interval is 0.5×10^{-6} s^{-1}. (f)–(j) **Same as in (a)–(e)** but for La Niña years. Regions where the t-test is statistically significant at the 90% confidence level are shaded by color. The green line denotes the boundary of each domain.

The notable difference in observed TC genesis locations over the South Pacific between El Niño and La Niña years is found on either side of 165°E (Figs. 3a and 3f). A marked increase in the number of TC genesis over the ESP is observed in the early TC season, whereas marked reduction of the TC genesis frequency is found over the AUS, in El Niño years than average (Figs. 4a and 4b). The difference that is apparent over the northern part of the ESP appears to be related to the change in the geographic distribution of low-level vorticity over the South Pacific along with changes in vertical wind shear and SST, whereas the change in TC frequency over the southern part of the ESP seems to be attributable mostly to that in vertical wind shear and anomalous descending, associated with ENSO (Fig. 3) [14]. The change in TC frequency over the AUS domain is apparently related mostly to the change in vertical wind shear, along with the change in the geographic distribution of low-level vorticity and SST, associated with ENSO (Fig. 3). The anomalous upper level convergence (divergence) linked to the low-level divergence (convergence) also appears to affect the reduction (enhancement) of the TC genesis frequency over the AUS domain in El Niño (La Niña) years (Figs. 3e and 3j).

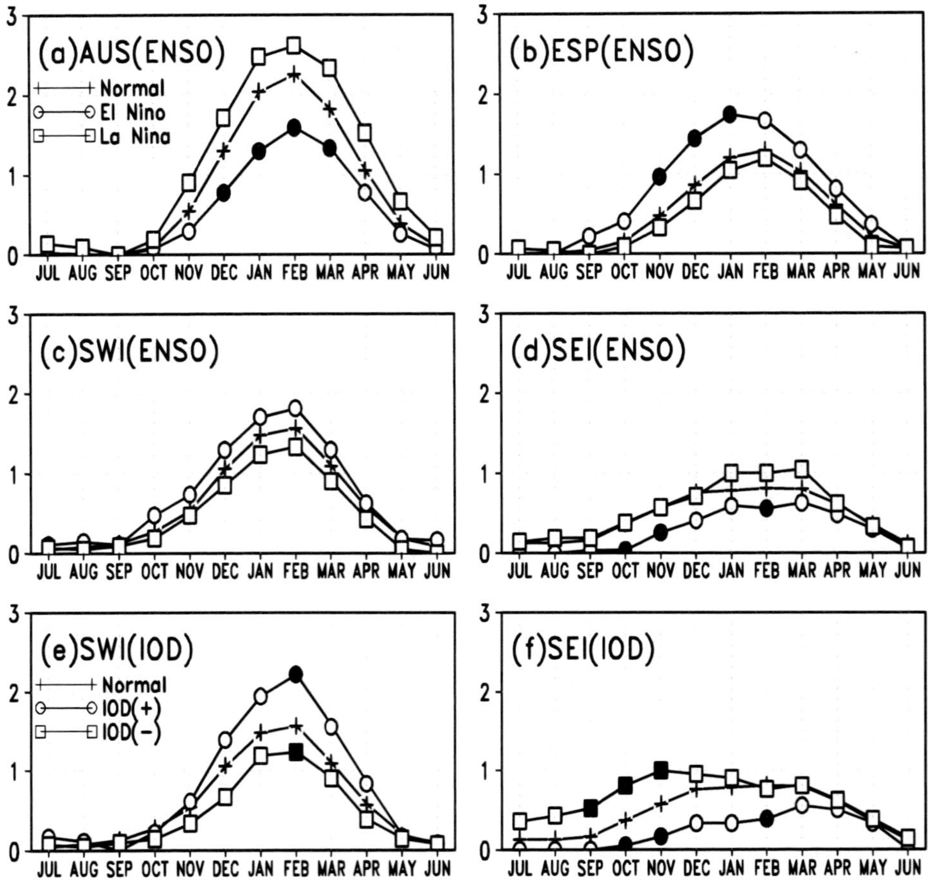

Figure 4. Seasonal evolution of observed three-months averaged TC number generated over the AUS (a), ESP (b), SWI (c), and SEI (d) domains. A solid line with a plus sign represents climatology, whereas a solid line with open circles (squares) represents composite for El Niño (La Niña) years. (e)–(f) Same as in (c)–(d) but for positive (negative) IOD years. Values where the t-test is statistically significant at the 90% confidence level are shown as closed circles (squares).

Figure 5 depicts the simulated TC genesis locations in the El Niño and La Niña years. They show a difference in model TC genesis locations over the South Pacific between El Niño and La Niña years, similar to the observations. In El Niño years, more model TCs tend to form over the northeastern part in the South Pacific when compared with La Niña years. The increase in model TC genesis over the northeastern part in the South Pacific is related to the change in low-level vorticity, along with decreased vertical wind shear and positive SST anomaly (Fig. 5). However, the change in vertical wind shear appears to be related to the difference in model TC genesis frequency over the southeastern part of the South Pacific between El Niño and La Niña years as primary contributors. In the Australian region, the changes in vertical wind shear and low-level convergence implied by the change in upper level divergence seem to contribute to the difference in TC genesis frequency between El Niño and La Niña years. These relations between TC genesis frequency and environmental conditions associated with ENSO in the model are consistent with the observed relations.

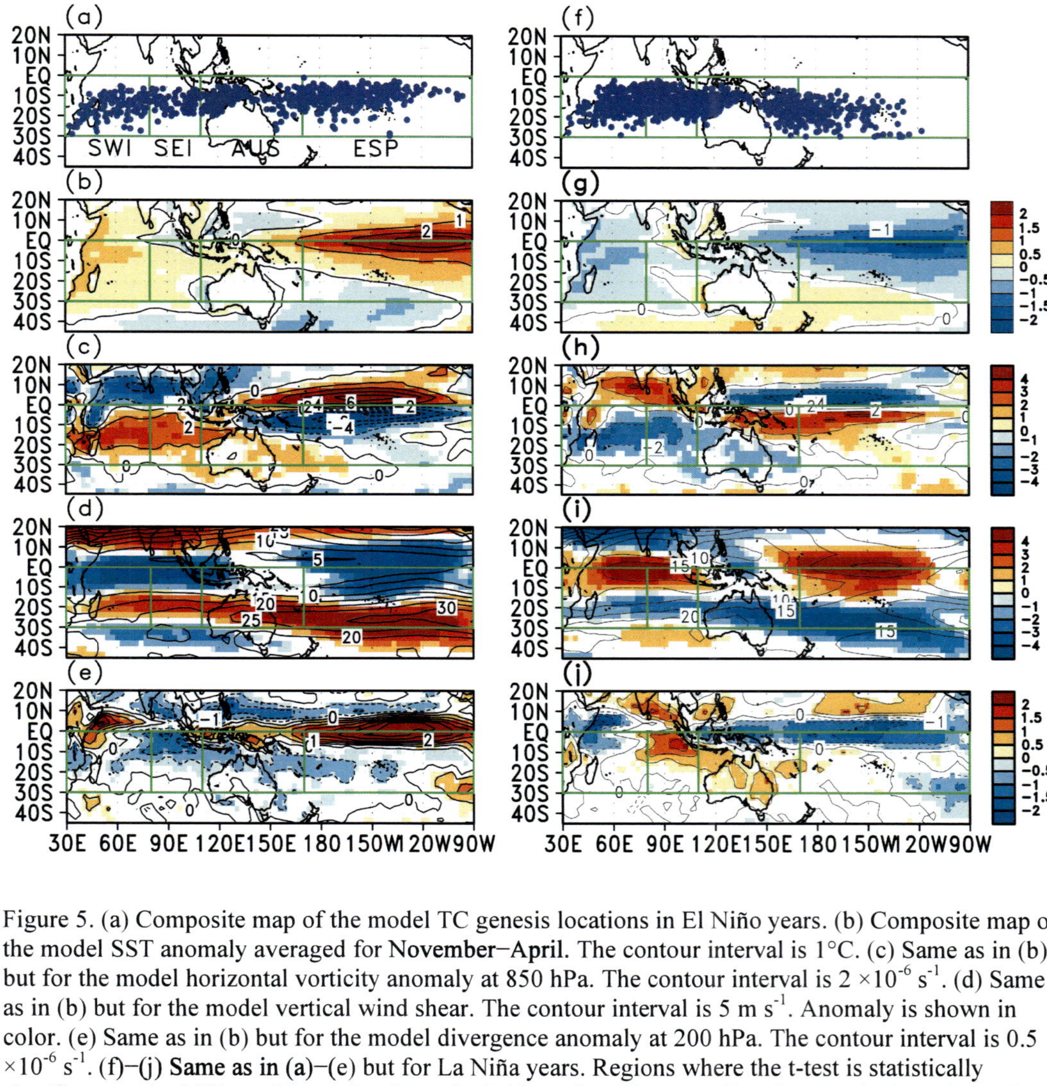

Figure 5. (a) Composite map of the model TC genesis locations in El Niño years. (b) Composite map of the model SST anomaly averaged for November–April. The contour interval is 1°C. (c) Same as in (b) but for the model horizontal vorticity anomaly at 850 hPa. The contour interval is 2×10^{-6} s^{-1}. (d) Same as in (b) but for the model vertical wind shear. The contour interval is 5 m s^{-1}. Anomaly is shown in color. (e) Same as in (b) but for the model divergence anomaly at 200 hPa. The contour interval is 0.5 $\times 10^{-6}$ s^{-1}. (f)–(j) Same as in (a)–(e) but for La Niña years. Regions where the t-test is statistically significant at the 90% confidence level are shaded by color. The green line denotes the boundary of each domain.

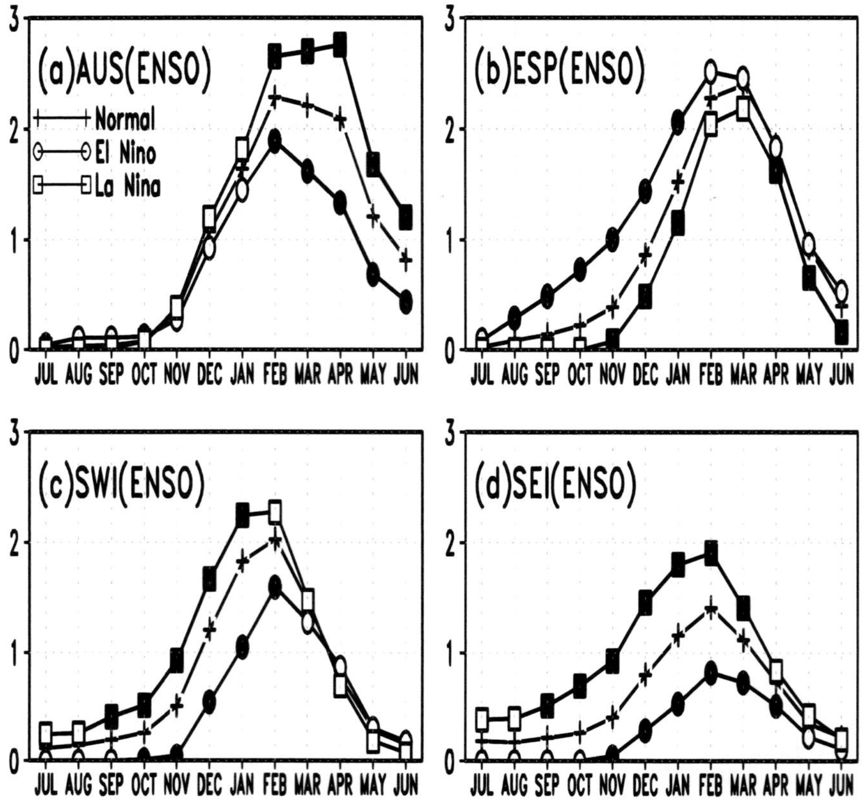

Figure 6. Seasonal evolution of model three-months averaged TC number generated over the AUS (a), ESP (b), SWI (c), and SEI (d) domains. A solid line with a plus sign represents climatology, whereas a solid line with open circles (squares) represents composite for El Niño (La Niña) years. Values where the t-test is statistically significant at the 90% confidence level are shown as closed circles (squares).

Figure 6 presents the composite seasonal evolution of model TC frequency over several domains in both El Niño and La Niña years. The boundary between the ESP and AUS domain is shifted to 170°E because the genesis locations of model TCs over the South Pacific in El Niño years extend more eastward (Figs. 5a and 5f) when compared with the observations (Figs. 3a and 3f). The mean model TC frequency in the ESP and AUS domains is much higher in the late TC season (Figs. 6a and 6b) when compared with the observations (Figs. 4a and 4b). This bias may be related partially to that in SST over the eastern part of the South Pacific (Fig. 2) accompanied by the model double ITCZ during the austral summer (not shown).

In El Niño years, an increase in the model TC frequency over the ESP is apparent in the early TC season, as was reported also for observations. Over the AUS domain, reduction (enhancement) of the model TC frequency in El Niño (La Niña) years is observed. These differences in the model between El Niño and La Niña years are generally similar to the observations, although the changes in the model TC genesis frequency over the AUS domain are overestimated during the austral fall (Figs. 4a and 6a).

Over the southern Indian Ocean, Ho *et al.* [25] and Kuleshov *et al.* [14] documented that the TC genesis locations seem to shift west in El Niño years when compared with La Niña years (Figs. 3a and 3f). However, no significant difference in the TC genesis frequency over the SWI domain is found in any TC season, although the analysis period or best track data used in their studies are different from those in the present analysis. Although marked reduction of TC genesis frequency is observed over the SEI domain in El Niño years (Fig. 4d), there are not any significant differences in TC genesis frequency over both the SWI and SEI domains in La Niña years (Figs. 4c and 4d). Xie *et al.* [26], on the other hand, documented that a relationship exists between TC activity over the southwestern Indian Ocean and local SST induced by either ENSO or IOD events that occur often independently with ENSO [27]. Therefore, we present the composite seasonal evolution of TC frequency over the southern Indian Ocean in both positive and negative IOD years in Figs. 4e and 4f. Because IOD events reach their peaks in September [27], we define positive (negative) IOD years when Dipole Model Index (DMI) averaged during August–October are more (less) than 0.9 times the standard deviation. Consequently, the following TC seasons are identified as positive IOD years (1972, 1977, 1982, 1994, 1997, and 2006) and negative IOD years (1971, 1975, 1989, 1992, 1996, 1998, and 2005).

A significant increase in TC genesis frequency is observed over the SEI in the early TC seasons of negative IOD years, while fewer TCs tend to form in positive IOD years than average (Fig. 4f). Because the eastern Indian Ocean SST anomaly during IOD events dissipates by the following January [27], this seasonal difference in TC genesis over the SEI between positive and negative IOD years is consistent with that of the local SST anomalies associated with IOD events. The significant difference in TC frequency over the SWI between positive and negative IOD years is also observed in February (Fig. 4e). The results suggest that DMI might become more useful index to predict seasonal TC genesis frequency in both the eastern and western parts of the southern Indian Ocean than ENSO indices. However, it is noteworthy that SST anomalies in other regions also play an important role in determining TC activity over the southwestern Indian Ocean [28].

Figure 7a displays the spatial distribution of regression coefficients of annual TC genesis frequency with DMI, showing the east-west dipole structure. The reduction (enhancement) of TC frequency over the SEI domain during positive (negative) IOD events is associated with the anomalous descent (ascent) in response to local SST cooling (warming) (Fig. 7m). The positive (negative) relative vorticity anomaly associated with anomalous easterlies (westerlies) over the SEI domain during positive (negative) IOD events is also unfavorable (favorable) for TC genesis in the region (Fig. 7g). Over the SWI domain, the anomalous ascent (descent) accompanied by local SST warming (cooling) is observed during positive (negative) IOD events, which is favorable (unfavorable) for TC genesis in the region (Fig. 7m). The change in vertical wind shear is not related to the difference in the TC genesis frequency over the southern Indian Ocean (Fig. 7j).

However, most of the IOD events co-occur with ENSO years during the analyzing period in this chapter. Therefore, one may argue whether the environmental conditions during the IOD events differ from those during ENSO years. To examine the contribution of individual ENSO and IOD events "as independent variables" to the TC frequency in the southern Indian Ocean, we here use a partial regression technique [29]. The partial regression coefficient gives the amount by which the dependent variable increases when one independent variable is

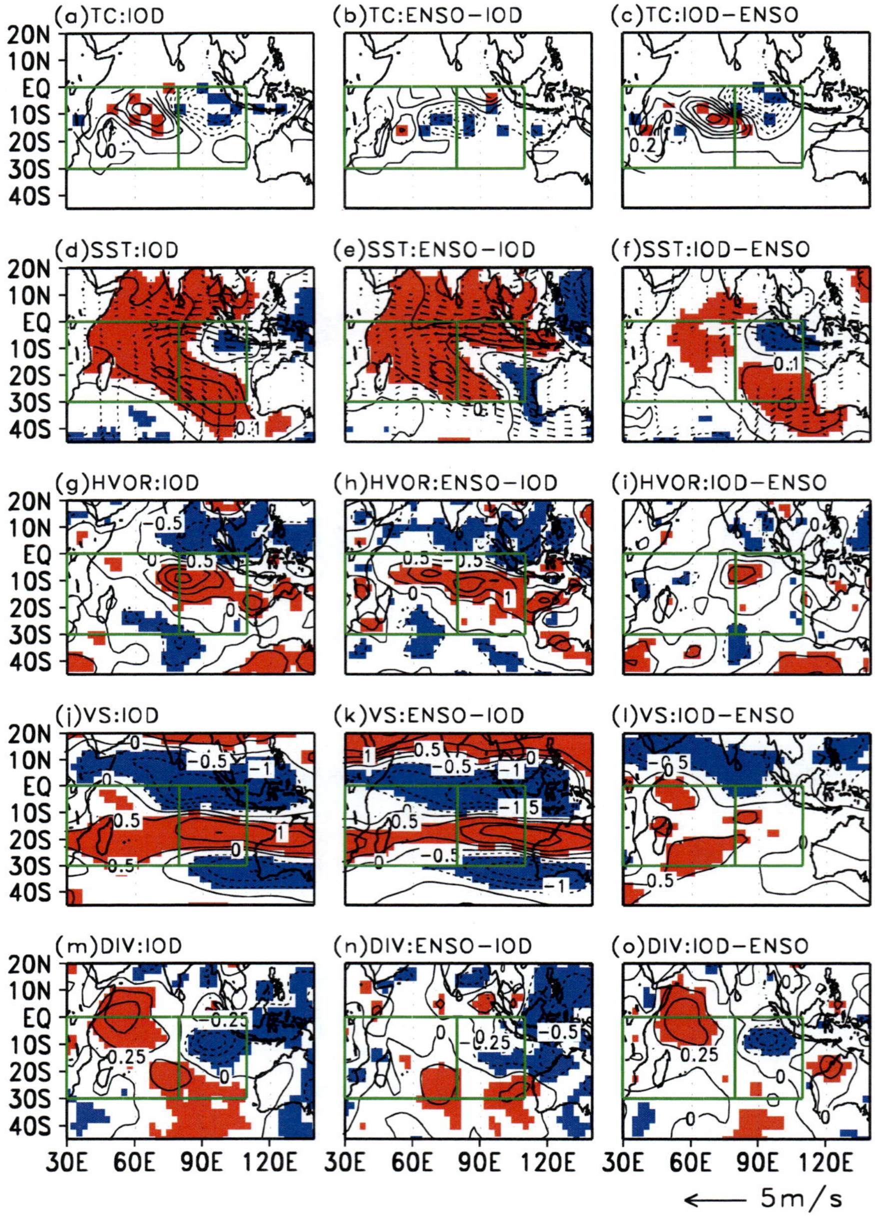

Figure 7. (a) Regression coefficients of the density of observed annual TC number per decade with August–October averaged DMI. The contour interval is 0.2 (b) Partial regression coefficients of the density of observed annual TC number per decade on October–December averaged Niño34 index independent of August–October averaged DMI. (c) Same as in (b) but on August–October averaged DMI independent of October-December averaged Niño34 index. (d)–(f) Same as in (a)–(c) but for observed SST and 850 hPa wind anomalies averaged for September–March of the following year. The contour interval is 0.1°C. (g)–(i) Same as in (d)–(f) but for observed horizontal vorticity anomaly at 850 hPa. The contour interval is 0.5×10^{-6} s^{-1}. (j)–(l) Same as in (d)–(f) but for observed vertical wind shear. The contour interval is 0.5 m s^{-1}. (m)–(o) Same as in (d)–(f) but for observed divergence anomaly at 200 hPa. The contour interval is 0.25×10^{-6} s^{-1}. Positive (negative) values where the t-test is statistically significant at the 90% confidence level are shaded as red (blue) color. The green line denotes the boundary of the SWI and SEI domains.

increased by one unit and all the other independent variables are held constant. The partial regression coefficient $PR_{BY,A}$ of the variable Y with the independent variable A, where the linear relationship with the independent variable B is removed, is calculated by

$$PR_{BY,A} = \frac{r_{BY} - r_{AY} r_{AB}}{1 - r_{AB}^2} \frac{\sigma_Y}{\sigma_B},$$

where r_{BY} is the correlation coefficient between B and Y, r_{AY} is the correlation coefficient between A and Y, and r_{AB} is the correlation coefficient between A and B. σ_Y and σ_B are the standard deviation of variable Y and B respectively.

Figure 7c shows the spatial distribution of partial regression coefficients of annual TC genesis frequency with DMI after taking away the effect of ENSO. The spatial distribution shows the east-west dipole structure. On the other hand, the spatial distribution of partial regression coefficients of annual TC genesis frequency with the Niño34 index after taking away the effect of IOD indicates that the TC genesis frequency tends to reduce (enhance) over the central part of the southern Indian Ocean during El Niño (La Niña) years in absence of IOD events (Fig. 7b). The nearly zonal uniformly reduction (enhancement) of TC genesis frequency over the southern Indian Ocean is associated with the positive (negative) relative vorticity anomaly extending from the southwestern Indian Ocean to the southeastern Indian Ocean (Fig. 7h). Thus, the geographic feature of the change in TC genesis frequency over the southern Indian Ocean during IOD events differs from that in ENSO years in absence of IOD events.

In contrast to the South Pacific and the Australian region, the model includes severe bias in the TC genesis frequency over the southern Indian Ocean, particularly, over the SWI domain in response to ENSO. The model TC frequency over the SWI increases (decreases) in La Niña (El Niño) years (Fig. 6c), whereas the observed TC frequency tends to reduce (enhance) (Fig. 4c). The atmospheric response to the northeast of Madagascar in the model is remarkably different from the observations (cf. Figs. 3 and 5). The negative (positive) low-level vorticity anomalies are found there in El Niño (La Niña) years, associated with anomalous upper-level convergence (divergence) in the model, which are opposite to the observations.

The seasonal difference in the model TC frequency over the southern Indian Ocean between positive (negative) IOD events is very similar to that between El Niño (La Niña) years (not shown). The spatial distribution of partial regression coefficients of annual TC genesis frequency with DMI after taking away the effect of ENSO in the CGCM shows the east-west dipole structure (Fig. 8c), whereas the corresponding spatial distribution for ENSO shows the nearly zonal uniformly change (Fig. 8b). These features in the CGCM are similar to those in the observation (Figs. 7b and c). However, the spatial distribution of regression coefficients of annual TC genesis frequency with DMI displays that the TC genesis frequency tends to reduce (enhance) over the entire southern Indian Ocean during positive (negative) IOD events in the CGCM (Fig. 8a). This discrepancy is because the amplitudes of the model Niño34 SST anomaly are twice as large as observed ones [18, 19] and most of the IOD events co-occur with ENSO years in the CGCM. Thus, the CGCM apparently overestimates the impact of ENSO on TC activity over the southern Indian Ocean. The relation of TC frequency over the southern Indian Ocean with IOD events should be examined in future with outputs of high-resolution models that correctly simulate the relation between ENSO and IOD.

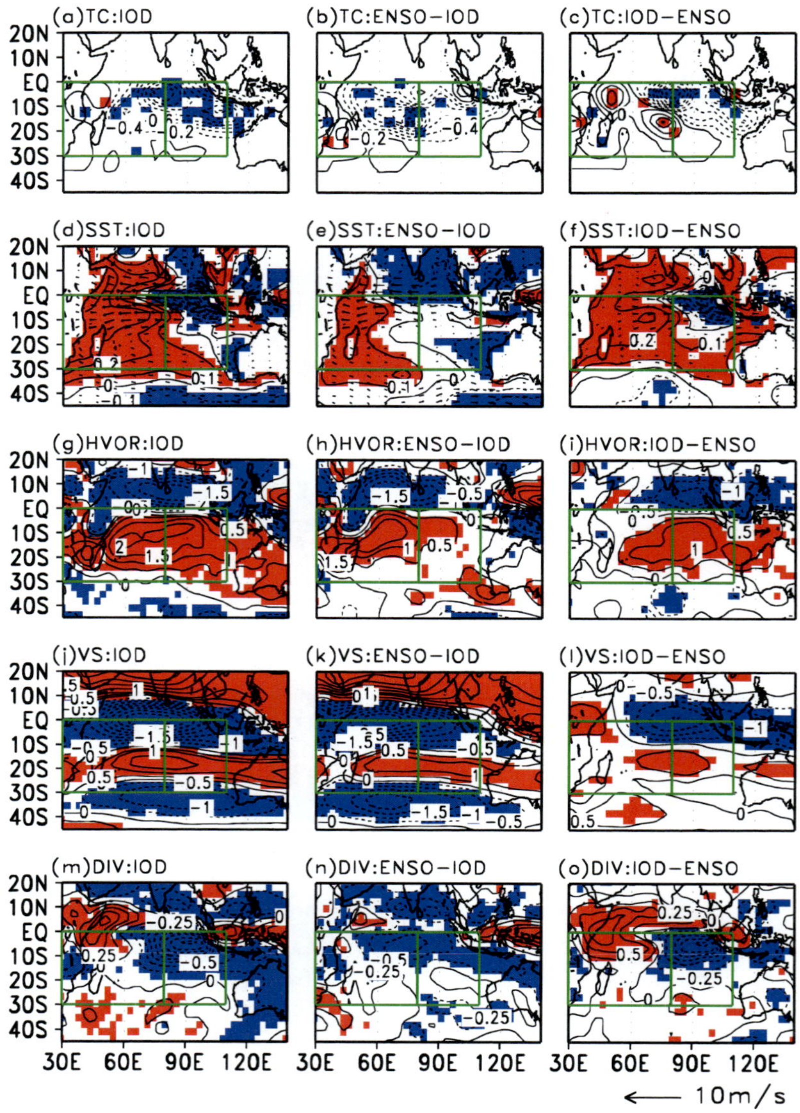

Figure 8. Same as in Fig. 7, but for the model.

Track

Next, we present the relation between ENSO and TC tracks in the observation and model, in Fig. 9. To examine the change in TC tracks in response to ENSO, here we use TC days as obtained by counting the hours of TC existing in each 4° × 5° latitude–longitude grid box. A notable change in TC days is found over the South Pacific. In association with an increase in observed TC genesis frequency in the northeastern part of the South Pacific in El Niño years, more TCs are observed to pass in the equatorward region from 20°S (Fig. 9a). However, the TCs observed to pass on the western and eastern coasts of Australia decrease (increase) in El

Niño (La Niña) years. The model well captures the changes in TC days associated with ENSO over the South Pacific and around Australia, although a remarkable difference in changes of TC days associated with ENSO exists between those of the model and the observations over the southern Indian Ocean (Fig. 9e).

We also present the difference in observed TC days between ENSO and IOD events. IOD events do not always occur with ENSO [27], but most positive (negative) IOD events co-occur in El Niño (La Niña) years. Consequently, the relation of TC days with IOD is similar to that with ENSO (Fig. 9b).

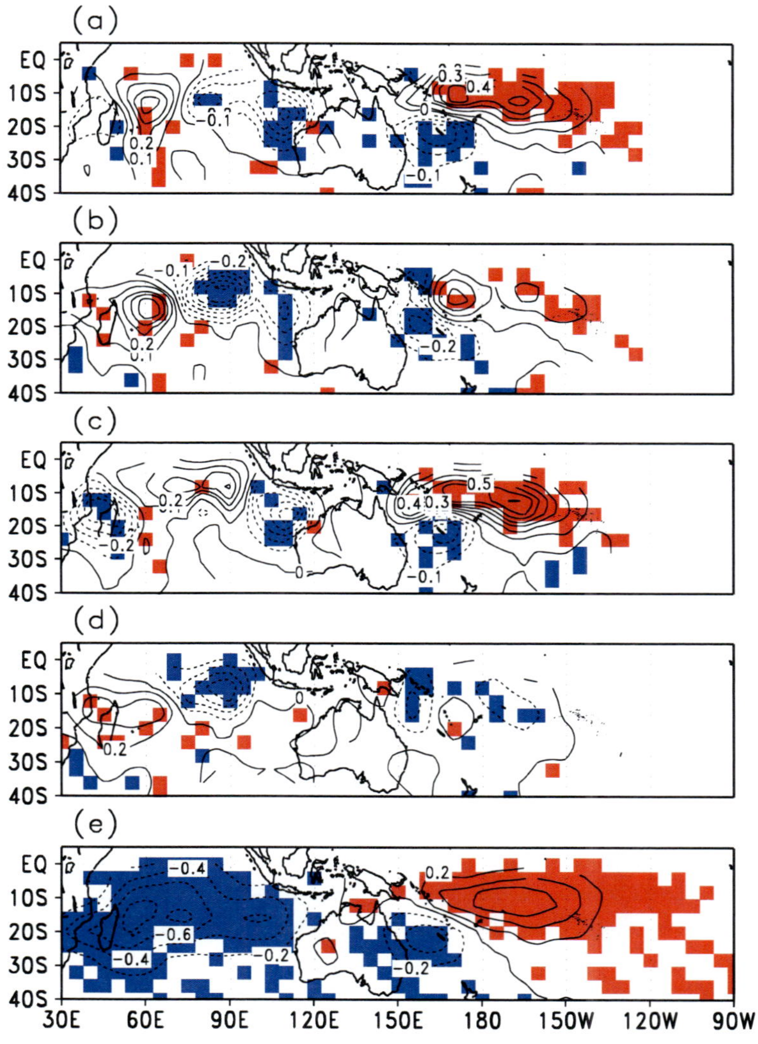

Figure 9. (a) Regression coefficients of TC days with October–December averaged Niño34 index. (b) Same as in (a) but with August–October averaged DMI. (c) Partial regression coefficients of TC days on October–December averaged Niño34 index independent of August–October averaged DMI. (d) Same as in (c) but on August–October averaged DMI independent of October–December averaged Niño34 index. (e) Same as in (a) but for the model. The contour interval is 0.2 days. Positive (negative) values where the t-test is statistically significant at the 90% confidence level are shaded as red (blue) color.

To assess the contribution of individual ENSO and IOD events "as independent variables" to the TC days, we again use a partial regression technique [29]. The spatial distribution of partial regression coefficients of TC days with DMI after taking away the effect of ENSO shows the east–west dipole structure. The reduction (enhancement) of TC days over the SEI domain during positive (negative) IOD events is associated with the anomalous descent (ascent) in response to local SST cooling (warming) (Fig. 9d). The opposite change is observed over the SWI domain. It is noteworthy that the TC days tend to increase (decrease) around Madagascar during positive (negative) IOD events (Fig. 9d) whereas they tend to decrease (increase) during El Niño (La Niña) years in absence of IOD events (Fig. 9c). This may cause the ambiguous relation between ENSO and TC activity over the southwestern Indian Ocean [28].

Intensity

The relation of the TC activity over the South Pacific and around Australia with ENSO in the model shows general agreement with the observational evidence. Therefore, we now examine the impact of ENSO on TC intensity. Here we define weak and intense model TCs, respectively, as model TCs with maximum surface winds less and more than 33 m/s. This threshold is based on the Saffir-Simpson scale. It is noted that there are no TCs with maximum surface winds more than 43 m/s, belonging to the category 2 on the Saffir-Simpson scale, because the horizontal resolution of the model has no ability to simulate such strong TCs. However, using the same threshold, Iizuka and Matsuura [18] obtained the relation between ENSO and TC intensity in the western North Pacific that is similar to the observed one. For observation, weak TCs are defined as TCs reaching maximum surface winds below category 3, whereas TCs belonging to categories 3–5 are defined as intense TCs.

Figure 10 shows scatter diagrams for the observed annual number of total TCs, weak TCs, intense TCs, and the ratio of the annual number of intense TCs to total TCs, against Niño34 index, over the AUS and ESP domains. Observations show an increase (decrease) in the annual TC frequency over the ESP domain in El Niño (La Niña) years, but the annual TC number is lower (higher) over the AUS domain, as reported in a previous study [15]. Similar relations have been found for intense TC numbers. However, there are not significant relations for the ratio of the annual number of intense TCs to total TCs over both the AUS and ESP domains (Figs. 10d and 10h). To confirm the relations in observations, we next examine the corresponding correlations in the model. The correlation coefficients in the model are similar to the observed ones (Fig. 11). This supports the observed relation that the changes in environmental conditions associated with ENSO contribute to TC genesis frequency over the South Pacific, but that they do not affect the ratio of the annual number of intense TCs to total TCs.

This differs from the response of TC activity over the western North Pacific to ENSO [12, 13]. In El Niño years, the mean TC genesis locations over the western North Pacific shift southeastward in association with low-level vorticity anomalies accompanied by anomalous westerlies induced by warm SST anomalies in the central and eastern equatorial Pacific. These changes are similar to those observed over the South Pacific. When comparing the South Pacific and the North Atlantic, however, the increase in vertical shear over the region with more TC tracks in El Niño years (east of Japan) is relatively small (not shown). This small change may be the reason why TCs over the western North Pacific are more intense and have longer lifetimes than normal during El Niño years.

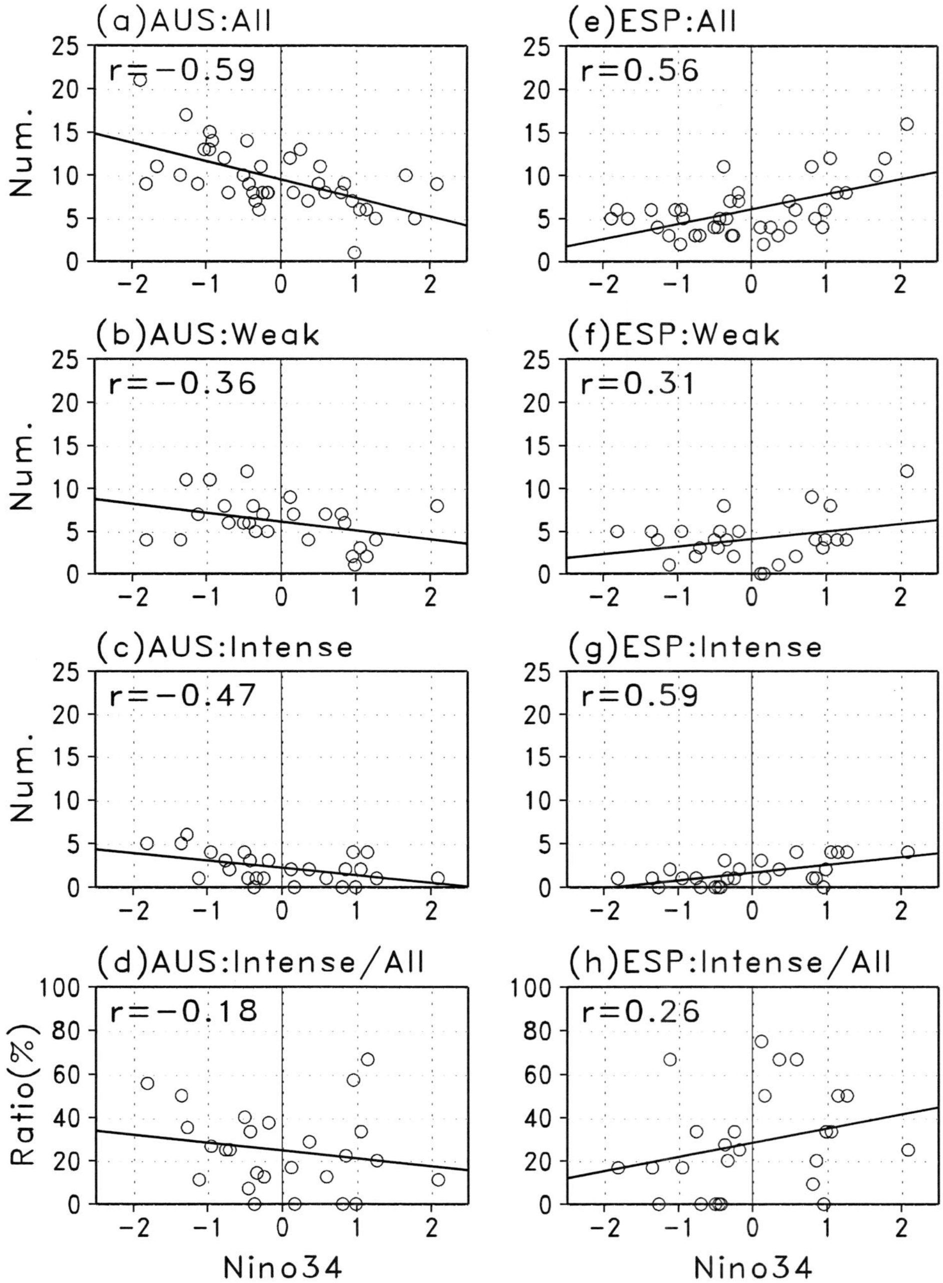

Figure 10. Scatter diagrams for observed annual number of TC (a), weak TC (b), intense TC (c), and ratio of intense TC to total TC number (d), generated over the AUS, with October−December averaged Niño34 index. (e)−(h) Same as in (a)−(d) but for ESP domain. Linear regression lines are shown as solid lines.

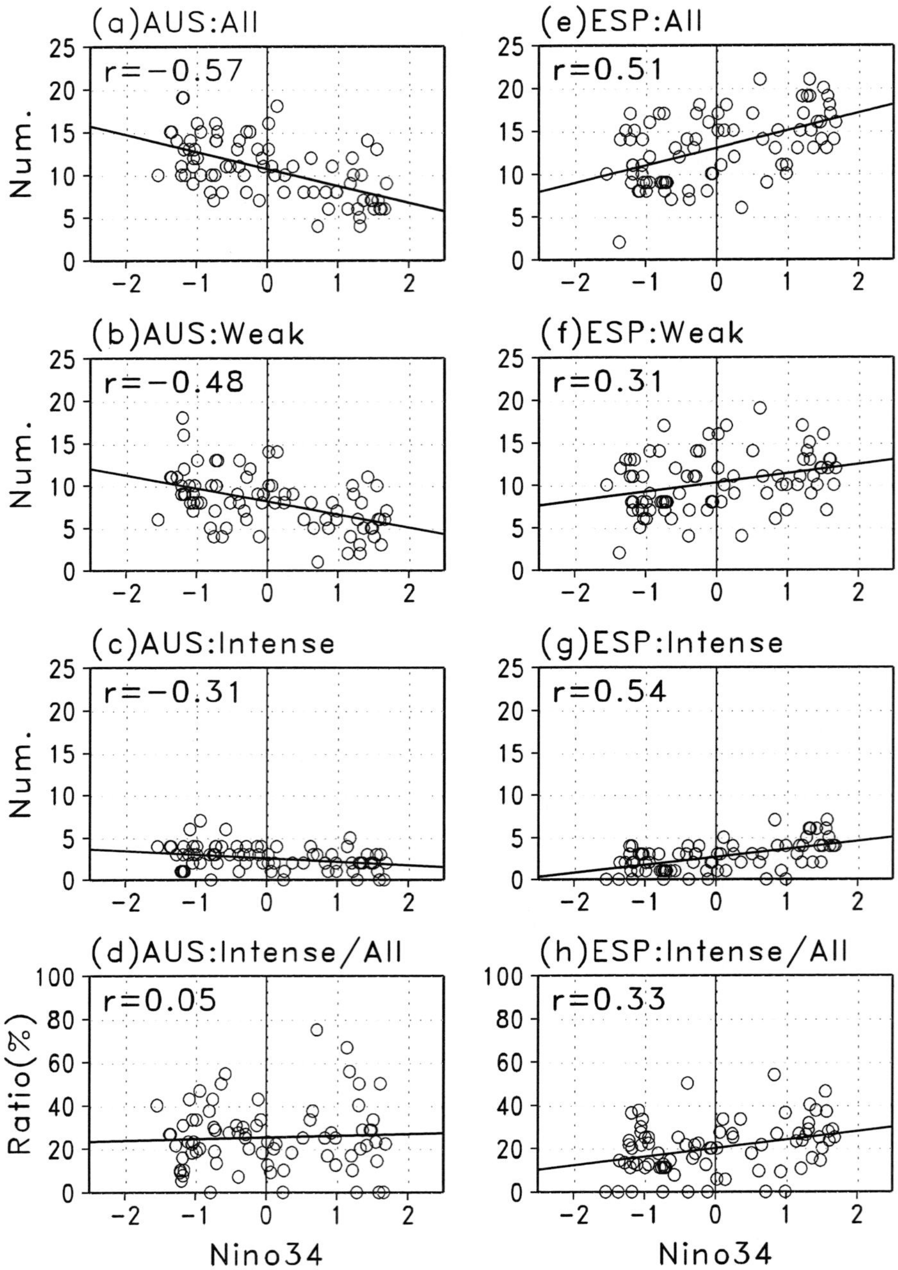

Figure 11. Same as in Fig. 10, but for the model.

Thus, the vertical wind shear may be attributed to an increase in the annual intense TC frequency on the interannual time scale, although warm SSTs play an important role in the increase in intense TC frequency under global warming [1–4].

We also examine the relation between the observed annual total of TCs, weak TCs, intense TCs, along with the ratio of the annual number of intense TCs to total TCs, against the Niño34 index as well as DMI, over the southern Indian Ocean. Because of the severe bias in the TC activity over the southern Indian Ocean in the model, we present only the relation based on observed data. Significant correlation at the 95% confidence level is found only in the annual number of total TCs and weak TCs over the SEI against the Niño34 index (Fig. 12).

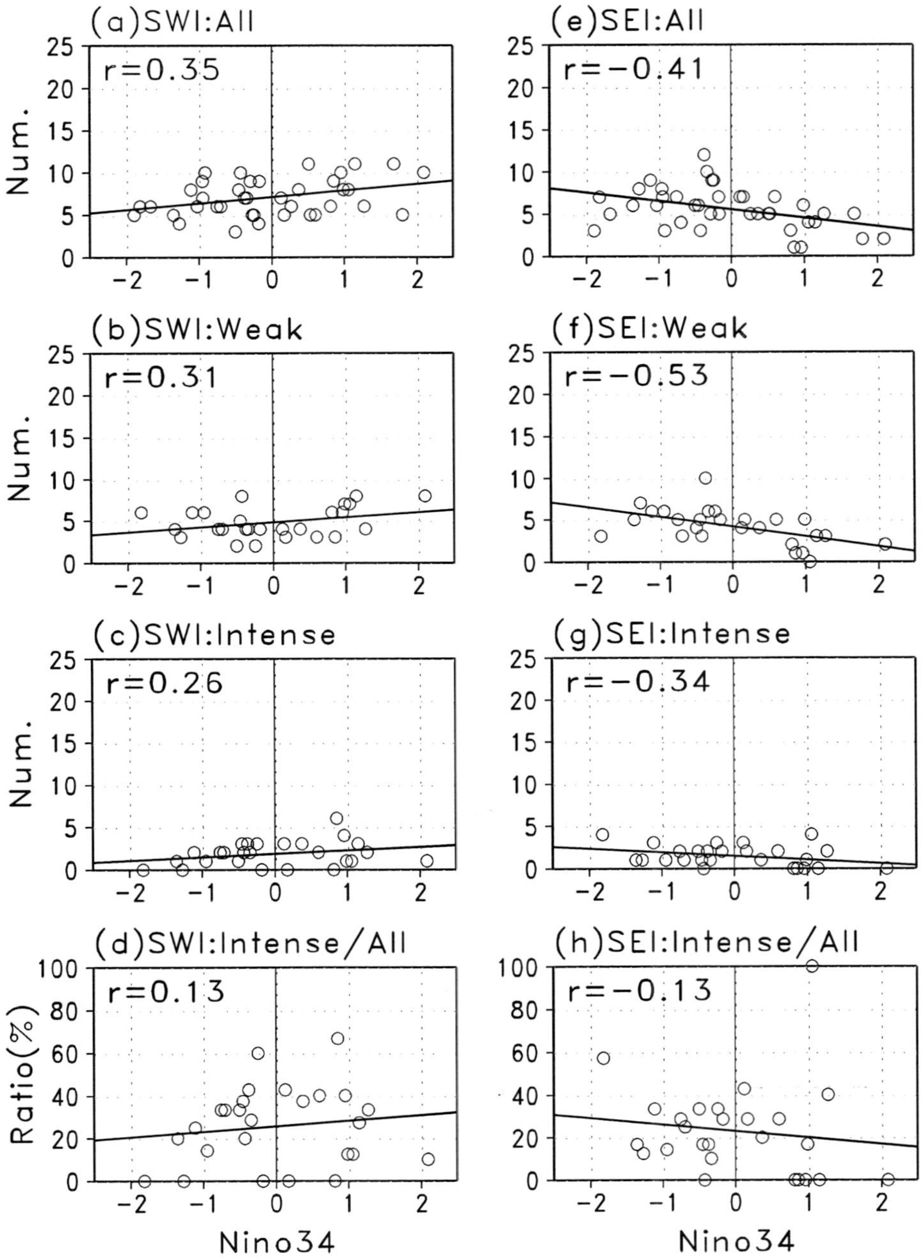

Figure 12. Scatter diagrams for observed annual number of TC (a), weak TC (b), intense TC (c), and ratio of intense TC to total TC number (d), generated over the SWI, with the October−December averaged Niño34 index. (e)−(h) Same as in (a)−(d), but for SEI domain. Linear regression lines are shown as solid lines.

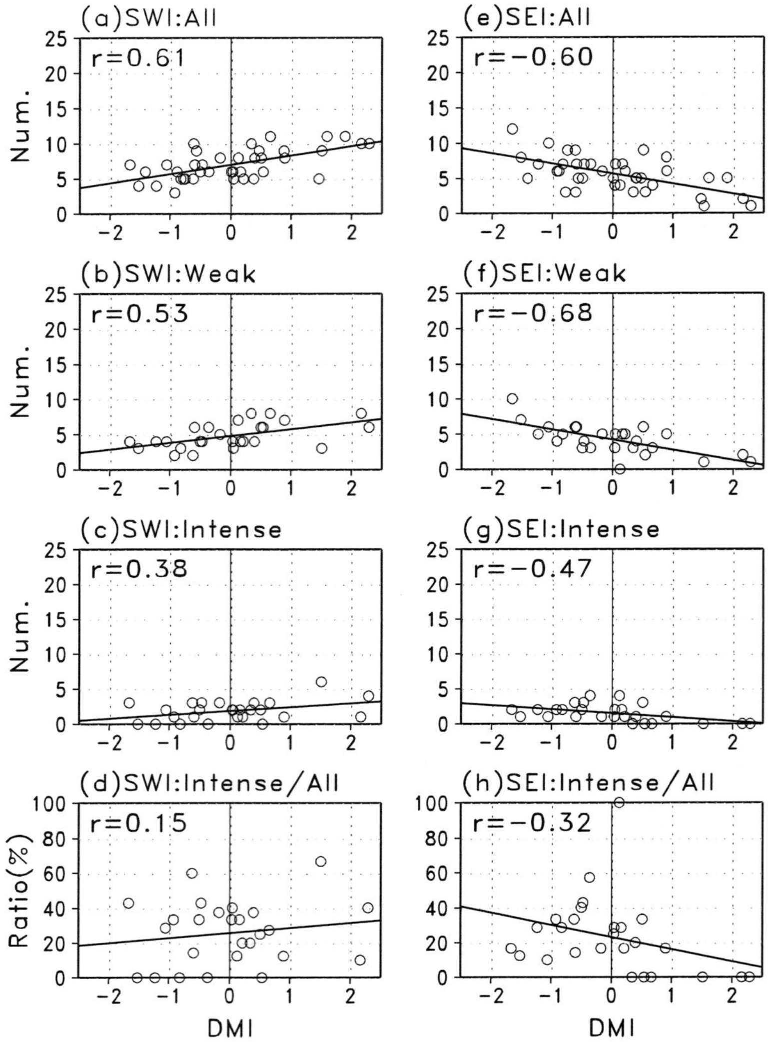

Figure 13. Same as in Figure 12, but with August−October averaged DMI.

However, significant correlation at the 95% confidence level is found in the annual number of total TCs, weak TCs, and intense TCs over the SEI against DMI (Fig. 13), which suggests that the DMI might become more useful index than ENSO indices to predict seasonal TC genesis frequency statistically over the eastern part of the southern Indian Ocean. The DMI also becomes a more useful predictor for annual TC frequency over the SWI domain than ENSO indices (cf. Figs. 12a and 13a). However, there are no significant relations for the ratio of the annual number of intense TCs to total TCs over both the SWI and SEI domains (Figs. 13d and 13h). Consequently, the changes in environmental conditions associated with IOD events seem not to affect TC in the development of their intensity. This is partly because a concurrent increase in vertical wind shear suppresses TCs to develop and sustain their intensity when more TCs are generated over the southwestern Indian Ocean in association with local SST warming [28]. However, further assessment of the relation between TC activity over the southern Indian Ocean and IOD must be undertaken using additional data.

SUMMARY

In this chapter, we described the impact of El Niño / Southern Oscillation (ENSO) on the activity of tropical cyclones (TCs) in the Southern Hemisphere using the best track data, along with the outputs simulated in a high-resolution (T213) coupled ocean-atmosphere general circulation model (CGCM). Both the observations and simulations show that, in El Niño (La Niña) years, the annual frequency of TCs in the northeast quadrant of the South Pacific increases (decreases), although it decreases (increases) in the southwest quadrant in the vicinity of the eastern coast of Australia. The increase (decrease) in TCs in the northeast quadrant is related to negative (positive) low-level vorticity anomalies associated with the anomalous westerlies (easterlies) induced by warm (cold) sea surface temperature anomalies in the central and eastern equatorial Pacific. An increase (decrease) in vertical wind shear associated with El Niño (La Niña) events contributes to the reduction (enhancement) of TC genesis frequency in the southeast quadrant of the South Pacific. The change in TC genesis frequency around Australia seems to be attributable to changes in the vertical wind shear and anomalous low-level convergence accompanied by anomalous upper-level divergence. These changes in the annual TC frequency over the South Pacific are accompanied by those in weak and intense TC frequency, but there are not significant relations for the ratio of the annual number of intense TCs to total TCs with ENSO. Therefore, the changes in environmental conditions associated with ENSO contribute to TC genesis frequency over the South Pacific, but they do not affect TCs in the development of their intensity.

We also found that the annual TC frequency observed over the eastern and western parts of the southern Indian Ocean is related more closely to the Indian Ocean Dipole (IOD) events than ENSO. During positive IOD events, the TC frequency generated in the early TC season is reduced, whereas an increase in TC frequency is identified during negative IOD events over the eastern part of the southern Indian Ocean in association with the anomalous descent in response to local SST cooling. The TC genesis frequency and associated TC days in the western part of the southern Indian Ocean tend to increase (decrease) in positive (negative) IOD events in association with the anomalous ascent (descent). However, there are not significant relations for the ratio of the annual number of intense TCs to total TCs because a concurrent increase in vertical wind shear suppresses TCs to develop and sustain their intensity.

Although this high-resolution CGCM is reasonably successful in simulating the relation between ENSO and TC activity over the South Pacific as well as the North Atlantic and western North Pacific, there are the discrepancies in the relation between ENSO and TC activity over the southern Indian Ocean simulated in the model. The model also fails to simulate the Madden–Julian Oscillation (MJO) which is known as an important mode for TC genesis, although the model simulates the convectively equatorial Rossby wave–like intraseasonal oscillation with sub-monthly time scales. Recently, a global cloud-resolving model was shown to be successful in simulating the MJO and associated TCs [30]. Thus, it is expected that the use of a finer-resolution global cloud-resolving model can be a powerful tool for elucidating TC activity on various time scales.

ACKNOWLEDGMENT

This research was supported partially by the Research Project of "Hot Spot of Midlatitude Air–Sea Interaction" and grants-in-aid (22310111) from the Japanese Ministry of Education, Culture, Sports, Science, and Technology, from the Global Environment Research Project Fund (S–5–3) of the Ministry of the Environment, Japan, and from the Research Project under the National Research Institute for Earth Science and Disaster Prevention.

REFERENCES

[1] K. A. Emanuel, *Nature* 326, 483 (1987).
[2] G. J. Holland, *J. Atmos. Sci.* 54, 2519-2541 (1997).
[3] T. R. Knutson, R. E. Tuleya, and Y. Kurihara, *Science* 279, 1018-1020 (1998).
[4] K. Oouchi, J. Yoshimura, H. Yoshimura, R. Mizuta, S. Kusunoki, and A. Noda, *J. Meteorol. Soc. Japan*, 84, 259-276 (2006).
[5] K. A. Emanuel, *Nature* 438, 686-688 (2005).
[6] P. J. Webster, G. J. Holland, J. A. Curry, and H.-R. Chang, *Science* 309, 1844-1846 (2005).
[7] J. C. L. Chan, *Science* 311, 1713b (2006).
[8] W. M. Gray, *Mon. Wea. Rev.* 96, 669-700 (1968).
[9] P. S. Chu, *In Hurricanes and Typhoons: Past, Present and Future*, edited by R. J. Murnane and K.-B. Liu, pp. 297–332, Columbia Univ. Press, New York (2004).
[10] W. M. Gray, *Mon. Wea. Rev.* 112, 1649-1668 (1984).
[11] L. J. Shapiro, *Mon. Wea. Rev.* 115, 2598-2614 (1987).
[12] B. Wang, and J. C. L. Chan, *J. Clim.* 15, 3252-3265 (2002).
[13] S. J. Camargo, and A. H. Sobel, *J. Clim.* 18, 2996-3006 (2005).
[14] Y. Kuleshov, F. C. Ming, L. Qi, I. Chouaibou, C. Hoareau, and F. Roux, *Ann. Geophys.* 27, 2523-2538 (2009).
[15] N. Nicholls, *Mon. Wea. Rev.* 107, 1221-1224 (1979).
[16] M. C. Wu, K. H. Yeung, and W. L. Chang, *EOS* 87, 48, 537-538 (2006).
[17] F. Vitart, *Quart. J. Roy. Meteor. Soc.* 132, 647-666 (2006).
[18] S. Iizuka, and T. Matsuura, *Clm. Dyn.* 30, 815-830 (2008).
[19] S. Iizuka, and T. Matsuura, *In Hurricanes and Climate Change*, edited by J. B. Elsner and T. H. Jagger, pp. 323–328, DOI: 10.1007/978-0-387-09410-6_17, Springer (2009).
[20] E. Kalnay, M. Kanamitsu, R. Kistler, W. Collins, D. Deaven, L. Gandin, M. Iredell, S. Saha, G. White, J. Woollen, Y. Zhu, A. Leetmaa, B. Reynolds, M. Chelliah, W. Ebisuzaki, W. Higgins, J. Janowiak, K. C. Mo, C. Ropelewski, J. Wang, R. Jenne, D. Joseph, *Bull. Amer. Meteorolo. Soc.* 77, 437–471 (1996).
[21] T. M. Smith, R. W. Reynolds, T. C. Peterson, and J. Lawrimore, *J. Clim.* 21, 2283-2296 (2008).
[22] J. L. McBride, and R. Zehr, *J. Atmos. Sci.* 38, 1132-1151 (1982).
[23] M. Sugi, A. Noda, and N. Sato, *J. Meteorol. Soc. Japan* 80, 249-272 (2002).
[24] R. E. Basher, and X. Zheng, *J. Climate* 8, 1249-1260 (1995).

[25] C. H. Ho, J. H. Kim, J. H. Jeong, and H. S. Kim, *J. Geophys. Res.* 111, D22191, doi:10.1029/2006JD007289 (2006).
[26] S.-P. Xie, H. Annamalai, F. A. Schott, and J. P. McCreary, *J. Climate,* 15, 864–878 (2002).
[27] N. H. Saji, B. N. Goswami, P. N. Vinayachandran, and T. Yamagata, *Nature* 401, 360–363 (1999).
[28] M. R. Jury, *Meteorol. Atmos. Phys.* 51, 101-115 (1993).
[29] N. H. Saji, B., and T. Yamagata, *Clim. Res.* 25, 151-169 (2003)
[30] H. Fudeyasu, Y. Wang, M. Satoh, T. Nasuno, H. Miura, and W. Yanase, *Geophys. Res. Lett.* 35, L22808, doi:10.1029/2008GL036003 (2008).

In: Cyclones: Formation, Triggers and Control
Editors: K. Oouchi and H. Fudeyasu
ISBN: 978-1-61942-976-5
©2012 Nova Science Publishers, Inc.

Chapter 3

SYNOPTIC AND CLIMATIC ASPECTS OF TROPICAL CYCLOGENESIS IN WESTERN NORTH PACIFIC

Tim Li*

International Pacific Research Center and Department of Meteorology, University of Hawaii at Manoa, Honolulu, Hawaii, US

ABSTRACT

In this chapter I first describe various precursor synoptic signals prior to tropical cyclone genesis in the western North Pacific. Then I discuss the origin of summertime synoptic-scale wave trains in the western Pacific and the energy source and dispersion characteristics of Pacific easterly waves. Next I present two cyclogenesis modeling results, focusing on the energy dispersion of a preexisting tropical cyclone and the genesis efficiency of initial mid-level versus bottom vortex. The climatologic aspect of tropical cyclogenesis is discussed, with a focus on the large-scale control of the atmospheric intraseasonal oscillation (ISO) and El Nino-Southern Oscillation (ENSO). A methodology is developed to rank key cyclogenesis parameters at different basins. Finally I describe the robust signals of future TC projection in the North Pacific from high-resolution global model simulations.

1. INTRODUCTION

Tropical cyclone (TC) genesis is a process through which random cumulus-scale cloud systems are well organized under favorable large-scale conditions and develop into a warm-core, cyclonic system with sustainable winds. Due to the lack of reliable data over open oceans and complicated scale interactions involved, our understanding of TC genesis remains very limited.

The western North Pacific (WNP) is the most frequent TC genesis region in the globe. The summer mean circulation in the WNP has some unique features. For example, the WNP is the region of the greatest warm pool in boreal summer, with averaged sea surface

* E-mail address: timli@hawaii.edu

temperature (SST) being greater than 29°C. The low-level circulation is characterized by a confluence zone between the monsoon westerly and the trade easterly. A meridional shear line along 150°E, 5°N and 120°E, 20°N separates the westerly to the south and the easterly to the north. The maximum low-level convergence and cyclonic vorticity overlap with intense convection, forming the WNP monsoon trough. Such a mean flow feature differs distinctively from that in the North Atlantic (NATL).

While the summer mean circulation provides favorable environmental conditions for cyclogenesis, it is synoptic-scale disturbances that trigger individual cyclogenesis events. Recent advances in satellite observations make it possible to reveal detailed synoptic-scale circulation patterns associated with tropical cyclogenesis. In section 2, synoptic-scale precursor disturbances prior to the formation of TC in the WNP will be described based on the analysis of satellite and NCEP-NCAR reanalysis data. Different from the NATL where the major perturbation source is Africa easterly waves, precursor perturbations in the WNP include TC energy dispersion (TCED) induced Rossby wave trains, northwest-southeast oriented synoptic wave trains (SWT), and Pacific easterly waves (PEW).

The origins of the synoptic wave trains and Pacific easterly waves are discussed in section 3. This is followed by the numerical modeling of cyclogenesis in section 4. First processes through which TCED induced Rossby wave train interacting with a monsoon gyre leads to the formation of a new TC are discussed. Next, given an environmental mid-level or bottom vortex initially, how the TC-scale vortex is developed and organized from random cumulus-scale clouds is further diagnosed with cloud-resolving model simulations. The genesis efficiency under the environmental mid-level and bottom vortex is compared.

From a climatologic aspect, what determines the TC genesis frequency in WNP? Can the same set of parameters determine the cyclogenesis in both the WNP and NATL? In section 5 the large-scale control of TC genesis by the atmospheric ISO and ENSO is discussed first. A methodology is then introduced to quantitatively measure the relative importance of genesis parameters in the NATL and WNP.

How Pacific TC frequency changes under global warming is addressed in section 6 through the diagnosis of two high-resolution global atmospheric general circulations models (AGCMs), forced with different future SST warming patterns. Both the models project a marked west-east shift in the North Pacific, with a large decrease (increase) of TC genesis number in the western (central) Pacific. The mechanism responsible for the distinctive regional TC changes is further discussed.

2. Precusor Synoptic Signals Prior to Cyclogenesis

The structure and evolution characteristics of tropical perturbations prior to TC formation in the WNP during 2000-2001 summers were investigated through the analysis of QuikSCAT, TMI and other observational data. Thirty-four TC genesis events were identified in the boreal summers of 2000 and 2001. Among 34 TCs, six (20%) were associated with TCED, eleven cases (30%) were associated with SWT (including those originated from equatorial mixed Rossby-gravity waves), and seven cases (20%) were associated with PEW. In the following I describe these and additional genesis scenarios.

A) Cyclogenesis in the Wake of a Preexisting TC Due to TCED

The first type synoptic-scale disturbance associated with TC genesis in the WNP is Rossby wave energy dispersion from a preexisting TC (TCED). A vortex is subject to Rossby wave energy dispersion even in a simple barotropic framework. While a vortex moves northwestward due to the mean flow steering and the planetary vorticity gradient, Rossby waves emit energy southeastward [1-5]. As a result, a Rossby wave train with alternating anticyclonic and cyclonic vorticity perturbations forms in its wake [6-7]. A study of three-dimensional (3D) Rossby wave energy dispersion showed that TCED-induced Rossby wave train has a baroclinic structure and that the upper-level wave train develops rapidly and triggers downward energy propagation that enhances the lower level Rossby wave train [8].

Cyclogenesis due to TCED was suggested previously [9-12] without a detailed description of the wave train structure and evolution. Using the satellite data, Li et al. showed the Rossby wave train pattern and its connection with cyclogenesis [13]. Among the 34 cases identified in the summers of 2000-2001, six cyclogenesis cases are associated with TCED. Figure 1 illustrates an example. It shows the evolution of the 3-8 day filtered QuikSCAT wind fields from 1 August to 9 August 2000. Letter "A" in Figure 1 represents the location of Typhoon Jelawat that formed on August 1. During the first 2-3 days, due to its weak intensity, Jelawat did not generate a visible Rossby wave train in its wake. During its northwestward journey, the intensity of Jelawat increased. With its steady intensification, the Rossby wave train became more and more evident in its wake. On 6 August, a clear wave train was observed. This Rossby wave train had a zonal wavelength of about 2500 km, oriented in a northwest-southeast direction.

The most remarkable characteristic of this wave train is that it had a large meridional wavelength. A half meridional wavelength was about 4000 km on 6 August. The meridional wavelength of the wave train was greatly reduced in subsequent days, leading to generation of a new TC on 9 August. Note that this new TC Ewiniar (represented by letter "B" in Figure 1) formed in the cyclonic vorticity center of the Rossby wave train.

The vertical profile of the Rossby wave train produced by TCED was examined with 3-8-day filtered NCEP-NCAR reanalysis data at different levels [14]. There was no clear wave pattern at 200hPa, but the wave train was evident at middle and lower troposphere with little vertical tilt. The cyclonic circulation region of the wave train coincided well with the region of concentrated cloud liquid water, whereas the anticyclonic circulation region of the wave train was cloud-free.

It was noted that some TCs did not reproduce a Rossby wave train in their wakes. Whether or not a TC generates a Rossby wave train depends on TC size or intensity [15,16], its radial wind profile [17], and its environmental flow such as vertical shear [18]. Whether or not a new TC can form in the Rossby wave train depends on the background flow the wave train is embedded in. A larger environmental low-level cyclonic vorticity and/or a larger environmental low-level convergence may favor cyclogenesis in the Rossby wave train [16, 19].

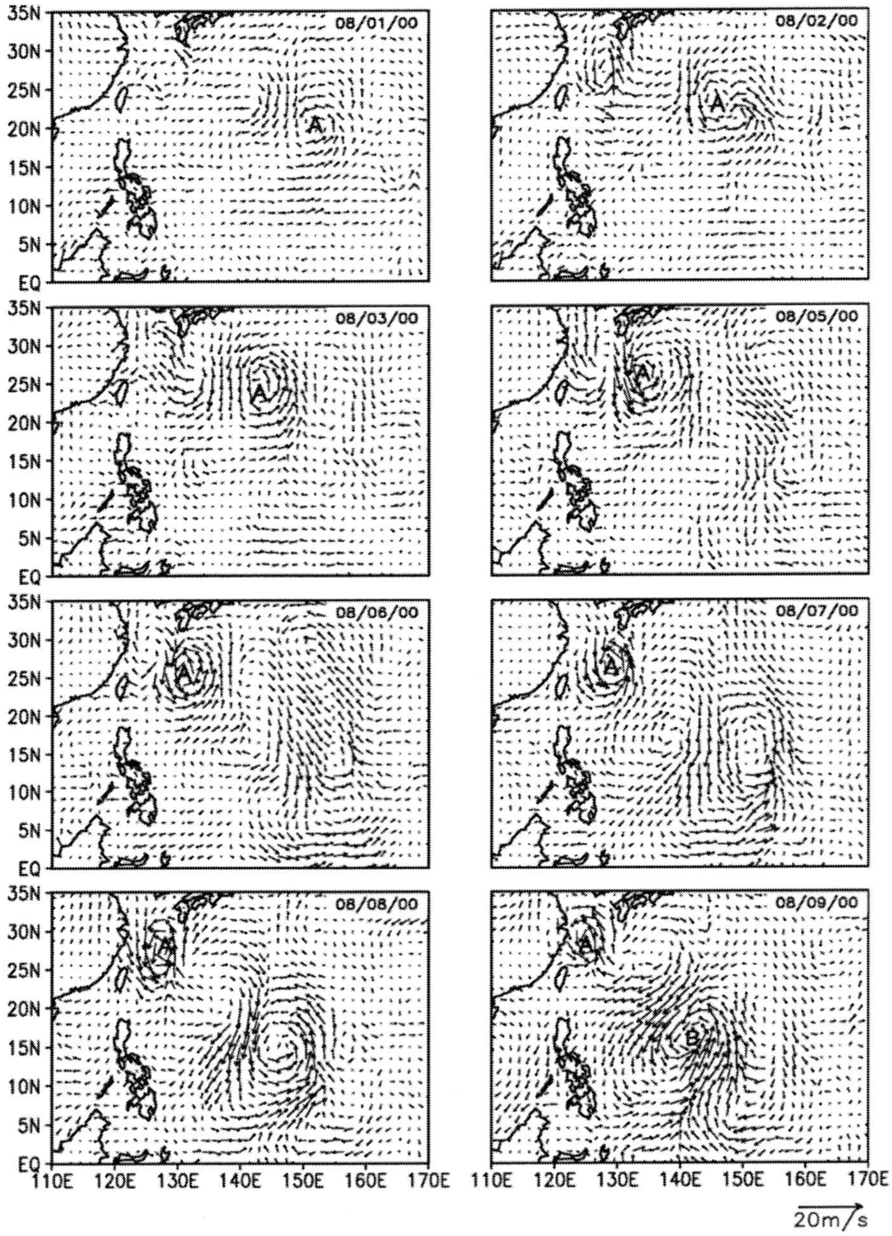

From [14], (c)American Meteorological Society, reprinted with permission.

Figure 1. Time sequences of synoptic-scale surface wind patterns associated with the Rossby wave energy dispersion of Typhoon Jelawat. "A" represents the center location of Jelawat that formed on 1 August 2000. "B" represents the center location of a new TC named Ewiniar that formed on 9 August 2000 in the wake of the Rossby wave train of Jelawat.

B) TC Genesis Associated with SWT

The second type of TC genesis happens within a synoptic wave train (SWT) that does not involve a preexisting TC. An observational study by Lau and Lau showed that the dominant synoptic-scale mode in the summertime WNP is a northwest-southeast oriented wave train, which has a typical wavelength of ~2500 km and propagates northwestward [20]. The intensity of the SWT increases as they move northwestward from the lower latitudes to the WNP.

During 2000 and 2001 summers, there were eleven TCs that formed in the SWT. An example is illustrated here for the formation of Typhoon Manyi, formed on 2 August 2001 (Figure 2). Three days prior to its genesis, a northwest-southeast oriented wave train developed in the WNP. This wave train had well-defined cyclonic and anticyclonic circulations, and covered a region between 130°E-160°E and 0°N-25°N. The wave train moved slowly northwestward. On 2 August, Manyi developed in the cyclonic vorticity region of the wave train.

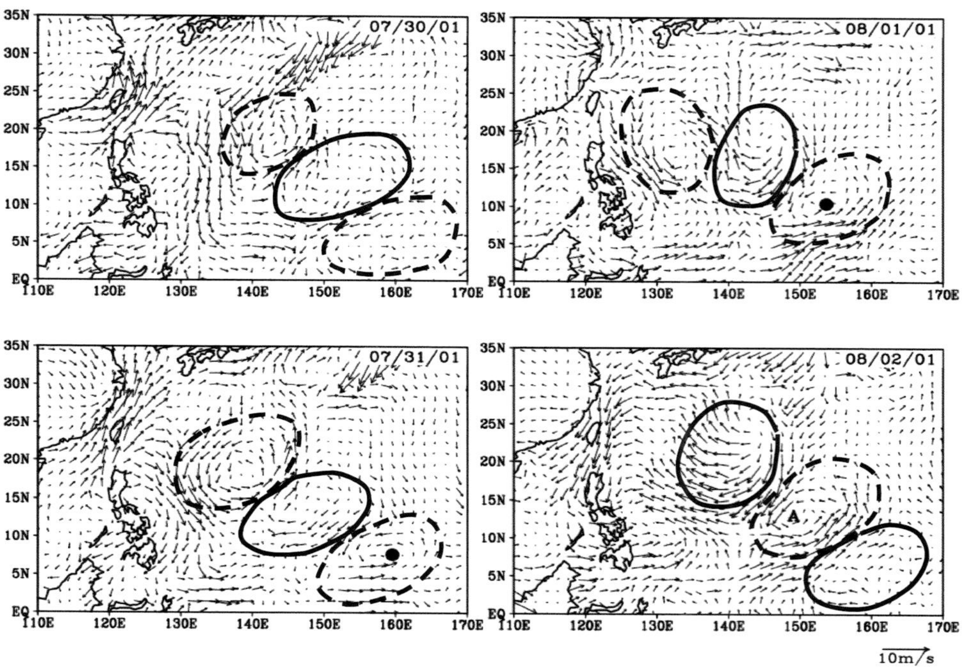

From [14], (c)American Meteorological Society, reprinted with permission.

Figure 2. Time evolution of 3-8-day filtered QuikSCAT surface wind patterns associated with synoptic wave train. The dots indicate the center location of cyclonic circulation prior to genesis of typhoon "Man-yi"; "A" indicates the center location of typhoon "Man-yi" on the day of genesis. Cyclonic/anticyclonic circulation is marked with dashed/solid circle.

Among the eleven SWT-induced cyclogenesis cases, four were associated with equatorial MRG waves. The MRG wave has an equatorially anti-symmetric structure with pronounced clockwise cross-equatorial flow. An example is given for Typhoon Sepat (see Figure 7 of [14]). An equatorial MRG wave appeared on 24 August 2001 near the date line. As its phase propagates westward, the group velocity of the MRG wave was eastward. This led to the development of the second MRG wave to its east, which had counter-clockwise flow. Meanwhile, the original MRG wave deviated to the north of the equator. This led to the development of a northwest-southeast oriented wave train in the WNP. The wave train was clearly seen in the 3-8-day filtered wind field on 26 August. As the wave train continued to develop, a new TC named Sepat formed in the cyclonic vorticity region of the wave train on 27 August. Such a transition process from an equatorial MRG wave to off-equator tropical depression (TD)-type disturbances is consistent with previous studies [21,22].

C) TC Genesis Associated with Pacific Easterly Waves

The third type of cyclogenesis in the WNP is associated with the development of Pacific easterly waves (PEW). Different form African easterly waves, the energy source of the PEW comes from mid-latitude Jet (a specific discussion of this subject will be given in section 3).

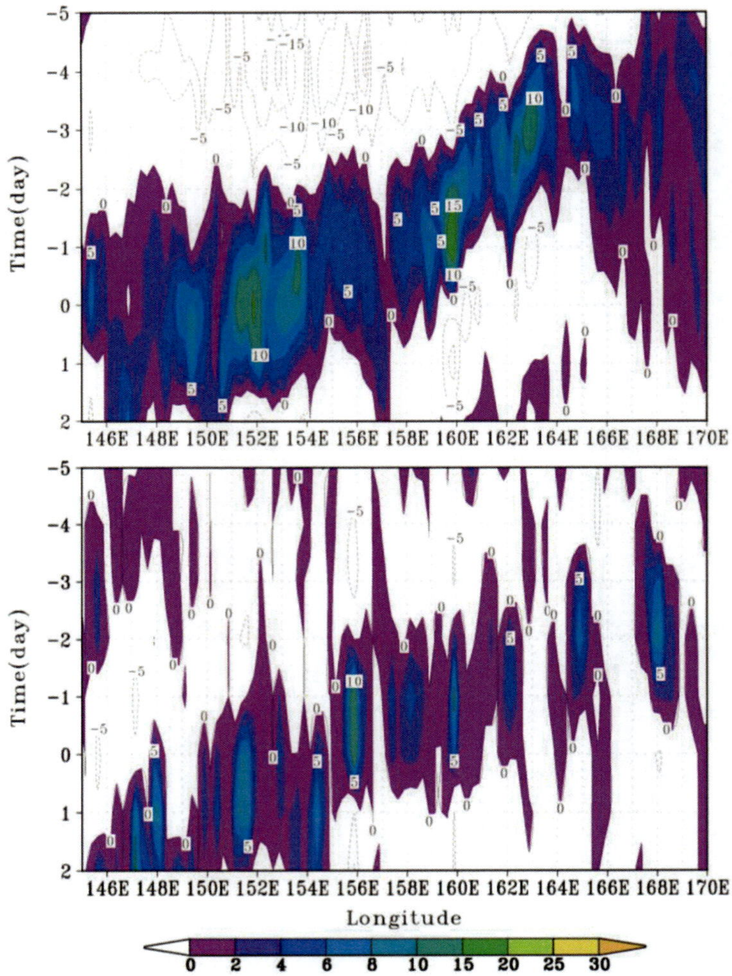

From [14], (c)American Meteorological Society, reprinted with permission.

Figure 3. Time-longitude profiles of the surface kinetic energy $(u^2+v^2)'/2$ (upper panel, units: m^2s^{-2}) and precipitation rate (lower panel, units: mm day^{-1}) along 25°N. The horizontal axis is longitude and the vertical axis is time in days. Typhoon Kong-rey formed at 150.2°E, 25°N on 22 Jul 2001 (corresponding to day 0). Both panels show clear westward-propagating signals prior to the birth of Kong-rey.

During 2000-2001 summers, seven PEW-induced cyclogenesis cases were identified. Only cyclogenesis cases that were associated with clear westward propagation signals in both the perturbation kinetic energy and rainfall fields derived from the TMI data were considered as PEW-induced cyclogenesis cases.

An example of PEW induced TC genesis is given for Typhoon Kong-rey, which formed on 22 July 2001. Figure 3 shows the time-longitude cross-section of the perturbation kinetic energy and the precipitation rate along the latitude 25°N where Kong-rey formed. Both the precipitation and the perturbation energy showed clear westward propagation prior to the genesis of Kong-rey. The wave signals could be traced back 4 to 5 days prior to Kong-rey's formation. The phase speed of the easterly wave was estimated around 4-5 longitude degrees per day.

The vertical cross-section of the easterly wave derived from 3-8-day filtered NCEP-NCAR reanalysis data showed that this easterly wave was quite deep, and had an equivalent barotropic structure [14]. The wave amplitude had two maxima at 600-700hPa and 200-300hPa respectively. This vertical profile was consistent with previous observational studies of easterly wave structures [23].

D) Precursor Signals in the Middle or Upper Troposphere

The previous three types of genesis scenarios are primarily related to low level precursor signals. It was found that some TCs did not have significant surface precursor signals; rather they had precursor signals in the middle or upper troposphere.

One such example is Typhoon Usagi, which formed on 9 August 2001. Figure 4 shows that this cyclogenesis event was associated with preexisting cloud activity. Three days prior to the genesis of Usagi, clouds represented by TMI cloud liquid water content had already appeared over the region where Usagi formed. However, from the 3-8-day filtered QuikSCAT wind fields, there was no clear cyclonic wind perturbation at the surface.

A further analysis of the NCEP final analysis product showed that 2-3 days prior to the TC genesis, a mid-level cyclonic vortex had already set up. The mid-level vortex was accompanied with a weak warm core above and a cold core below, and eventually led to the cyclone development near the surface. This mid-level vortex triggering process has been suggested by Simpson et al. [24], although specific processes that lead to the downward development of cyclonic vorticity are not clear. Numerical simulations in section 4 will further explore this cyclogenesis process.

Some TC genesis events in the WNP may be related to upper tropospheric forcing. Colon examined the evolution of 200hPa flows and found that a poleward flow aloft, such as those in the eastern side of troughs in the westerlies or those in the western side of anticyclones, is a favorable environmental condition for the development of low-level perturbations [25]. Sadler developed a synoptic model to describe the role of the Tropical Upper Tropospheric Trough (TUTT) [26]. He suggested that TUTT might have three effects on TC development. Firstly, the accompanying sub-equatorial ridge on the south side of TUTT lies over the low-level low that may decrease the vertical shear; secondly, a divergent flow on the south and east side of the cyclonic circulation in TUTT may increase the ventilation aloft to help the development of a low-level low; thirdly, a channel to large-scale westerlies may be established for efficient outflow of the heat released by increased convection in the developing depression.

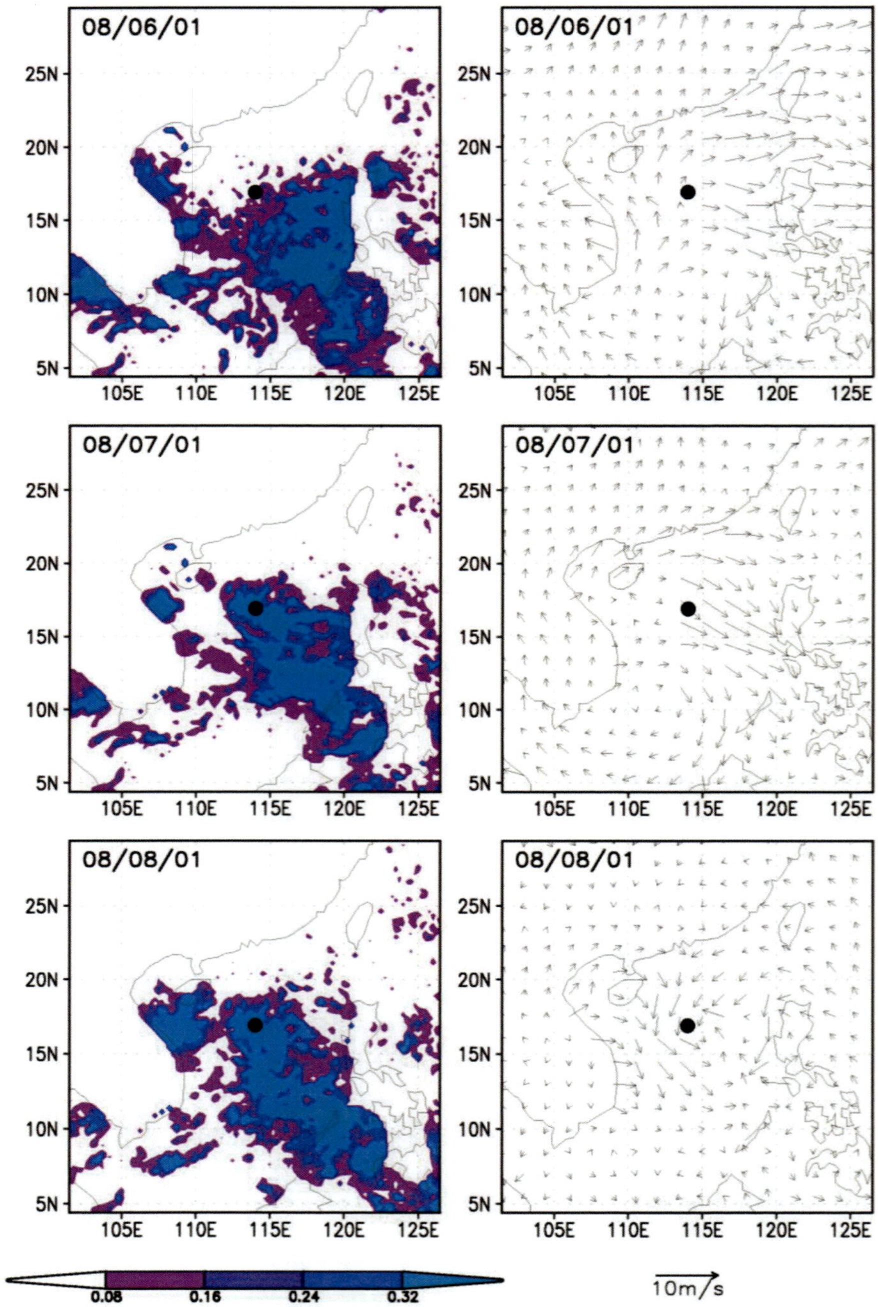

From [14].

Figure 4 Time evolution of TMI cloud liquid water (left) and 3-8-day filtered QuikSCAT surface wind (right) prior to the genesis of TC Usagi. Usagi formed on 9 Aug 2001. The dots indicate the position where Usagi formed.

During 2000-2001 summers, three genesis events were associated with the upper tropospheric forcing [14]. An example is Typhoon Nari, which formed on 6 September 2001. Two days prior to its formation, an upper-level trough is located approximately 20 degrees west of the genesis location. One day later, this trough moved eastward 10 longitude degrees. In the mean time, this upper-level trough deepened, and the southwesterly flow on the east side of the trough became stronger. On the genesis date, this trough was within 5 degrees to the northwest of Nari's genesis position. As a result, strong southwesterlies dominated in this area, and an outflow channel was set up in upper levels north of Nari. This led to the enhancement of low-level disturbances and was possibly responsible for Nari's genesis.

3. Origin of WNP Synoptic-Scale Wave Trains and Easterly Waves

Dominant synoptic-scale variability in the WNP is a northwest-southeast oriented synoptic-scale wave train [20]. The wave train extends toward the southeast with a wave-like pattern that consists of alternating regions of cyclonic and anticyclonic circulations, propagating northwestward. The wave train has a typical wavelength of ~2500 km and a time scale of 6-10 days.

What causes the SWT in the WNP? Li hypothesized that the SWT is an unstable mode of the summer mean flow in the presence of the convection-circulation-moisture (CCM) feedback [27]. To test this hypothesis, an anomaly atmospheric general circulation model (AGCM) was employed. The model was modified from the original version of Princeton global spectrum atmospheric model [28]. It was linearized by a specified 3D basic state (see the appendix of [27] for the detailed derivation of the model and the treatment of circulation-dependent perturbation heating), similar to that of Ting and Yu [29]. Previously this anomaly model was used in studying an asymmetric atmospheric response to equatorially symmetric forcing [30] and MJO initiation in the western Indian Ocean [31].

The model used sigma $(\sigma = p/p_s)$ as its vertical coordinate, and vertically there were 5 evenly distributed sigma levels, with a top level at σ=0 and a bottom level at σ=1. The horizontal resolution was T42. Rayleigh friction was applied to the momentum equations, with the damping rate of 1 day^{-1} taken in the lowest model level $(\sigma = 0.9)$ to mimic the planetary boundary layer (PBL), linearly decaying to 0.1 day^{-1} at the level of $\sigma = 0.7$. Newtonian cooling with an e-folding time scale of 10 days was applied to the temperature equation at all the model levels. Because the focus of the study was on tropical perturbations, a strong damping of 1 day^{-1} was applied in the perturbation momentum and temperature equations over higher latitude regions (beyond 40°N and 40°S). A realistic summer (JJA) mean state of $\overline{u}, \overline{v}, \overline{T}, \overline{P}_s$ was prescribed as the basic state, taken from the long-time mean of the NCEP-NCAR reanalysis.

Initially a small perturbation with a zonal wavenumber-9 structure was introduced. In the presence of both the summer mean flow and the convection-circulation-moisture feedback, the model captured the growth of the most unstable mode in WNP. The fastest growing mode in the model had a typical observed synoptic-scale wave train structure in WNP (see Figure

5a). The wave train, oriented in the northwest-southeast direction, had alternative cyclonic and anti-cyclonic circulation regions and a zonal wavelength of 2500 km. The preferred geographic location for the perturbation growth appeared in the WNP, primarily due to high moist static energy (MSE) in the region. This is because the background MSE condition would lead to the strongest CCM feedback, given the same perturbation vorticity. The unstable mode, with an eastward tilting vertical structure, propagated northwestward.

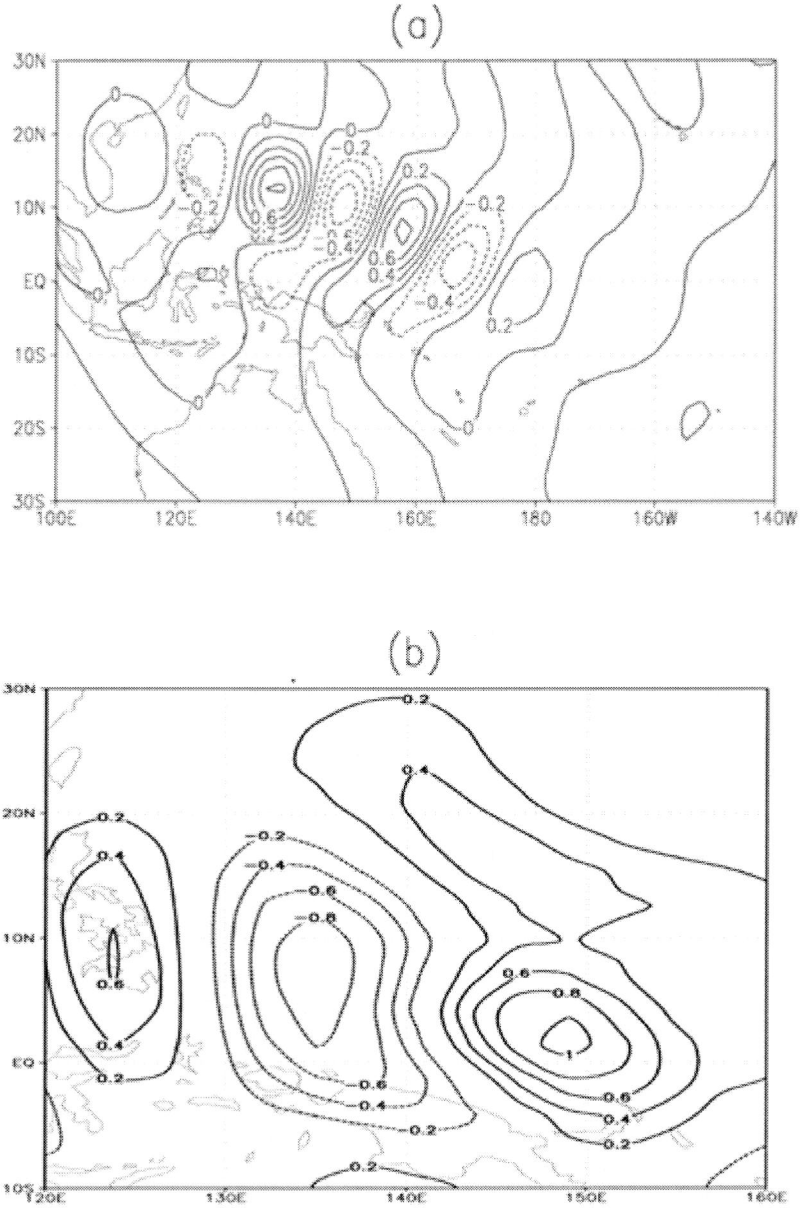

From [27], (c)American Meteorological Society, reprinted with permission.

Figure 5. Horizontal patterns of normalized meridional wind fields at σ =0.7 (contour interval: 0.2) associated with (a) the most unstable mode in the presence of the CMC feedback and (b) the least damped mode in the lack of the CMC feedback.

In the absence of the CCM feedback, the initial perturbation did not grow; rather it decayed. This indicated that the climatologic summer mean flow alone (without involving of the CCM feedback) was unable to destabilize tropical perturbations. Thus, the summer mean flow only supported a weakly damped mode in the tropics. This differs fundamentally from mid-latitude baroclinic instability waves, which gain energy directly from the mean westerly jet. The numerical result was consistent with the observational study by Lau and Lau, who pointed out that the primary heat source for the SWT was the latent heating associated with deep convection [32].

Even though the climatologic summer mean flow alone was unable to lead to the perturbation growth, it did play a role in determining the preferred length scale, wave train structure and propagation characteristics. Figure 5b illustrates the meridional wind structure of the least damped mode in the absence of the CCM feedback. A northwest-southeast oriented wave train pattern emerged. The wave train had alternative cyclonic and anti-cyclonic circulation regions, with a wavelength of about 2500 km. The weakly damped mode propagated toward the west and northwest.

Further sensitivity experiments with different initial conditions indicated that the final structure of the unstable mode was not sensitive to the initial conditions. In all cases the model reproduced the most unstable modes that had a synoptic wave train structure similar to that shown in Figure 5a. This supported the hypothesis that the summertime synoptic wave train in the WNP was a result of instability of the summer mean flow in the presence of the CCM feedback. Any types of perturbations under such a background condition could grow and develop into the wave train pattern.

To reveal the energy accumulation mechanism and dispersion characteristics of Pacific easterly waves, Tam and Li conducted an observational analysis using NCEP-NCAR reanalysis data [33]. A phase-independent wave-activity flux of Takaya and Nakamura [34] was applied to examine the wave activity and energy accumulation associated with PEW. The convergence of the wave activity flux was found in the region of 135-160E, 10-25N. Such convergence could lead to the amplification of Rossby waves, and its pattern was in agreement with the region of positive growth of the observed perturbations.

Negative zonal group velocity associated with the PEW was found everywhere in the WNP, indicative of Rossby wave energy moving to the west at low level. The zonal wavelength of the PEW was estimated. It was found that the zonal wavenumber, k, increased as the wave moves westward. The zonal scale contraction led to the convergence of the group velocity (hence wave activity). There was accumulation of wave activity in the low levels associated with a wave packet of the PEW. The largest contribution to the wave activity accumulation came from the convergence of intrinsic group velocity, which was attributed to the contraction of wavelength toward the west and the change of the horizontal tilt of waves. By comparison, convergence of the mean flow contributed less to the wave activity accumulation.

It was found that the phase relationship between convection and vorticity depended on longitudinal positions. Signals of enhanced convection were in phase with positive low-level vorticity anomalies west of about 150°E, but to the east the vorticity led the convection by one quarter of a wavelength. This indicated that there were two different dynamical regimes along the 'storm tracks' of synoptic-scale activity. In the eastern portion or at the entrance of the storm tracks, heat flux associated with synoptic-scale eddies was found to be equatorward (i.e. $v'T'<0$) in the mid troposphere. This could be seen in composite maps of circulation as well as eddy covariance statistics.

Associated with the negative heat flux signals was downward injection of wave activity from the upper levels. In the upper troposphere, southward wave activity from the extratropics was able to penetrate into the Tropics near the dateline. Figure 6 is a schematic diagram illustrating 3D energy propagation characteristics associated with PEW. The result suggested that the energy source of Pacific easterly waves originated from mid-latitude Pacific Jet. This hypothesis was supported by a case study in which upper-level activity was able to initiate a synoptic-scale wave train with strong surface circulation features. PV anomalies intruded into the Tropics due to wave breaking were seen to cause downward development, leading to low-level disturbances which subsequently moved westward and grew.

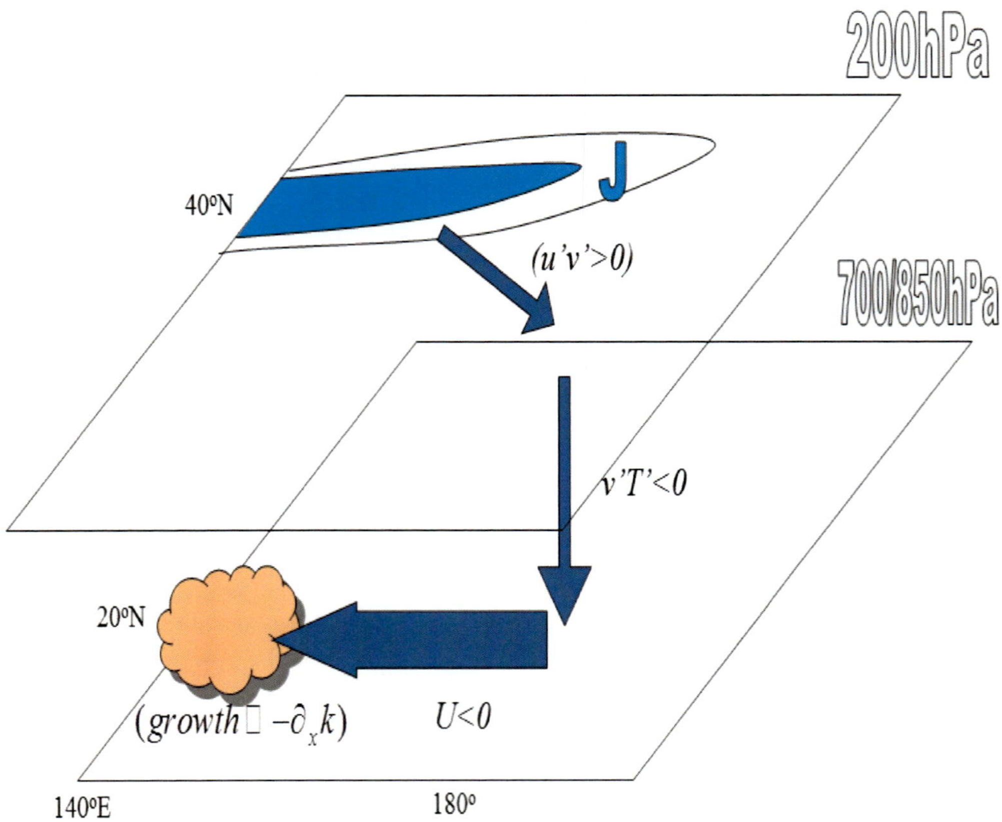

Figure 6. A schematic diagram illustrating the southward wave energy propagation from the mid-latitude Pacific Jet to tropical central Pacific, downward energy propagation in the tropical central Pacific, and westward wave activity flux in the lower troposphere over the western North Pacific. Both wave scale contraction and confluent mean flows contribute to the energy accumulation of easterly waves in the maximum perturbation growth region (135E-160E, 10-25N).

4. Numerical Simulations of Tropical Cyclogenesis

A) TCED Induced Cyclogenesis

The cyclogenesis associated with TC energy dispersion were simulated in a 3D model. The atmospheric model used in the study was the triply nested movable mesh primitive equation model (TCM3). A detailed description of the model and experiment design can be found in Li et al. [19]. An initial vortex was spun up on a beta-plane for 10 days. The environmental temperature and moisture profiles were the same as those in Wang [35] and represented the mean summertime conditions over the western Pacific. At the end of the 10-day integration, a 3-D Rossby wave train pattern developed. To study the mean flow effect, an idealized monsoon gyre (MG) pattern was specified. This MG had a size of 3000 km and a central minimum pressure 2-3 hPa lower than the surrounding environment. The maximum wind of the MG was about 5 m s^{-1} at a radius of 300 km from the center. Both the pattern and amplitude of the specified MG were quite close to the observed composite [36].

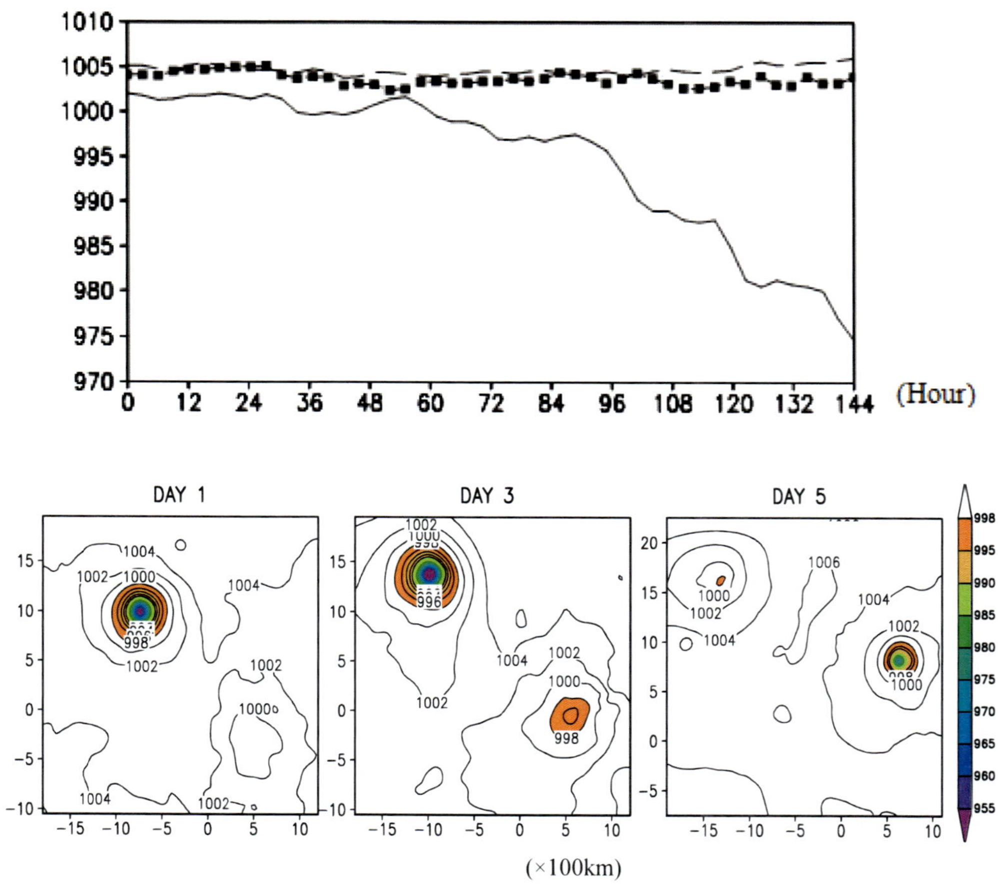

From [19], (c)American Meteorological Society, reprinted with permission.

Figure 7. Time evolution of minimum sea-level pressure fields (upper panel) for Exp1 (dashed line), Exp2 (dotted line) and Exp3 (solid line). The lower panels show the surface pressure (unit: hPa) patterns at day 1, 3 and 5 in Exp3.

The following three experiments were conducted. In the first experiment (Exp1), the model was initialized with the Rossby wave train only. This experiment was designed to examine whether the wave train alone can develop into a TC in a resting environment. In the second experiment (Exp2), how the background MG evolved with time was examined, without the involvement of the TCED-induced Rossby wave train. In the third experiment (Exp3), the Rossby wave train was superposed on the MG as the initial condition; it was designed to examine whether the wave train-MG interaction could lead to cyclogenesis.

The simulation results (Figure 7) showed that while the wave train alone or the MG alone did not cause cyclogenesis, the superposition of wave train and MG led to the generation of a new TC. This result suggested that a proper mean flow pattern was essential for the TCED-induced wave train to develop into a TC. The simulated new vortex exhibited realistic TC characteristics, including a tilted eyewall, a warm core, and a vigorous in-up-and-out mean secondary circulation with mean subsidence in the eye region.

Figure 8 shows the evolution of simulated vorticity, relative humidity, vertical velocity and rainfall rate (averaged over a 120 km by 120 km domain centered at the maximum low-level vorticity) in Exp3. Two important features in the vorticity field were worth noting. First, maximum vorticity generation occurred in the PBL, and with the TC development the cyclonic vorticity gradually penetrated from the PBL into the upper troposphere. Secondly, during the initial development stage, the PBL vorticity experienced an oscillatory growth. For instance, the vorticity grew rapidly between hours 60 and 64, slowed down or even decayed afterwards, and re-intensified at hours 72-78. The averaged rain rate also exhibited a clear oscillatory evolution with periods ranged from 9 to 15 hours. This oscillation feature was clearly seen in the vorticity or rainfall field averaged even over a larger domain (e.g., 240 km by 240 km).

The oscillatory vorticity development was closely related to the evolution of vertical motion and relative humidity fields. It can be seen from Figure 8 that there are alternations of updraft and downdraft, corresponding well to convectively "active" and "break" phases. While the ascending motion associated with penetrated convection led to PBL convergence and thus a positive vorticity tendency, the downdraft led to PBL divergence and thus a negative vorticity tendency. A vorticity budget analysis indicated that the generation of the PBL vorticity in the model was primarily attributed to the divergence term. The vertical motion regulated the moisture profile through vertical moisture transport, leading to the alternative change of relative humidity in mid-troposphere.

The oscillatory development found in the model was validated by observed rainfall and cloud-top temperature evolutions [19]. The diagnosis of the model output showed that the oscillatory development was attributed to the discharge and recharge of the PBL moisture and its interaction with convection and circulation. The moisture-convection feedback regulated the TC development through controlling the atmospheric stratification, raindrop-induced evaporative cooling and downdraft, PBL divergence, and vorticity generation. On one hand, ascending motion associated with deep convection transported moisture upward and led to the discharge of PBL moisture and a convectively stable stratification. On the other hand, the convection-induced raindrops evaporated, leading to mid-level cooling and downdraft. The downdraft further led to dryness and a reduction of equivalent potential temperature. This reduction along with the recharge of PBL moisture due to surface evaporation led to re-establishment of a convectively unstable stratification and thus new convection.

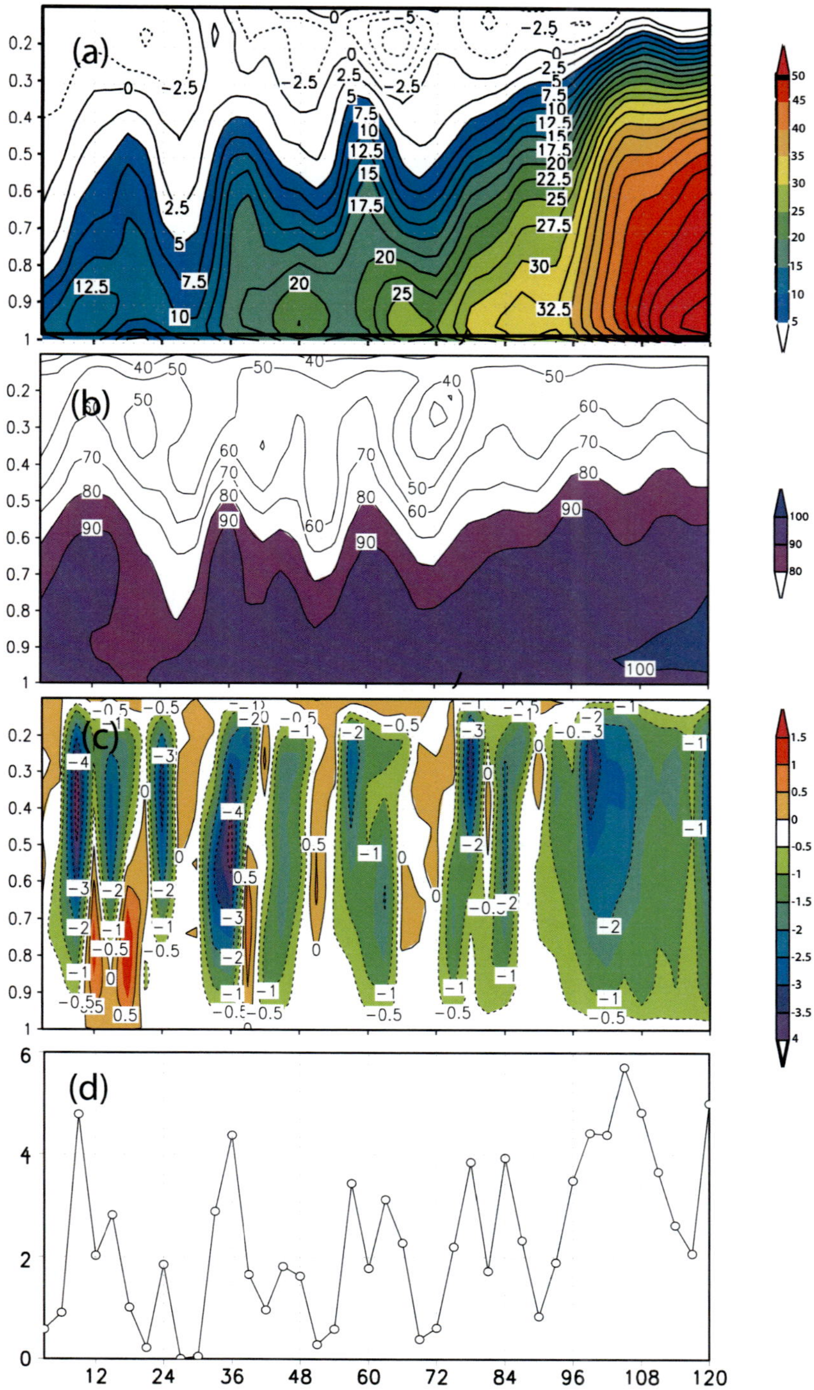

From [19], (c)American Meteorological Society, reprinted with permission.

Figure 8. Time evolution of (a) vorticity (unit: $10^{-5}\ s^{-1}$), (b) relative humidity (unit: %), (c) vertical motion (unit: $10^{-5}\ s^{-1}$) and (d) rainfall rate (unit: mm/hr) averaged over a 120 km by 120 km domain centered at the low-level maximum vorticity in Exp3. The horizontal axis is time (unit: hours) and the vertical axis in (a)-(c) is the sigma level.

B) Cyclogenesis Efficiency: Mid-Level vs. Bottom Vortex

TC genesis may occur either in an environment with a near bottom vortex (EBV) or an environment with a mid-level vortex (EMV). Which scenario leads to greater genesis efficiency? A cloud-resolving WRF model was used to address this question. The model was quadruply nested. The mesh sizes in all four domains were 181×181 with horizontal grid sizes of 54, 18, 6, and 2 km, respectively. There were 27 levels in the vertical. The Kain-Fritch convective scheme [37] was applied in the two outer meshes and an explicit microphysics scheme [38] was used in the finer meshes. Initial water vapor mixing ratio and other thermodynamic variables were assumed horizontally homogeneous and had the vertical profiles of typical January mean observations at Willis Island, northeast of Australia [39]. The model was set on an *f*-plane centered at 15°N, and a quiescent environment with a constant sea surface temperature (SST) of 29°C was specified. For more detailed model description, readers are referred to [40].

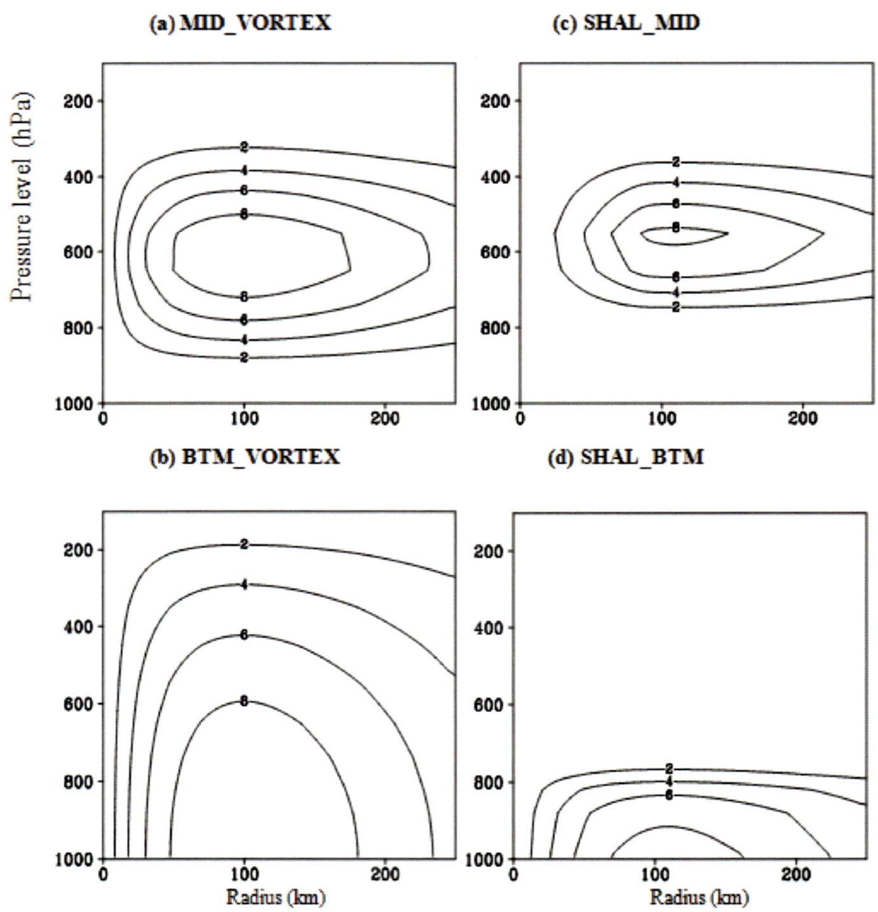

From [40].

Figure 9. The vertical-radial cross section of tangential velocity (ms^{-1}) of the initial vortex in (a) MID_VORTEX, (b) BTM_VORTEX, (c) SHAL_MID and (d) SHAL_BTM.

Four experiments were designed. The first experiment (MID_VORTEX) mimicked a mid-level precursor condition, in which initial vortex had a maximum vorticity at 600 hPa, with a maximum wind speed of 8 m s^{-1} at a radius of 100 km and a size of 500 km radius where the wind vanished. The vorticity gradually decreased both upward and downward, and vanished at the surface. In the second experiment (BTM_VORTEX), an initial maximum precursor perturbation with a maximum wind speed of 8 ms^{-1} was located at the surface. Figure 9a and Figure 9b show the vertical-radial cross section of the tangential wind of these initial vortexes. The third and the fourth experiment contained a shallow mid-level vortex (SHAL_MID) and a shallow bottom vortex (SHAL_BTM), respectively, to identify the PBL effects and to better separate the mid-level and the bottom vortices. Their vertical wind profiles are shown in Figure 9c and Figure 9d.

In all the four experiments, after integration of several days, a TC with realistic dynamic and thermodynamic structures formed. The simulated TC characteristics included a tilted eyewall, a vigorous in-up-and-out secondary circulation, and subsidence within the eyewall. The maximum tangential wind speed reached to 60 ms^{-1} at a radius of about 20 km. Strong inflow appeared in the PBL, with maximum inflow just outside the radius of the maximum tangential wind. A broad outflow layer appeared in the upper troposphere outside the eyewall.

A marked difference among these experiments was the timing of a rapid pressure drop, suggesting that each event possessed distinct genesis efficiency. Here the genesis time was defined as time when the maximum surface wind reached 17.5 ms^{-1} – a criterion for the tropical storm strength. Figure 10 shows the time evolutions of the corresponding maximum surface wind speed from the four experiments. Based on the definition above, the genesis time occurred at hour 33 for BTM_VORTEX, hour 66 for MID_VORTEX, hour 78 for SHAL_BTM, and hour 114 for SHAL_MID, respectively.

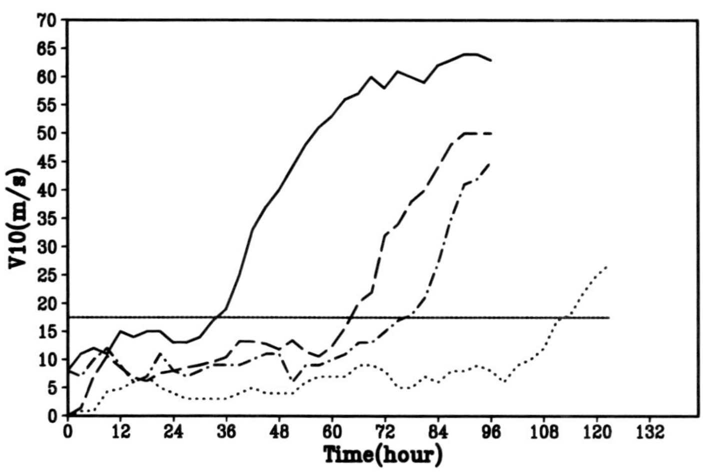

From [40].

Figure 10. Time evolution of the simulated maximum surface wind speed in BTM_VORTEX (solid line), MID_VORTEX (dashed line), SHAL_BTM (dot-dashed line) and SHAL_MID (dotted line). The horizontal dotted line in the bottom panel indicated the threshold of TC genesis when the wind speed reaches 17.5 ms^{-1}.

Figure 11 presents the vertical cross section of relative vorticity and temperature anomaly fields averaged over a 100km×100km domain centered at the minimum sea level pressure (MSLP) from the four experiments. In MID_VORTEX, the initial mid-level vorticity maximum coincided well with a cold (warm) core below (above) 600 hPa (top left panel of Figure 11), due to the thermal wind balance relationship. During the first 30 hours, the mid-level cyclonic vorticity gradually weakened and the "cold core" structure became less distinct. The mid-level vortex re-intensified and created a maximum vorticity center at 500 hPa at hour 48. Associated with this re-establishment of the mid-level vortex was the development of a cold-warm-core couplet and the mid-level convergence. The temperature anomaly (defined as the departure from the mean temperature over the innermost domain) showed a clear cold (warm) anomaly below (aloft) 600 hPa. A weak divergence appeared around 800 hPa before hour 60, implying descending motions and a relative dry layer between 600 and 800 hPa. The thermodynamic structure at hours 48-60 bore the characteristics of stratiform precipitation regime [41], indicating that stratiform cloud was dominated at this time.

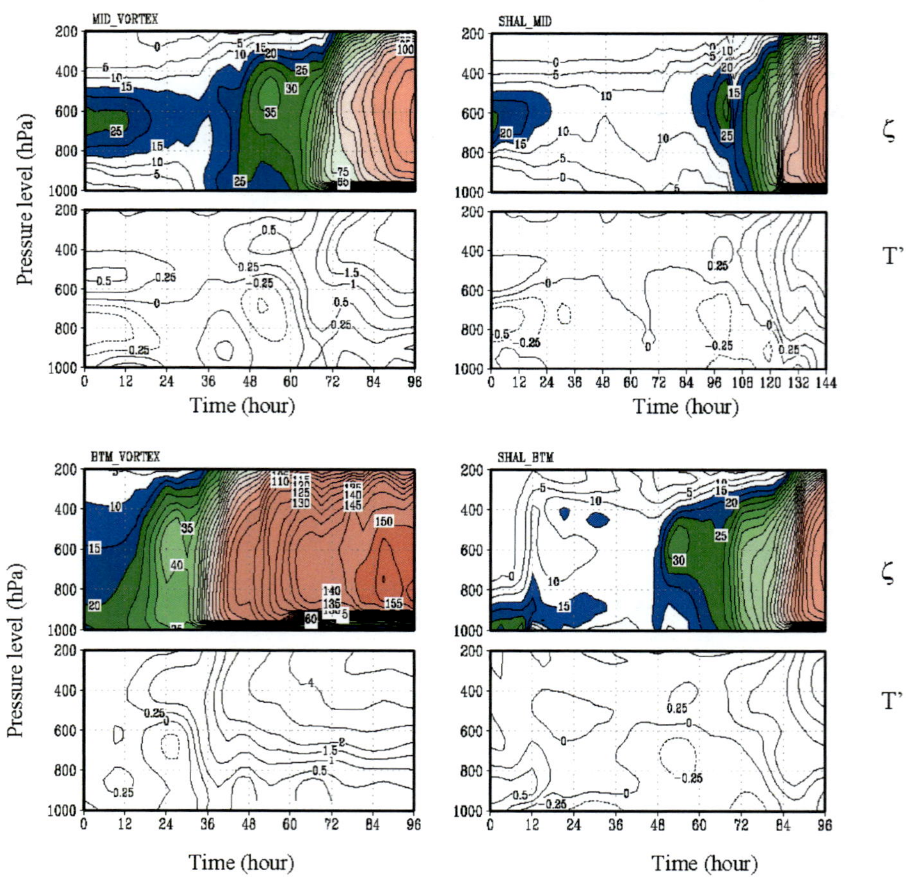

From [40].

Figure 11. Time-vertical (unit: hPa) cross section of relative vorticity ($1\times10^{-5}\ s^{-1}$) and temperature anomaly (K) averaged over a 100 km×100km domain centered at the MSLP in MID_VORTEX (top left), BTM_VORTEX (bottom left), SHAL_MID (top right) and SHAL_BTM (bottom right).

The similar plots for BTM_VORTEX were depicted in the bottom left panel of Figure 11. In this case, a rapid pressure drop occurred after hour 33, indicating faster TC development compared to MID_VORTEX. Although initially the maximum vorticity appeared at the low level, a mid-level vorticity enhancement was observed at hour 30. Associated with this mid-level vorticity enhancement was a cold-warm-core couplet (i.e., a cold core at lower level and a warm core above, bottom left panel of Figure 11). The temperature and vorticity profiles at that time were associated with a mid-level convergence and downdrafts below, reflecting stratiform cloud regime. Although the TC genesis was slower in the two shallow vortex cases, the vorticity and temperature evolution characteristics in SHAL_MID and SHAL_BTM (right panels of Figure 11) bore many similarities to their counterparts with deeper vorticity profiles.

A common feature among the EBV and EMV experiments was the formation of a mid-level maximum vorticity prior to TC genesis irrespective where the initial vorticity maximum located. A comparison of the BTM_VORTEX and MID_VORTEX experiments indicated that the former had greater genesis efficiency. A possible cause of this genesis efficiency difference was attributed to the initial column integrated absolute vorticity. Note that the former had greater column-integrated (from 1000 to 200 hPa) absolute vorticity than the latter. Physically, it was argued that vertically integrated absolute vorticity could affect vorticity segregation [42]. To examine additional factors that affected genesis efficiency, two shallow vortex cases (SHAL_BTM and SHAL_MID) were further considered, in which the vertically (1000-200 hPa) integrated absolute vorticity were almost same. The simulation results showed that the cyclogenesis time in SHAL_BTM occurred at hour 78, which was much faster than the genesis time (hour 114) in SHAL_MID. Thus the numerical model results indicated that given the same column integrated absolute vorticity, a bottom vortex was more efficient in cyclogenesis than a mid-level vortex. The difference was likely attributed to the effect of surface fluxes and PBL processes.

A common development characteristic among the four experiments was the occurrence of upscale cascade from randomly distributed cumulus-scale (~5 km) vertical hot towers (VHTs) to an organized meso-vortex-scale (~50-100 km) system (MVS). Sporadic cumulus-scale (~5 km) convective cells developed shortly after the integration started, due to the conditionally unstable environmental condition and the surface evaporation over the ocean. While the cumulus-scale (~5 km) VHTs were randomly triggered initially, they were gradually organized and merged into a MVS through merging around the environmental vorticity center. This warm-core MVS differed from conventional mid-latitude mesoscale convective system that has a distinct cold-core structure. As shown in Figure 12, in MID_VORTEX, several convective cells moved closer and merged into a larger vortex with the horizontal scale of about 50 km at 48 hour. A similar evolution feature was seen in BTM_VORTEX, although cumulus-scale VHTs and a MVS developed earlier than in MID_VORTEX.

The vorticity segregation proposed by Schecter and Dubin [42] is a possible mechanism for the VHTs merging. That is, cyclonic (anticyclonic) vorticity anomalies move up (down) the ambient vorticity gradient. As a result, cyclonic vorticity anomalies tend to move inward

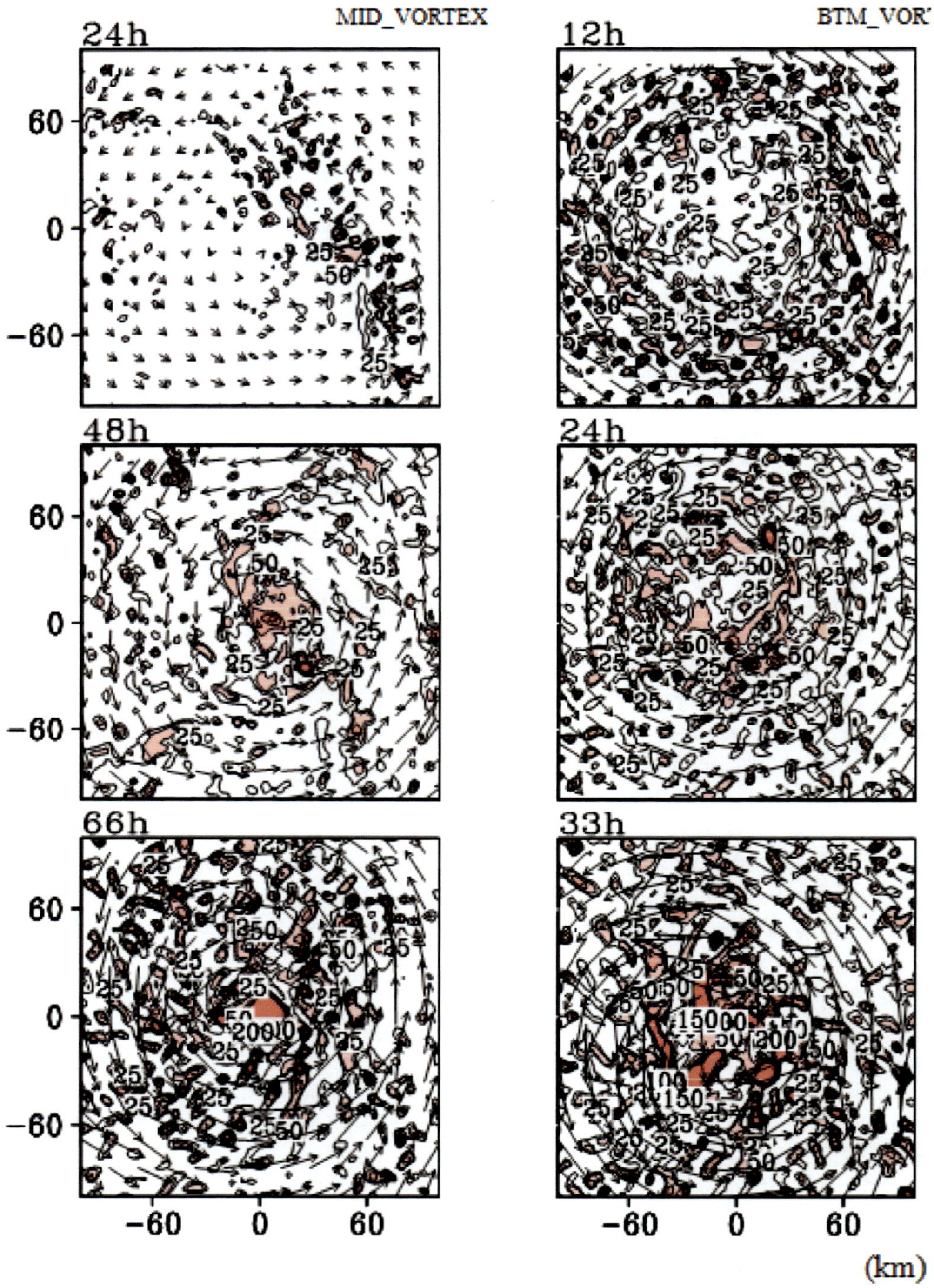

From [40].

Figure 12. Horizontal patterns of low-level (900hPa) wind field (vector) and relative vorticity (greater than $3\times10^{-4}s^{-1}$ are shaded) in MID_VORTEX (left) and BTM_VORTEX (right).

toward the vortex centre, while anticyclonic anomalies tend to move outward away from the centre. This vorticity segregation as well as vortex merger or axisymmetrization [43-45] could cause upscale cascade from cumulus-scale (~5 km) VHTs to a MVS. As seen from all experiments, the upscale cascade was the first key step toward cyclogenesis. Once the MVS system formed, the adjacent convective cells would be further absorbed, or organized into outer spiral rainbands.

Another common development feature was the setup of a deep moist layer prior to cyclogenesis. For instance, in MID_VORTEX, there was a steady increase of the moist layer in the core region during hour 24-60. The 90% RH layer thickened from 900 hPa at hour 24 to about 500 hPa at hour 60. It was the establishment of this near-saturation air column that signified the next development stage: deepening of cyclonic vorticity and a rapid drop of minimum sea level pressure. Such a deepening of the moist layer appeared clearly in the other three experiments.

5. Climatic Control of Cyclogenesis

In addition to a marked annual cycle, TC genesis frequency in the WNP experiences a significant intraseasonal variation [46-50]. Several studies investigated how the atmospheric intraseasonal oscillation (ISO) modulates TC activity [51-53]. For example, Maloney and Dickinson suggested that the modulation is through enhancing or suppressing the tropical depression (TD)-type disturbances in lower troposphere [53]. The enhancement of the TD-type disturbances is attributed to either Rossby wave accumulation caused by large-scale convergence [54,55] or barotropic energy conversion during ISO westerly phases [52,53].

The large-scale control of the ISO on the WNP cyclogenesis in 2000 and 2001 summers was examined by Fu et al. [14]. The pattern and the characteristics of the ISO were represented by the 20-70-day filtered OLR and 850hPa zonal wind fields. Both the wind and OLR fields were widely used to represent ISO activity in many previous studies. Observational analyses showed that the ISO was characterized by enhanced (suppressed) convection and westerly (easterly) in 850hPa zonal wind in its active (suppressed) phase. From a thermodynamic perspective, enhanced convection provided a greater large-scale low-level moisture convergence that favored the development of tropical depressions. From a dynamic perspective, the ISO westerly anomaly helped increase the cyclonic vorticity to the north and created favorable environment for tropical depression development. As a result, more TCs formed in the active phase of ISO.

The time series of area-averaged (5°N-25°N, 110°E-160°E) OLR perturbations are shown in the left panels of Figure 13. There were two ISO events in the WNP each year. The active phase (negative OLR perturbations) occurred from 28 June to 15 July and from 6 August to 5 September in 2000, and from 16 June to 6 July and 8 August to 26 August for year 2001. Figure 13 also shows area averaged OLR and zonal wind perturbations on the 34 cyclogenesis date in 2000 and 2001. It appears that most of TC geneses events in 2000 and 2001 were coincided with ISO active phases.

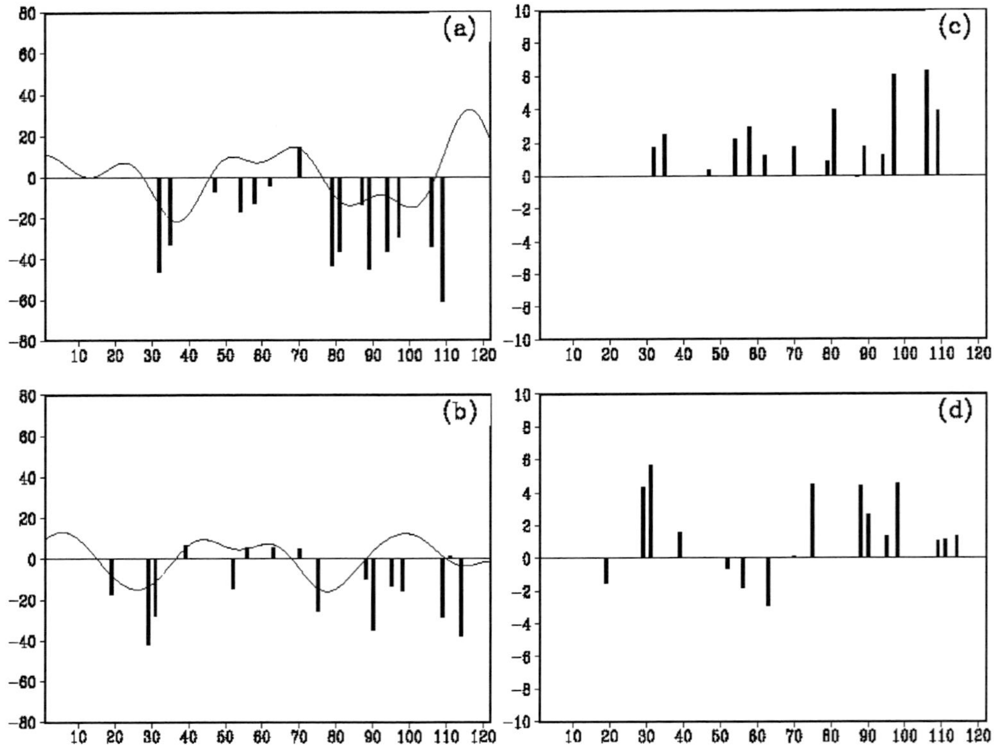

From [14], (c)American Meteorological Society, reprinted with permission.

Figure 13. The 20-70-day filtered OLR (unit: W/m^2) (left panels) and 850mb zonal wind (unit: m/s) (right panels) in 2000 (upper panels) and 2001 (lower panels) as a function of time (From 1 Jun to 30 Sep, totally 122 days). Each bars represents the OLR or zonal wind perturbations at the location of TC genesis when TCs occur. The curves in the left panels are time series of 20-70-day filtered OLR averaged in the domain (5°N-25°N, 110°E-160°E).

The relationship between the barotropic energy conversion and the TC genesis location during active and suppressed ISO phases are shown in Figure 14. Positive conversion appeared in both the ISO active and suppressed phases, indicating that synoptic-scale disturbances always obtained kinetic energy from the total background mean (including ISO) flow [56]. Regions of enhanced barotropic conversion favored the development of synoptic seed disturbances, which could further grow into TCs. It was noted that TCs tended to form in the positive barotropic energy conversion regions. The amplitude and spatial pattern of the energy conversion varied with the ISO phase. During the ISO active phase, larger energy conversion occurred over the WNP south of 25°N (Figure 14a). During the ISO suppressed phase, positive conversion with smaller amplitude appeared over the WNP (Figure 14b). The TC genesis number in the WNP was approximately proportional to the barotropic energy conversion. For example, the TC genesis number during the ISO active phase was 43, 1.5 times greater than that during the ISO suppressed phase (27). A similar ratio appeared in the barotropic energy conversion amplitude.

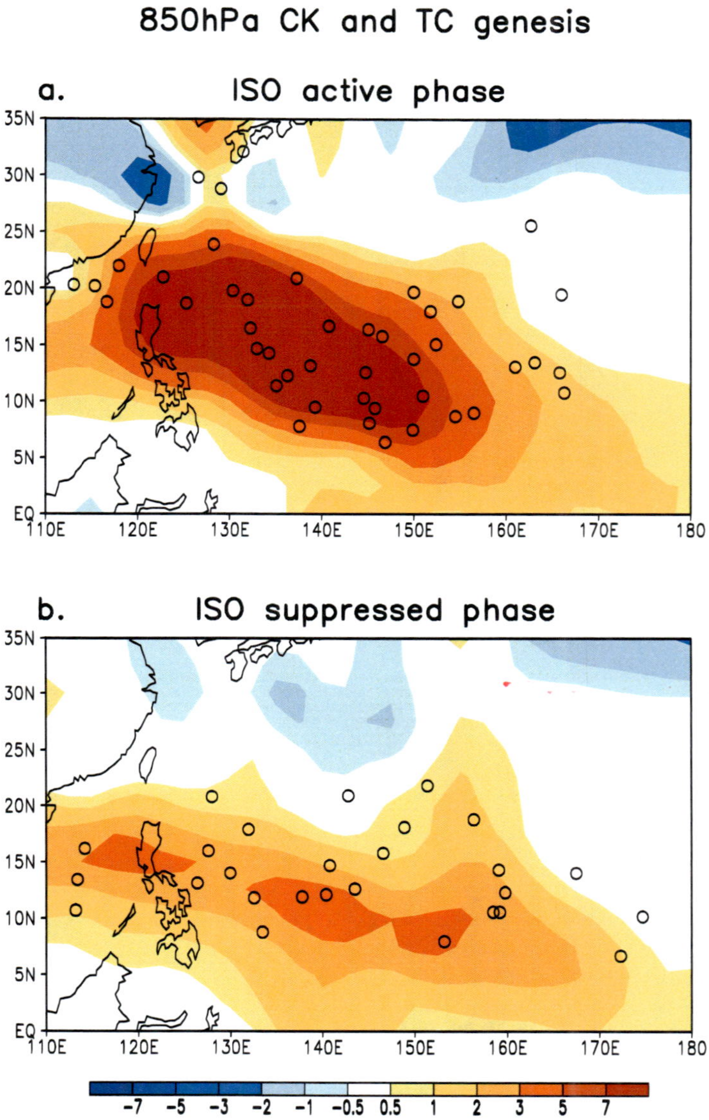

From [56], (c)American Meteorological Society, reprinted with permission.

Figure 14. Composites of 850-hPa barotropic energy conversion field (shading, unit: $10^{-5}\,m^2s^{-3}$) and TC genesis location (circle) during the ISO (a) active and (b) suppressed phase.

What determined the interannual variation of TC genesis number in the WNP? A calculation of the simultaneous correlation between JJAS TC number and Nino3.4 index during 1979-2010 showed a positive value, but the simultaneous correlations were not statistically significant, from both Hadley and NOAA SST datasets (Table 1). Lagged correlations of JJAS TC number with the Nino3.4 index in the preceding winter (DJF 0) and in the succeeding winter (DJF +1), however, were both statistically significant at the 95% level. The correlation with the Nino3.4 in the preceding winter was negative. This indicates that TC frequency decreased (increases) during El Nino (La Nina) decaying summer. The correlation with the Nino3.4 in the succeeding winter was positive, indicating that TC frequency increased (decreased) during El Nino (La Nina) developing summer.

Table 1. Correlation coefficients between JJAS TC number in the WNP and Nino3.4 index at preceding winter (DJF 0), concurrent summer (JJAS), and succeeding winter (DJF+1) during 1979-2010

		DJF 0	JJAS	DJF+1
Raw SST	Hadley	**-0.36**	0.25	**0.37**
	NOAA	**-0.37**	0.28	**0.35**
Detrended SST	Hadley	**-0.36**	0.26	**0.37**
	NOAA	**-0.35**	0.29	**0.36**

The first raw shows the correlation coefficients using raw SST data from the Hadley Center and NOAA.

The second raw indicates the correlation coefficients using the linearly-detrended SST fields from the same data sources.

Bold numbers indicate correlation coefficients significant at the 0.05 level.

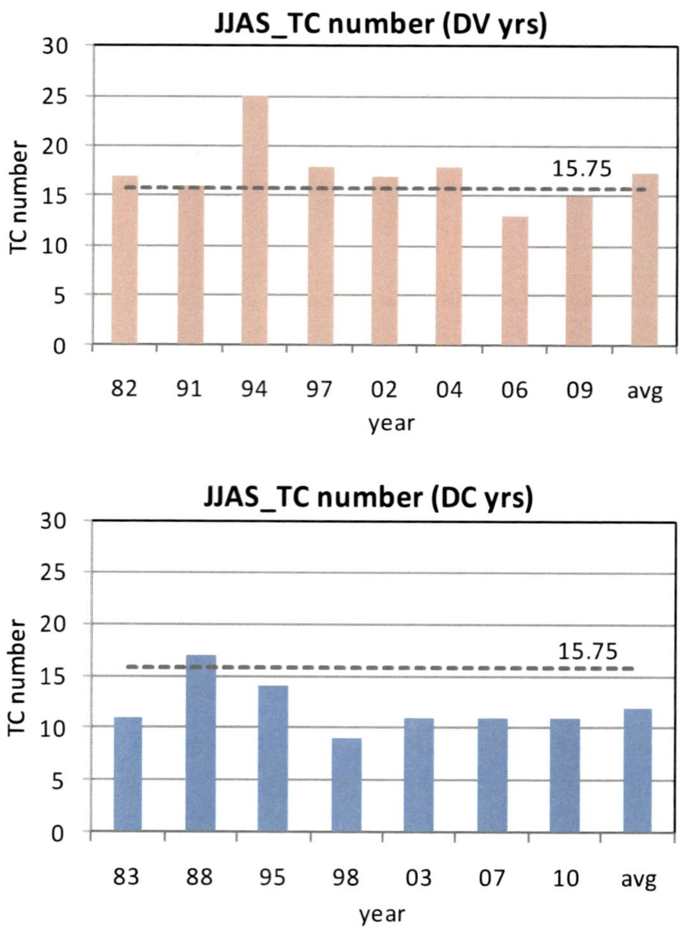

Figure 15. JJAS tropical cyclone number in the WNP during the El Niño developing years (upper panel) and dacaying years (bottom panel). The dashed line with number indicate the climatological TC number for 1979-2010.

The observed TC number at each of El Nino developing and decaying summers is shown in Figure 15. The average TC number during 1979-2010 was 15.75. Six out of eight El Nino developing summers had more than normal TC frequency, with exceptional cases in 2006 and 2009. Six out of seven El Nino decaying summers (except in 1988) were characterized by less than normal TC frequency in the WNP. The ENSO-phase-dependent feature was quite consistent with Iizuka and Matsuura, who analyzed the ENSO-TC relationship in a coupled atmosphere-ocean general circulation model [57].

What caused the ENSO-phase-dependent cyclogenesis feature? During the El Nino developing summer, equatorial westerly anomalies appeared in the central Pacific, which strengthened cyclonic shear and led to the southeastward shift of the monsoon trough. As seen from total and anomalous 850-hPa vorticity fields (Figure 16), a maximum low-level vorticity belt extended southeastward. In the anomaly field, a large-scale negative vorticity anomaly occupied the WNP typhoon genesis region. The southeastward shift of the monsoon trough led to more frequent TC genesis in the southeast quadrant of the WNP [58]. During the El Nino decaying summer, a large-scale anomalous anticyclone appeared in the WNP (bottom panel of Figure 16). This led to a weakening of the WNP monsoon trough. The ENSO-phase-dependent background flow condition caused distinctive behaviors in atmospheric ISO, with a stronger (weaker) ISO variability during the El Nino developing (decaying) summer [59]. Thus the change of the large-scale background condition (including both the mean flow and ISO variability) was responsible for the opposite TC changes during ENSO developing and decaying summers.

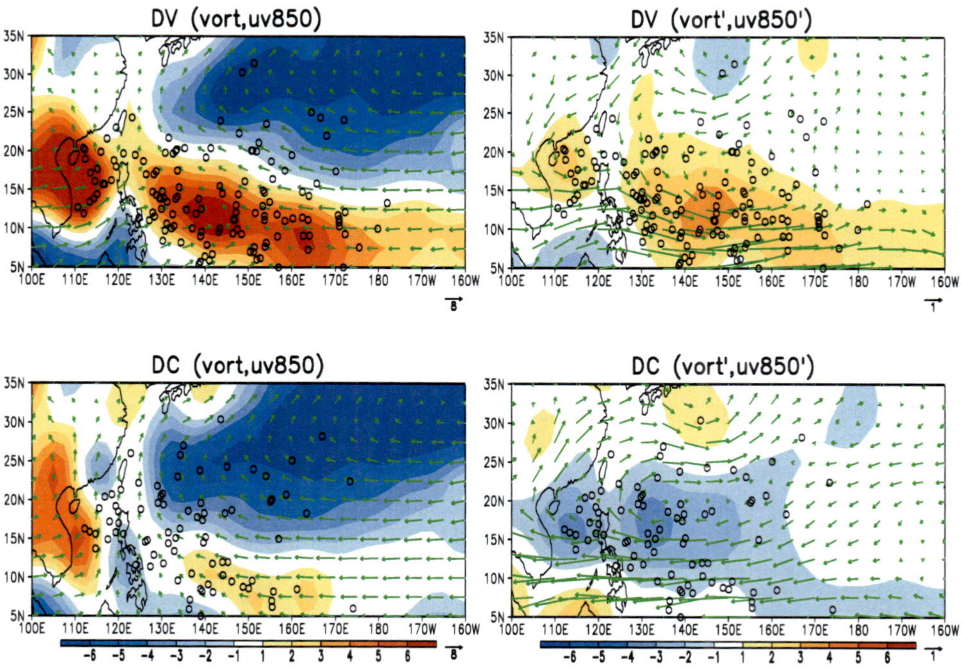

Figure 16. Left: Composites of 850-hPa vorticity (shading, unit: 10^{-6} s^{-1}), wind field (vector, unit: m s^{-1}) and TC genesis location (circle) for the El Niño developing years (upper) and dacaying years (bottom). Right: Same as left panel except that the shadings and vectors represent the anomalous vorrticity and wind fields, respectively.

Previous studies proposed a set of universal TC genesis parameters [60, 61]. Given distinctive mean flow and disturbance characteristics in the NATL and WNP, it is likely that factors controlling cyclogenesis differ. Peng et al. and Fu et al. investigated the differences of developing versus non-developing tropical disturbances using global daily analysis fields of the Navy Operational Atmospheric Prediction System (NOGAPS) from year 2003 to 2008 [62, 63]. A box difference index (BDI) was introduced to quantitatively measure the differences between the developing and non-developing disturbances. The definition of the index is as below:

$$BDI = \frac{M_{DEV} - M_{NONDEV}}{\sigma_{DEV} + \sigma_{NONDEV}}$$

where M_{DEV} and σ_{DEV} (M_{NONDEV} and σ_{NONDEV}) represent the mean and standard deviation of a variable for the developing (non-developing) cases.

To illustrate what a BDI value means, they took relative humidity as an example and considered a simple case that the standard deviations of the developing and non-developing groups were same. Figure 17 shows what it looks like in a box-and-whiskers figure when BDI is 0, 0.5 and 1. It is obvious that BDI=0 implies that there is no difference between the two groups. BDI=1 denotes a case in which the developing and non-developing groups are well separated, whereas BDI=0.5 implies that the two groups are partially separated.

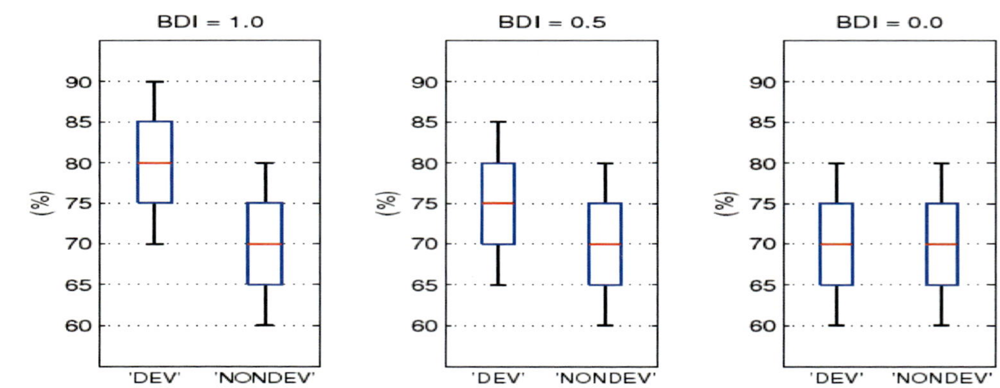

From [62], (c)American Meteorological Society, reprinted with permission.

Figure 17. Idealized box-and-whiskers figures for BDI=1.0, 0.5 and 0. The red line denotes the mean value of a genesis parameter (say, relative humidity at 500 hPa), the blue box represents the standard deviation of the parameter within the developing (denoting as 'DEV') and non-developing (denoting as 'NONDEV') disturbance groups. Whiskers (black line) represent the minimum and maximum of the samples.

With the aid of the BDI, Peng et al. and Fu et al. objectively evaluated how important a genesis parameter was in distinguishing developing and non-developing disturbances [62, 63]. Larger BDI amplitude implies a greater possibility to differentiate the developing and non-developing groups. Using this objective method, they ranked key genesis parameters at the NATL and WNP basins.

Table 2 lists top ranked genesis parameters in the NATL and the WNP based on their BDI calculations. It is interesting to note that the top three parameters in the NATL belong to the thermodynamic factor. For example, the vertically integrated specific humidity from 925hPa to 400hPa ranks the top of all the parameters, while the rain rate and SST follow as the second and the third most important parameters in NATL. In contrast, major genesis parameters in the WNP (except the rain rate) belong to the dynamic factors. Therefore, the BDI may provide a basis for selecting predictors for a statistical TC genesis forecast model at each basin.

Table 2. List of top four genesis parameters in the NATL and the WNP based on the BDI rank. "T" and "D" in the bracket denote the thermodynamic factor and the dynamic factor respectively

Basin / BDI rank	NATL	WNP
1	925 - 400hPa water vapor content (**T**)	800hPa maximum relative vorticity (**D**)
2	Rain rate (**T**)	Rain rate (**T**)
3	SST (**T**)	1000-400hPa averaged $\partial u / \partial y$ (**D**)
4	700hPa maximum relative vorticity (**D**)	1000-500hPa averaged divergence (**D**)

6. PROJECTED FUTURE TC CHANGES IN NORTH PACIFIC

How global warming will affect TC activity is a hotly debated topic [64-66]. With the increase of the global SST and surface moisture, it was anticipated that more TCs would develop. However, many climate models simulated a global decreasing trend of TC frequency [67-72]. One explanation of the decreasing trend was attributed to an increase of atmospheric static stability. This is because the global warming leads to a larger increase of air temperature in the upper troposphere than in the lower troposphere; as a result, the atmosphere becomes more stable, which suppresses the TC frequency [67, 72]. Does the same mechanism affect the regional TC change?

Li et al. investigated the cause of regional TC changes in the Pacific under global warming based on two high-resolution global AGCMs [73]. The first AGCM used was ECHAM5 [74] at a horizontal resolution of T319 (about 40-km grid). SST, the lower boundary condition of the model, was derived from a lower-resolution (T63) coupled version of the model (ECHAM5/MPI-OM) [75], which participated in the fourth assessment report of intergovernmental panel for climate change (IPCC-AR4). Two different climate change scenarios (20C3M and A1B) were applied. In 20C3M scenario, increasing historical greenhouse gases in 20th century were prescribed as a radiative forcing. In A1B scenario,

In: Cyclones: Formation, Triggers and Control
Editors: K. Oouchi and H. Fudeyasu
ISBN: 978-1-61942-976-5
©2012 Nova Science Publishers, Inc.

Chapter 4

VARIABILITY OF TYPHOON TRACKS AND GENESIS OVER THE WESTERN NORTH PACIFIC

Hisayuki Kubota*
Research Institute for Global Change,
Japan Agency for Marine-Earth Science and Technology
(RIGC, JAMSTEC), Yokosuka, Japan

ABSTRACT

The variability of typhoon tracks and genesis over the Western North Pacific (WNP) is reviewed in terms of intraseasonal to interdecadal time-scales. Intraseasonal variability of typhoon activity is related to the phase of Madden-Julian Oscillation and to shorter periods of 10-30 day westward propagating disturbances. Modulations of typhoon activity associated with the El Niño Southern Oscillation are found in typhoon genesis location, landfall numbers, intensity, life span, and typhoon rainfall; the mechanism is mainly due to the relocation of monsoon trough with the eastward shift of the Walker circulation. Pacific-Japan pattern of the Asian summer monsoon, Indian Ocean warming, and Quasi-Biennial Oscillation also influence the interannual variability of typhoon activity. Interdecadal variability of typhoon activity is found in the WNP: however there are many discrepancies among various studies on this topic. Short period of typhoon track dataset and difference between typhoon track datasets make it difficult to draw a unified picture of typhoon variability. The recovery of historical typhoon track data, which has not been used in the typhoon studies, will be a great tool for studying long-term variability.

INTRODUCTION

Typhoon can cause disasters in countries around the Western North Pacific (WNP) due to strong wind, heavy rain and powerful storm surge. When the first instrumental observation was conducted in the 19th century in East Asia, it was high priority to identify typhoon before

* E-mail address: Kubota@jamstec.go.jp

carbon dioxide concentration was increased at a rate 1% per year till it reached 720 ppm and was then kept constant. A 'time-slice' method [76] was applied, in which the high-resolution AGCM was forced by SST during two 20-year periods (1980-1999 and 2080-2099). The two periods are hereafter referred to as 20C and 21C, respectively. The second model was the GFDL high-resolution (50-km) atmospheric model (HiRAM2.1). For detailed model description and its simulation of the annual, inter-annual and decadal variations of TC activity, readers are referred to [77]. HiRAM2.1 was forced by an ensemble SST warming pattern in 21C derived from 18 IPCC AR4 models. It is worth mentioning that the two AGCMs have distinctive physical parameterization schemes in convection and radiation.

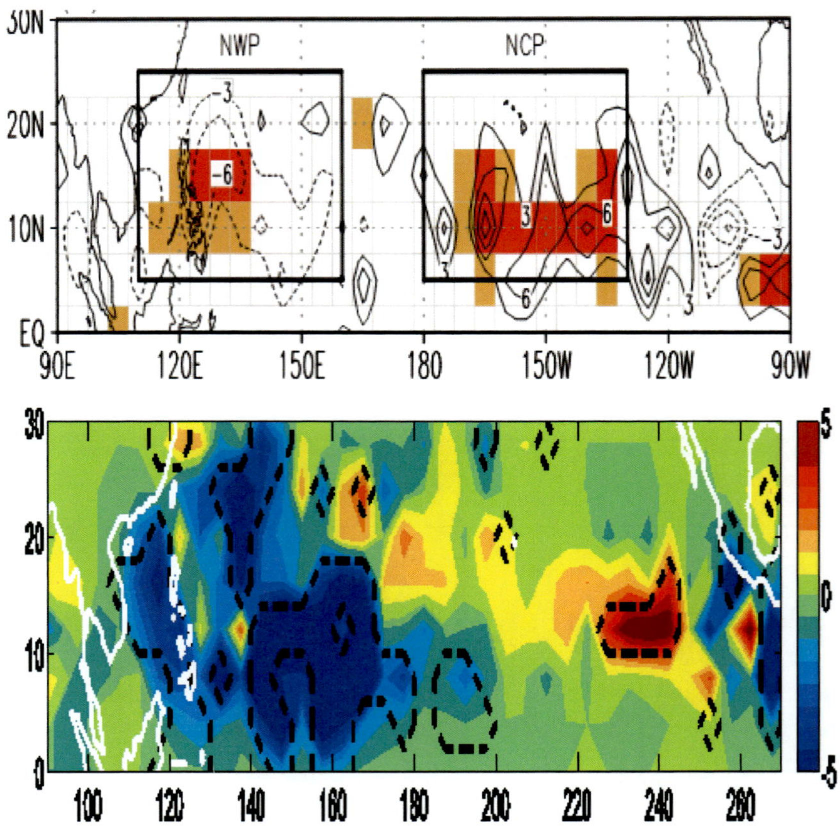

From [73], Copyright [2010] American Geophysical Union, reproduced by permission of American Geophysical Union.

Figure 18. Differences (21C minus 20C) of TC genesis number at each 2.5°x2.5° box for a 20-year period derived from the ECHAM5 T319 (top) and HiRAM2.1 (bottom). In the top panel, red (orange) shaded areas indicate 95% (90%) confidence level. In the bottom panel, dashed line denotes the 95% confidence level.

Figure 18 shows the difference of TC genesis frequency between the 20C and 21C simulations. In 20C, TCs form primarily over the western and eastern Pacific, similar to the distribution of the observed genesis locations. In 21C, however, more TCs shift their genesis locations to the Central Pacific. As seen from the difference map, there are two notable TC decrease and increase regions over the Pacific. One is over the WNP and the other the central North Pacific (CNP). In the ECHAM5 simulation, the numbers of TCs in 21C decreases by 31% over the WNP but increases by 65% over the CNP. Similarly, HiRAM2.1 projects a 30% decrease of TC genesis number in the WNP but a 19% increase in the CNP. Thus both the high-resolution AGCMs with distinctive model physics and different SST warming patterns simulate two opposite TC trends in the WNP and CNP. Further HiRAM2.1 simulations with different future SST warming patterns derived from GFDL CM2.0 and CM2.1, HADCM3, HADGEM1, ECHAM5, CCCMA, MRIGCM, and MIROCHI confirmed that the west-east shift of TC genesis location in the North Pacific is indeed a robust signal (Ming Zhao, personal communication).

To understand the cause of this west-east shift, Li et al. diagnosed the dynamic and thermodynamic conditions in northern summer (July – October) over the WNP (5-25°N, 110°E-160°E) and CNP (5-25°N, 180-130°W) regions, respectively. The change of atmospheric stability parameter was examined. The result showed that the upper-level air temperature increased at a greater rate than that at lower levels in both the regions. As the static stability is measured by the vertical gradient of the potential temperature, the result implies that the atmosphere becomes more stable under the global warming in both the regions. Thus, the static stability change cannot explain the opposite trends of TC frequency between the WNP and CNP.

A further analysis revealed that the fundamental cause of the opposite TC trends lies in the change of dynamic conditions. As we know, TCs originate from the tropical disturbances such as synoptic wave trains and easterly waves. The 21C simulation showed an increased variability of synoptic-scale disturbances over the CNP region but a decreased synoptic activity over the WNP region. Here the strength of the synoptic-scale disturbances was represented by the variance of the 850-hPa vorticity field filtered at a 2-8 day band using Lanzcos digital filter [78]. Both the models showed a remarkable decrease of the synoptic-scale variance over WNP but an increase of the variance in CNP. Thus, it is likely that the decreasing trend in the WNP is caused by the reduced synoptic-scale activity whereas the increasing trend in the CNP is caused by the strengthening of synoptic disturbances.

The contrast of the synoptic-scale activity between the WNP and CNP possibly resulted from the change of the background vertical wind shear and low-level divergence (Figure 19). Previous theoretical and modeling works pointed out that the easterly shear and low-level convergence of the background mean flow favored the development of tropical disturbances [79-80, 27]. In the 20C simulation, the western Pacific and Indian monsoon regions have a prevailing easterly wind shear in association with large-scale convective heating. This easterly wind shear becomes weaker in the warming climate (Figure 19a). The weakening of the easterly wind shear corresponds to a weakened western Pacific monsoon, as seen from Figure 19b. The weakening of the easterly shear, along with low-level divergence to its east (Figure 19c), further suppresses the development of tropical disturbances in the WNP. In contrast, the easterly wind shear and low-level convergence in the CNP are strengthened, and so is the precipitation. They favor the development of TCs in the CNP.

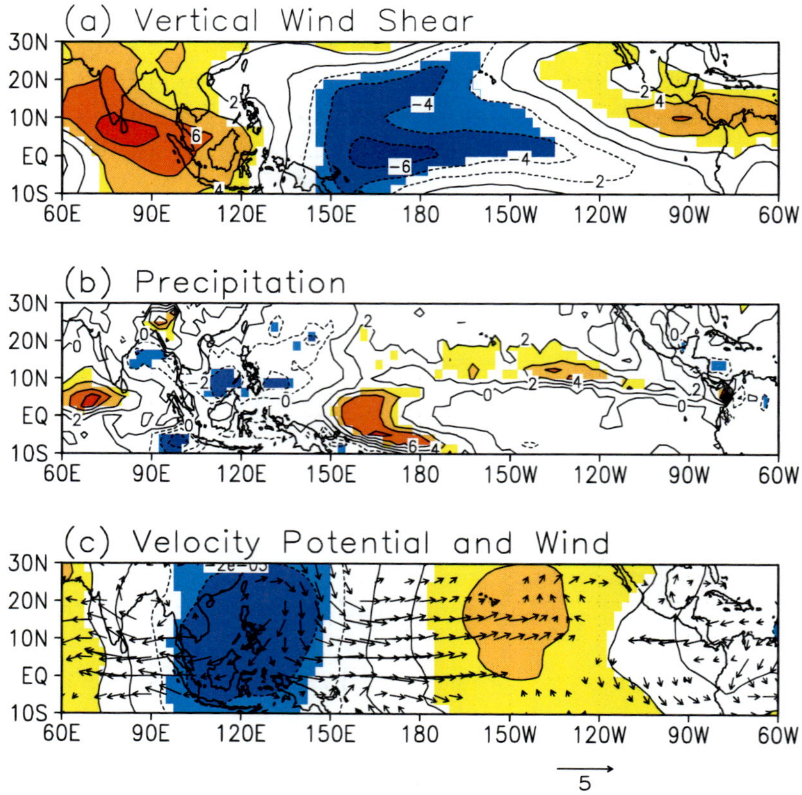

From [73], Copyright [2010] American Geophysical Union, reproduced by permission of American Geophysical Union.

Figure 19. Difference (21C – 20C) fields of (a) the vertical shear of zonal wind (200 hPa minus 850 hPa, unit: m s^{-1}), (b) precipitation (unit: mm day^{-1}), and (c) velocity potential (unit: s^{-1}) and wind (unit: m s^{-1}) at 850 hPa during northern summer (July-October). Areas that exceed the 95% confidence level (Student's t test) are shaded (for contour) and plotted (for vector).

It was argued that the change of the background vertical shear and the low-level divergence was closely related to changes of the trade wind in the tropics (Figure 19c). Many IPCC-AR4 models predicted an El Nino-like warming pattern in the Pacific under global warming [81], that is, a greater SST warming would occur in the tropical eastern and central Pacific compared to that in the tropical western Pacific. As a result, the zonal SST gradient was reduced across the tropical Pacific. The reduced zonal SST gradient would decrease the trade wind [82] and weaken the Walker circulation. The weakening of the trades would lead to the decrease of the boundary-layer convergence in the western Pacific monsoon region and the increase of the boundary convergence in the CNP (Figure 19c). The former would suppress the monsoon convective heating and decrease the easterly shear in the WNP, whereas the latter would strengthen the local boundary layer moisture convergence and convective heating and thus increase the easterly wind shear in the CNP.

CONCLUSION

This chapter summarizes some of recent progress in understanding tropical cyclogenesis in the WNP from both synoptic and climatic aspects. Observational analyses indicate that synoptic precursor perturbations at low levels in the WNP include TC energy dispersion induced Rossby wave trains, northwest-southeast oriented synoptic wave trains, and Pacific easterly waves. This feature differs from that in the NATL where dominant precursor signal is African easterly wave. In addition the near surface precursor signals, some cyclogenesis events are associated with a mid-level vortex or TUTT.

Dominant synoptic-scale variability in the WNP is a southeast-northwest oriented synoptic wave train. A theoretical study suggested that this wave train was originated from the instability of the summer mean flow in the presence of the convection-circulation feedback in the WNP. The structure and dispersion characteristics of the Pacific easterly wave were examined through the diagnosis of observational data. A wave activity diagnosis showed that the zonal scale contraction primarily contributes to the convergence of the wave activity, while the convergence of the mean flow contributes less to the wave activity accumulation. An analysis of group velocity in the upper troposphere and in the vertical direction showed that there was downward injection of wave activity from the upper levels and in the upper troposphere wave activity directed southward form the extratropics to the tropical central Pacific. The result indicated that the energy source of Pacific easterly waves originated from mid-latitude Pacific Jet.

The cyclogenesis associated with TCED was simulated in a 3D model. A new TC with realistic dynamic and thermodynamic structures formed in the wake of a pre-existing TC when a large-scale monsoon gyre was present. Maximum vorticity generation appeared in the PBL and the vorticity growth exhibited an oscillatory development. This oscillatory growth was also seen in the observed rainfall and cloud-top temperature fields. The diagnosis of the model output showed that the oscillatory development was attributed to the discharge and recharge of the PBL moisture and its interaction with convection and circulation. A cloud-resolving (2-km grid) model was used to investigate cyclogenesis efficiency in an environmental mid-level and bottom vortex. The simulation results showed that a bottom vortex was more efficient in cyclogenesis than a mid-level vortex, given the same column integrated absolute vorticity. Both the mid-level and bottom vortex experiments shared common development characteristics: 1) a transition from random cumulus-scale (~5 km) convective cells into an organized meso-vortex-scale (~50 km) system through upscale cascade, 2) a setup of a nearly saturated air column prior to a rapid drop of the central minimum pressure, and 3) a convective-stratiform phase transition.

TC frequency in the WNP is regulated by ISO activity. The number of TCs increases (decreases) during an active (suppressed) phase of ISO. Cyclogenesis activity is also El Nino-phase dependent. The TC genesis number increases (decreases) during El Nino developing (decaying) summer. A Box Difference Index was used to rank major genesis parameters that were capable of differentiating the developing and non-developing disturbances. It was found that whereas TC genesis in the NATL was more sensitive to the thermodynamic factors, dynamic factors played a major role in determining whether or not a tropical disturbance would develop in the WNP.

The future change of TC genesis location in the North Pacific under global warming was investigated with use of two global high-resolution models (ECHAM5 T319 and GFDL HiRAM2.1). A significant shift was found in the location of TCs from the western to central North Pacific. The shift to more cyclones in the central Pacific and less in the western Pacific was not attributable to a change in atmospheric static stability, but to a change in the variance of tropical synoptic-scale perturbations associated with a change in the background vertical wind shear and boundary layer divergence.

ACKNOWLEDGMENTS

The author appreciates two anonymous reviewers for constructive comments. Contributions by Drs. Melinda Peng, Pangchi Hsu, Bing Fu, Xuyang Ge, Francis Tam, MinHo Kwon, and Ming Zhao are greatly appreciated. This work was supported by ONR grants N000140810256 and N000141010774, NSF grant AGS-1106536, NOAA grant NA11OAR4310087, and by the International Pacific Research Center that is sponsored by the Japan Agency for Marine-Earth Science and Technology (JAMSTEC), NASA (NNX07AG53G) and NOAA (NA17RJ1230). This is SOEST contribution number 8484 and IPRC contribution number 816.

REFERENCES

[1] R. A. Anthes, Tropical cyclone: their evolution, structure and effects. *Meteor. Mongr.* No. 41, *Amer. Meteor. Soc.*, 208pp, (1982).
[2] G. R. Flierl, *J. Phys. Oceanogr.*,14, 47, (1984).
[3] G. R. Flierl and K. Haines, *Phys. Fluids*, 6, 3487, (1994).
[4] Z. Luo, *Acta Meteor. Sinica*, 8, 51, (1994).
[5] N. R. McDonald, *J.Fluid Mech.*, 361, 237, (1998).
[6] L.E. Carr III and R.L. Elsberry, NPS Tech. Rep. NPS-MR-94-002, 273 pp, (1994).
[7] L.E. Carr III and R.L. Elsberry, *Mon. Wea. Rev.* 123, 265, (1995).
[8] X. Ge, T. Li., Y. Wang, and M. Peng, *J. Atmos. Sci.*, 65, 2272, (2008).
[9] W. M. Frank, *Mon.Wea.Rev.*, 110, 572, (1982).
[10] N. E. Davidson and H. H. Hendon, *Mon.Wea.Rev.* 117, 1458, (1989).
[11] L. M. Briegel and W.M. Frank, *Mon. Wea. Rev.* 125, 1397, (1997).
[12] E. A. Ritchie and G. J. Holland, *Mon.Wea.Rev.*, 125, 1377, (1997).
[13] T. Li, B. Fu, X. Ge, B. Wang, and M. Peng, *Geophy. Res. Let.*, 30, 2122, (2003).
[14] B. Fu, T. Li, M.S. Peng, and F.Z. Weng, *Wea. Forecasting*, 22, 763, (2007).
[15] G. R. Flierl, M. E. Stern and J. A. Whitehead, Jr., *Dyn. Atmos. Oceans*, 7, 233, (1983).
[16] T. Li and B. Fu, *J. Atmos. Sci.* 63, 1377, (2006)
[17] L. J. Shapiro and K. V. Ooyama, *J. Atmos. Sci.*, 47, 170, (1990).
[18] X. Ge, T. Li, and X. Zhou, *Geophys. Res. Lett.,* 34, L23807, doi:10.1029/2007GL031867, (2007).
[19] T. Li, X. Ge, B. Wang, and Y. Zhu, *J. Atmos. Sci.*, 63, 1390, (2006).
[20] K.-H. Lau and N.-C. Lau, *Mon. Wea. Rev.,* 118, 1888, (1990).

[21] Y. N. Takayabu and T. Nitta, *J. Meteor. Soc. Japan*, 71, 221, (1993).
[22] M. Dickinson and J. Molinari, *J. Atmos. Sci.* 59, 2183, (2002).
[23] R. J. Reed and E. E. Recker, *J. Atmos. Sci.*, 28, 1117, (1971). R. J. Reed, D. C. Norquist, and E. E. Recker, *Mon. Wea. Rev.,* 105, 317, (1977).
[24] J. Simpson, E. Ritchie, G. J. Holland, J. Halverson, and S. Stewart, *Mon. Wea. Rev.*, 125, 2643, (1997).
[25] J. A. Colon and W.R. Nightingale, *Mon. Wea. Rev.* 91, 329, (1963).
[26] J. C. Sadler, *Mon. Wea. Rev.,* 104, 1266, (1976).
[27] T. Li, *J. Atmos. Sci.* 63, 1093, (2006).
[28] I. M. Held, and M. J. Suarez, *Bull. Amer. Meteor. Soc.*, 75, 1825, (1994).
[29] M. Ting and L. Yu, *J. Atmos.Sci.*, 55, 3565, (1998).
[30] B. Wang, R. Wu, T. Li, *J. Climate*, 16, 1195, (2003).
[31] X. Jiang and T. Li, *J. Climate*, 18, 3777, (2005).
[32] K.-H. Lau and N-C. Lau, *Mon. Wea. Rev.,* 120, 2523, (1992).
[33] C.-Y. Tam and T. Li, *Mon. Wea. Rev.*, 134, 1630, (2006).
[34] K. Takaya and H. Nakamura, *J. Atmos. Sci.,* 58, 608, (2001).
[35] Y. Wang, *Mon. Wea. Rev.,* 129, 1370, (2001).
[36] E. A. Ritchie and G. J. Holland, *Mon. Wea. Rev.*, 127, 2027, (1999).
[37] J. S. Kain and J. M. Fritsch, *J. Atmos. Sci.*, 47, 2784, (1990).
[38] Y.-L. Lin, R. D. Rarley, and H. D. Orville, *J. Appl. Meteor.*, 22, 1065, (1983).
[39] G. J. Holland, *J. Atmos. Sci.,* 54, 2519, (1997).
[40] X. Ge, T. Li and M. Peng, *J. Tropical Meteorology*, in press (2012).
[41] S. E. Yuter and R. A. Houze, Jr., *Mon. Wea. Rev.*, 123, 1941, (1995).
[42] D. A. Schecter and D. H. Dubin, *Phys. Rev. Lett.*, 83, 2191, (1999).
[43] E. A. Hendricks, M. T. Montgomery, and C. A. Davis, *J. Atmos. Sci.*, 61, 1209, (2004).
[44] M. T. Montgomery, M. E. Nicholls, T. A. Cram, and A. B. Saunders, *J. Atmos. Sci.*, 63, 355, (2006).
[45] K. J. Tory, M. T. Montgomery, and N. E. Davidson, *J. Atmos. Sci.*, 63, 3077, (2006).
[46] W. M. Gray, *Meteorology over the Tropical Oceans*, D. B. Shaw, ED., Royal Meteorological Society, 155, (1979).
[47] N. Yamazaki and M. Murakami, *J. Meteor. Soc. Japan,* 67, 791, (1989).
[48] D. L. Hartmann, M. L. Michelsen, and S. A. Klein, *J. Atmos. Sci.*, 49, 1277, (1992).
[49] H. Liebmann, H. Hendon, and J. D. Glick, *J. Meteor. Soc. Japan*, 72, 401, (1994).
[50] E. D. Maloney and D. L. Hartmann, *J. Climate*, 13, 1451, (2000).
[51] E. D. Maloney and D. L. Hartmann, *J. Atmos. Sci.*, 58, 2545, (2001).
[52] A. H. Sobel and E. D. Maloney, *Geophys. Res. Lett.*, 27, 1739, (2000).
[53] E. D. Maloney and M. J. Dickinson, *J. Atmos. Sci.,* 60, 2153, (2003).
[54] G. J. Holland, *Meteor. Atmos. Phys.*, 56, 57, (1995).
[55] A. H. Sobel and C. S. Bretherton, *J. Atmos. Sci.*, 56, 3106, (1999).
[56] P.-C. Hsu, T. Li, and C.-H. Tsou, *J. Climate*, 24, 927, (2011).
[57] S. Iizuka, and T. Matsuura, *Clim.Dyn.*, 30, 815, (2008).
[58] B. Wang and J. C. L. Chan, *J. Climate,* 15, 1643, (2002).
[59] A. Lin and T. Li, *J. Climate*, 21, 6304, (2008).
[60] W. M. Gray, *Mon. Wea. Rev.,* 96, 669, (1968).
[61] K. A. Emanuel and D. S. Nolan, *26th Conference on Hurricanes and Tropical Meteorology*, Am. Meteorol. Soc., Miami, FL, (2004).

[62] M. Peng, B. Fu, T. Li, and D. E. Stevens, *Mon.Wea.Rev.*, in press (2012).
[63] B. Fu, M. Peng, T. Li, and D. E. Stevens, *Mon.Wea.Rev.*, in press (2012).
[64] P. J. Webster, G. J. Holland, J. A. Curry, and H.-R. Chang, *Science,* 309, 1844, (2005).
[65] K. A. Emanuel, *Nature,* 436, 686, (2005).
[66] C. W. Landsea, B. A. Harper, K. Hoarau, and J. A. Knaff, *Science,* 313, 452, (2006).
[67] M. Sugi, A. Noda, and N. Sato, *J. Meteorol. Soc. Japan,* 80, 249, (2002).
[68] R. E. McDonald, and coauthors, *Clim. Dyn.,* 25, 19, (2005).
[69] A. Hasegawa, and S. Emori, *Scientific Online Letters on the Atmosphere,* 1, 145, (2005).
[70] J. Yoshimura, M. Sugi, and A. Noda, *J. Meteorol. Soc. Japan,* 84, 405, (2006).
[71] K. Oouchi, and coauthors, *J. Meteorol. Soc. Japan,* 84, 259, (2006).
[72] L. Bengtsson, K. I. Hodges, M. Esch, N. Keenlyside, L. Komblush, J. J. Luo, and T. Yamagata, *Tellus* 59A, 539, (2007).
[73] T. Li, M. Kwon, M. Zhao, J.-S. Kug, J.-J. Luo, and W. Yu, *Geophys. Res. Lett.*, 37, L21804, doi:10.1029/2010GL045124, (2010).
[74] E. Roeckner, and coauthors, Max Planck Institute for Meteorology Rep. 349, 127 pp, (2003).
[75] H. Jungclaus, and coauthors, *J. Climate,* 19, 3952, (2006).
[76] L. Bengtsson, M. Botzet, and M. Esch, *Tellus,* 48A, 57, (1996).
[77] M. Zhao, I. M. Held, S.-J. Lin, and G. A. Vecchi, *J. Climate*, 22, 6653, (2009).
[78] C. E. Duchon, *J. Appl. Meteoro.* 18, 1016, (1979).
[79] B. Wang and X. Xie, *J. Atmos. Sci.,* 53, 449, (1996).
[80] K. P. Sooraj, D. Kim, J.-S. Kug, S.-W. Yeh, F-F. Jin, and I.-S. Kang, *Clim. Dyn.*, 33, doi 10.1007/s00382-008-0483-6, (2008).
[81] S. Solomon, and coauthors, *Cambridge University Press for the Intergovernmental Panel on Climate Change,* (2007).
[82] R. S. Lindzen and S. Nigam, *J. Atmos. Sci.,* 44, 2440, (1987).

its arrival and to issue warning of approaching typhoon in order to save people's lives [1]. Seasonality of major typhoon tracks was investigated since the late 19th century [2]. Climatology of seasonal change of typhoon tracks was known as early as the 1920s [3, 4]. Besides seasonality of typhoon tracks, the variability of typhoon formation was focused on after the atmospheric oscillations such as the El Niño-Southern Oscillation (ENSO) and Madden-Julian Oscillation (MJO) were discovered [5-7].

Typhoon track dataset is necessary for typhoon studies. Typhoon track dataset over the WNP, however, has gaps during the World War II (WWII). Most of the typhoon studies used typhoon track dataset after the WWII. This chapter first focuses on the history of typhoon track dataset, and then reviews atmospheric intraseasonal to interdecadal variability over the WNP and its influence on typhoon activity in the region.

History of Detecting Typhoon Tracks Over the WNP

The study on the variability of typhoon activity began after the typhoon tracks were recorded over the WNP. Continuous instrumental weather observations over the WNP started at a meteorological station in Manila of the Philippines in 1865. The Manila Observatory was established by Spanish Jesuits [8]. The first remarkable typhoon passage was observed at Manila station around September 27 1865 (Figure 1) [9]. Barometer measured minimum pressure in early morning of Sep. 27. Counter-clockwise change of wind direction during the typhoon passage indicates a typhoon passed east or north of Manila. Typhoon tracks have been continuously recorded from 1879 [2]. By accumulating the typhoon track reports around the Philippine Archipelago, Algué [2] noted seasonal variability of typhoon track over the WNP using the typhoon track data from 1879 to 1894 (Figure 2). The cyclonic structure of typhoon and clouds and rain associated with typhoon system were investigated by Algué [10]. Hurricane-induced waves and swells were also reported in this study. Seasonal numbers of typhoon and main typhoon tracks were classified. Due to the limitation of the observation coverage in the WNP, the maximum numbers appeared in September instead of August in a recent analysis by Joint Typhoon Warning Center (JTWC) (Figure 3). The Manila Observatory was renamed as the Philippine Weather Bureau after 1901 by American meteorologists.

Jesuits missionary expanded to China and established Zi-Ka-Wei Observatory near Shanghai in 1872 [8]. Typhoon track analysis was performed by Forc [3], who used typhoon track data from 1893 to 1918. Compared to Manila station, Zi-Ka-Wei station was located much north near 31°N, therefore not only tropical cyclones (TCs) but also mid-latitude cyclones passed over this region. Froc [3] did not separate cyclones from typhoons in his study. Another independent meteorological agency established in Tokyo Japan in 1876. The Tokyo Meteorological Observatory started to draw isobar contours to make weather map since 1883. First remarkable typhoon was observed through a weather map on August 25 1884 (Figure 4). Typhoon was located on the west coast of Kyushu Island. The Tokyo Meteorological Observatory was renamed to the Central Meteorological Observatory of Tokyo Japan (CMOJ: formerly the Japan Meteorological Agency), which recorded typhoon tracks. Another independent meteorological agency was established in Hong Kong in 1884. The Royal Observatory Hong Kong also recorded typhoon tracks by collecting weather information through telegram. There was no typhoon definition to distinguish it from weak tropical disturbances at that time. Only the Philippine Weather Bureau used 750mmHg (about 1000hPa) for typhoon definition in its bulletins.

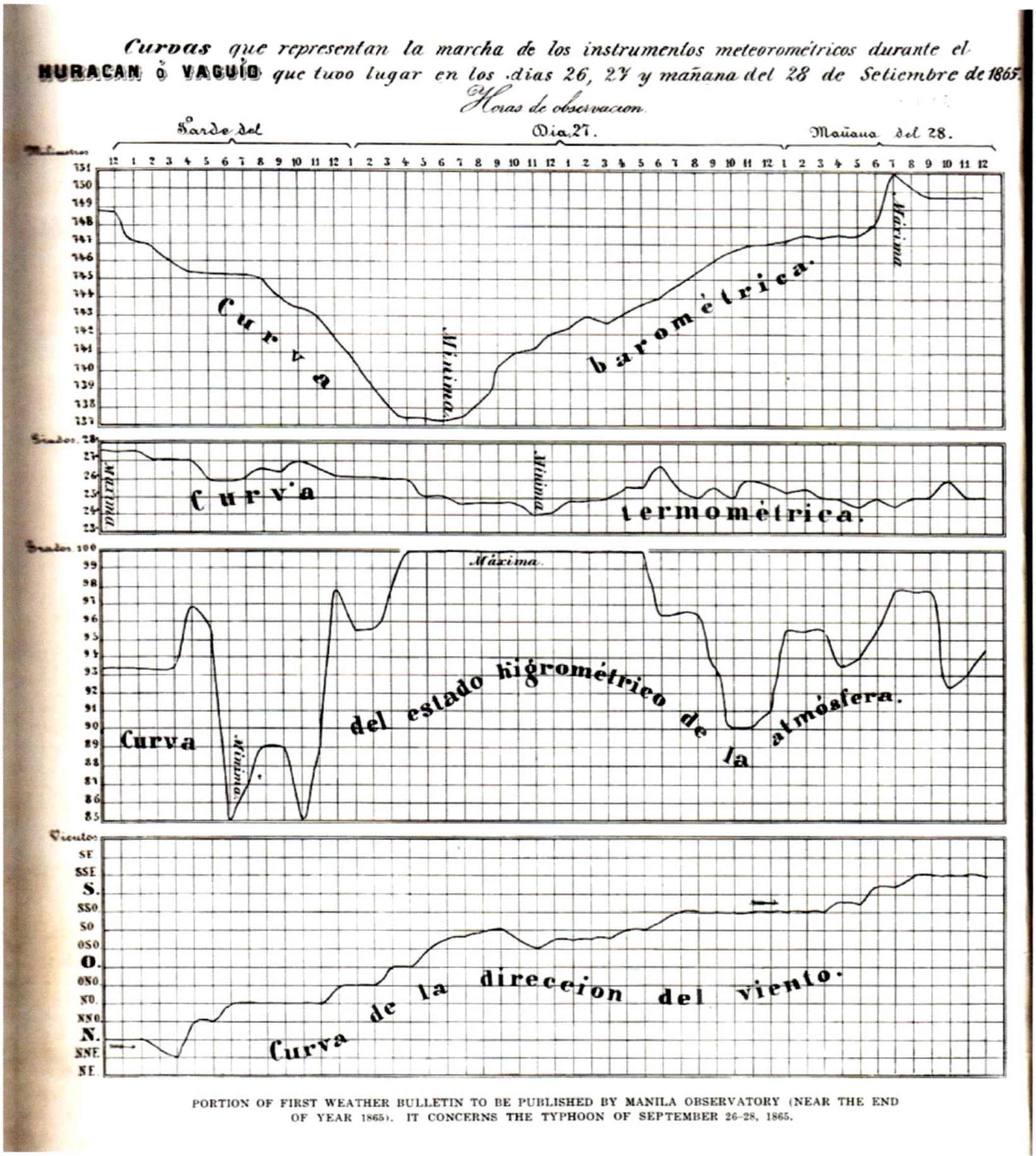

Figure 1. Time series of meteorological observations (pressure, temperature, humidity, and wind direction) in the Manila Observatory during the typhoon passage of September 26-28, 1865. After Depperman [9].

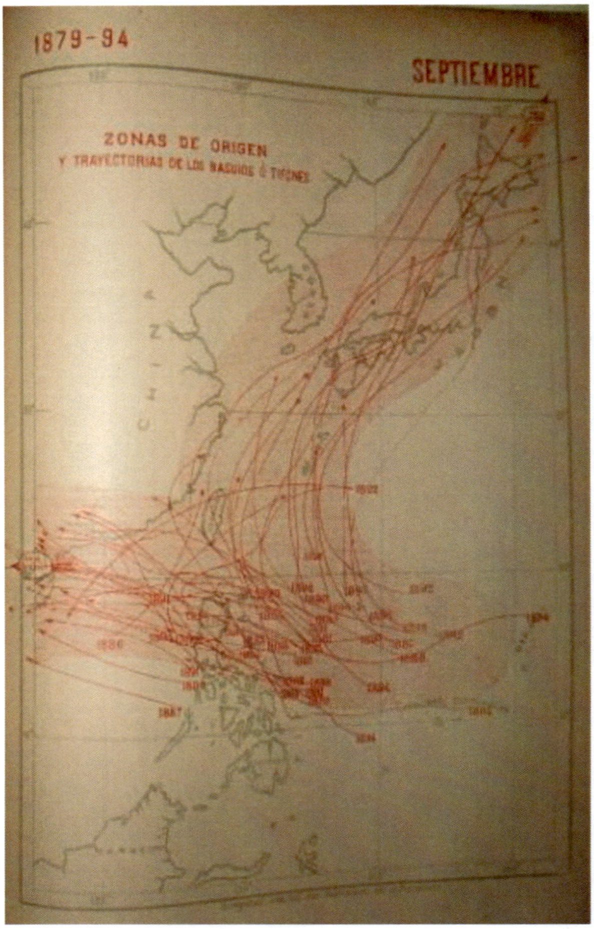

Figure 2. Typhoon tracks on September 1879-1894. After Algué [2].

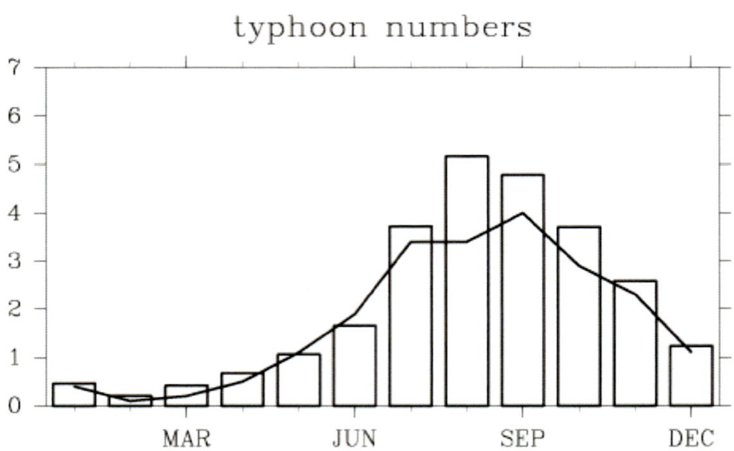

Figure 3. Monthly-mean number of typhoon over the WNP from 1945 to 2009 based on the JTWC dataset (boxes) and from 1880 to 1901 of the Philippine Weather Bureau. After Algué [10].

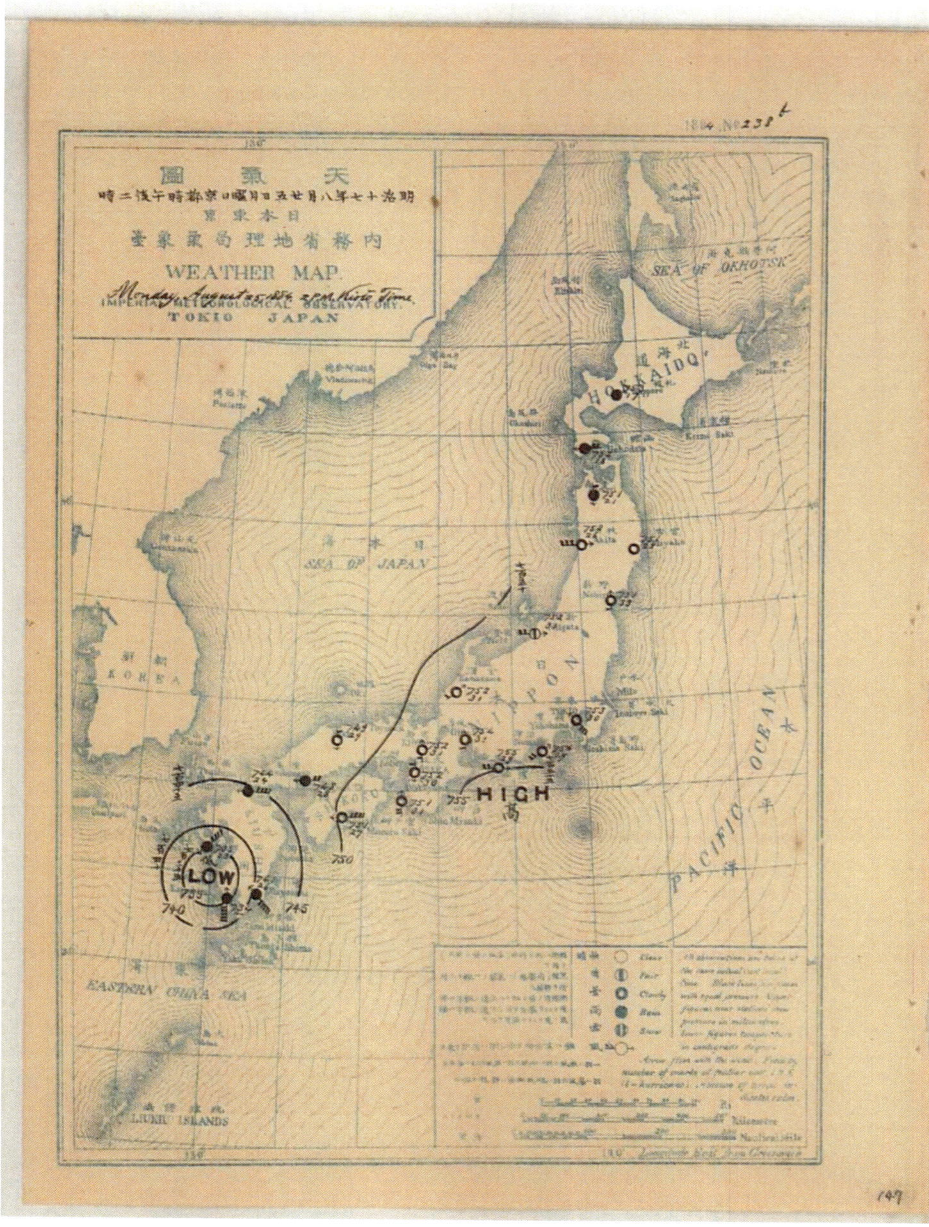

Figure 4.Weathermap of August 25, 1884. Provided by the JMA.

Typhoon study developed in the early 20th century at the Philippine Weather Bureau. Weather station grew to more than 300 stations in the Philippines by 1940. Fine surface structures of typhoon were investigated by Deppermann [9] who updated and corrected the surface structure of typhoon fromAlgué [10]. However the Philippine Weather Bureau stopped reporting typhoon tracks in its monthly bulletins in August 1940. From September 1940, Japan started reporting observations in the Philippines, Indochina Peninsula, Malaysia, Thailand, Indonesia, and New Guinea in the Monthly Bulletins of Southern Seas. CMOJ could trace typhoon tracks over a wide region near the dateline in the WNP from 1940 [11].

Aircraft reconnaissance of weather observations was performed by Japan in 1940, and aviation investigation report was published by the CMOJ. However after the WWII outbreak the coverage of typhoon track area over the WNP shrunk year after year (Figure 5) [12]. In 1941 typhoon tracks were detected up to 160°E. However in 1944 most typhoons in the data were located west of 150°E due to the battles that prevented data collection from the eastern part of the study region. It means the typhoon was generated and passed in the eastern part, however, observation was limited due to the battle with America.

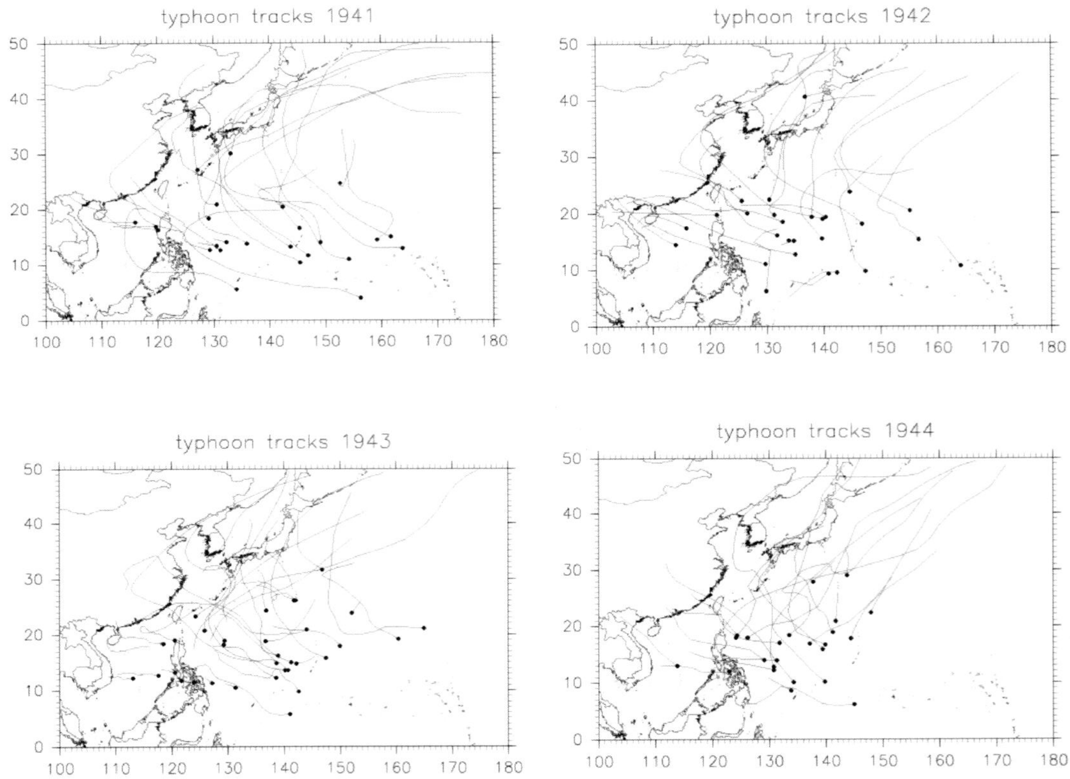

Figure 5. Typhoon tracks (lines) and genesis location (dots) over the WNP from 1941 to 1944 based on the CMOJ typhoon track dataset.

Americans began widespread aircraft reconnaissance over the WNP in 1945. From this year on, Americans started reporting the typhoon (tropical storm) best track data using new definition of exceeding 35kt of 1-minute average of the maximum wind speed near the storm center for TC, and the JTWC started tomonitor typhoon tracks from 1959. After the WWII, the Shanghai Typhoon Institute started reporting typhoon track dataset by a new typhoon definition, same as the JTWC's but using two-minute average data from 1949; the Japan Meteorological Agency (JMA) started from 1951, and the Hong Kong Observatory (HKO), from 1961. Both JMA and HKO used the same criteria of maximum wind speed for typhoon definition, namely for 10-minute average.

After the new typhoon best track data became popular, old typhoon track data was gradually forgotten. Gao and Zeng [13] combined four old typhoon track datasets from 1884 to 1955. Chin [14] published typhoon track data from 1884 to 1953 based on the collection of the four typhoon track datasets. However, no one used these datasets for a long time. Recently, climate change and global warming issues became a great concern; their influences on typhoon were highlighted. The demand for historical typhoon track data became high. The recovery of historical typhoon track data was then carried out. The Central Weather Bureau of Taiwan created a typhoon track dataset in the vicinity of Taiwan from 1897 to 1996 [15]. The HKO digitized the typhoon track locations in Chin [14] and interpolated the data into 6-hourly data. Kubota and Chan [16] recovered and digitized typhoon track data of the Philippine Weather Bureau. Now the International Best Track Archive for Climate Stewardship (IBTrACS) project is underway [17] to collect global long-term typhoon track data.

INTRASEASONAL VARIABILITY

Over the WNP, tropical convection activities vary with large-scale atmospheric condition. Their time scale can be separated according to their dynamical structures. The main processes are solar radiation and equatorial waves. Solar insolation promotes diurnal and seasonal time scales due to the Earth's rotation and revolution cycles. Equatorial waves are coupled with convections [18]. Off-equatorial rotational mode involves Rossby wave mode. Over the WNP, MJO is dominant by 30 to 60 day periods, with a structure of the combination of Kelvin and Rossby waves [19-21]. The MJO consists of large-scale convective systems and propagates eastward from the equatorial Indian Ocean to the central Pacific. During the active phase of MJO, strong westerly wind bursts are observed near the equator and cyclone pairs are observed in off-equatorial regions in both Northern and Southern hemispheres [22-24]. Some of the cyclones developed into typhoons. The relationship between MJO and typhoon activity was first examined by Liebmann et al. [25]. The number of typhoon increased during the active phase of MJO. However, the ratio of storms vs typhoons formed from depressions did not change with the phase of MJO. Even in the absence of MJO, typhoon number increased in the active phase of intraseasonal oscillation. During the Asian summer monsoon season, intraseasonal oscillation of convective activity modulates, and northward propagation can be seen over the WNP [26, 27]. Oscillations of convective activity with shorter periods of 10 to 30 days over the off-equatorial region have westward propagating characteristics and large amplitude during September and December [27-30]. Typhoon activity is also modulated by 20 to 25 day oscillation during this season [28]. Figure 6 shows the frequency of the TC within 10 degrees of radius of 12.5°N 142.5°E in Guam divided by eight bins of phases in 20-25 day oscillations. Highest pressure in Guam occurred in phase 3 and lowest pressure in phase 7. TC frequency has intraseasonal oscillation with maximum frequency during the lowest pressure in phase 7.

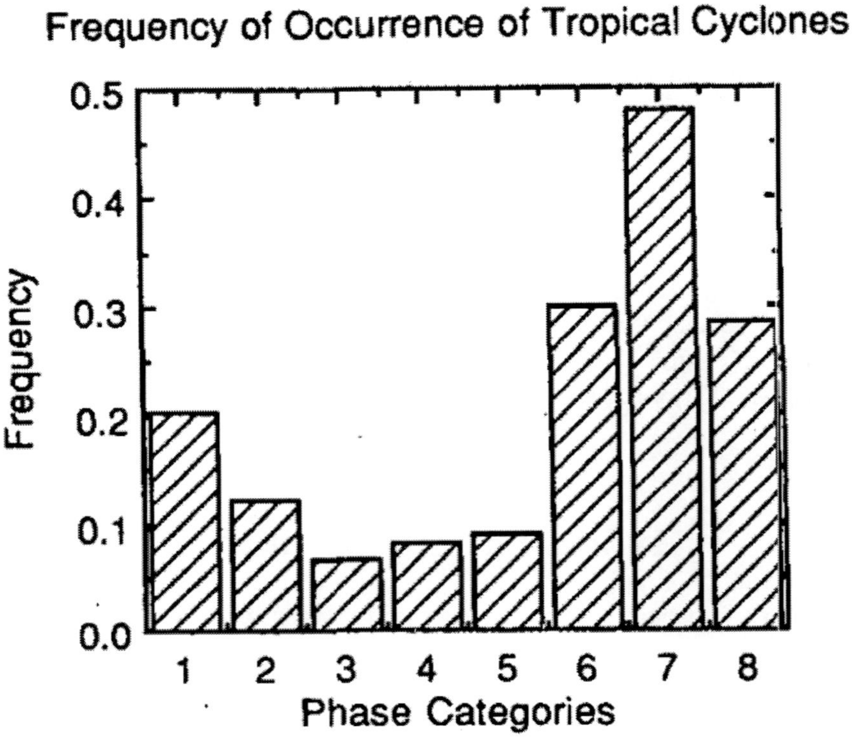

Figure 6. Histogram of tropical cyclone (TC) frequency within 10-degree radius of 12.5°N 142.5°E in eight phase bins of 20-25 day oscillations in Guam. Highest pressure in Guam occurred in phase 3 and lowest pressure in phase 7. After Hartmann et al. [28], with the permission of the American Meteorological Society.

Interannual Variability

El Niño Southern Oscillation

The most prominent atmospheric and oceanic interannual variability over the WNP is the ENSO. During El Niño, the warmest sea-surface temperature (SST) region shifts eastward and favorable condition for convection moves to the central Pacific, along with the eastward shift of the Walker circulation [31]. The change of typhoon formation during El Niño was found by Atkinson [32]. Chan [33] noted the number of typhoon was correlated to the Southern Oscillation Index (SOI), with the SOI leading about one year. It means typhoon number decreases in the year following the El Niño event. However Lander [34] disagreed that there was any correlation of annual typhoon number over the WNP with ENSO. Instead, he pointed out that the displacement of the typhoon genesis location was robustly associated with ENSO. During the developing phase of El Niño, typhoon genesis location shifted eastward [34-36]. Figure 7 shows the TC generated at EQ-20°N, 160°-180°E mostly during low SOI. And a north to south shift of typhoon genesis location was seen during the summer of El Niño [37]. These displacements of typhoon genesis locations can be explained by the eastward and southward relocations of monsoon trough during El Niño [34, 37]. Monsoon trough is consisted by moist southwesterly monsoon flow in the south and northeasterly trade wind in the north where the cyclonic rotation tends to form typhoons [38, 39]. Chia and Ropelewski[40] added 200-850 hPa vertical wind shear, western Pacific SST, and the position and strength of western Pacific subtropical high as components related to the displacement of typhoon genesis location associated with ENSO.

Typhoon landfall number in Southeast Asia usually increases during summer and autumn monsoon seasons in the climatology. During summer of the developing phase of El Niño, typhoon landfall number increases in Southeast Asia, which is associated with the southward shift of typhoon genesis location [41]. On the other hand, typhoon landfall number decreases in autumn in Southeast Asia including Philippines due to the eastward shift of typhoon genesis location [42-44].

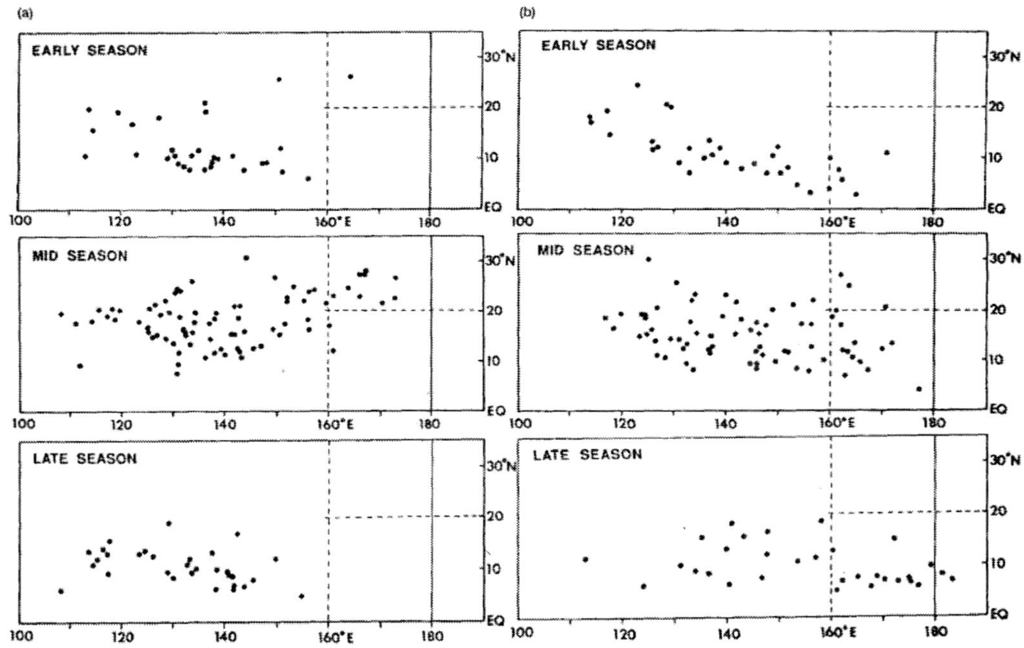

Figure 7. TC genesis locations in early season (from March to mid-July), mid season (from mid-July to October) and late season (from October to January) for the five highest SOI (a) and five lowest SOI (b) during 1970-1991. After Lander [34], with the permission of the American Meteorological Society.

Typhoon life span and intensity are also influenced by the displacement of typhoon genesis location associated with the ENSO. During El Niño developing year, typhoon genesis location shifts southeastward to near the date line. Figure 8 shows typhoon tracks in autumn of El Niño and La Niña year. Typhoon tends to curve northward in El Niño years [45]. Typhoon usually moves west-northwestward, therefore it travels a long distance in about seven days on average in El Niño years (only four days in La Niña years) over the warm pool region. This means typhoon can intensify while it travels a long span over a favorable condition for its development, and typhoon becomes more intense during El Niño [46, 47].

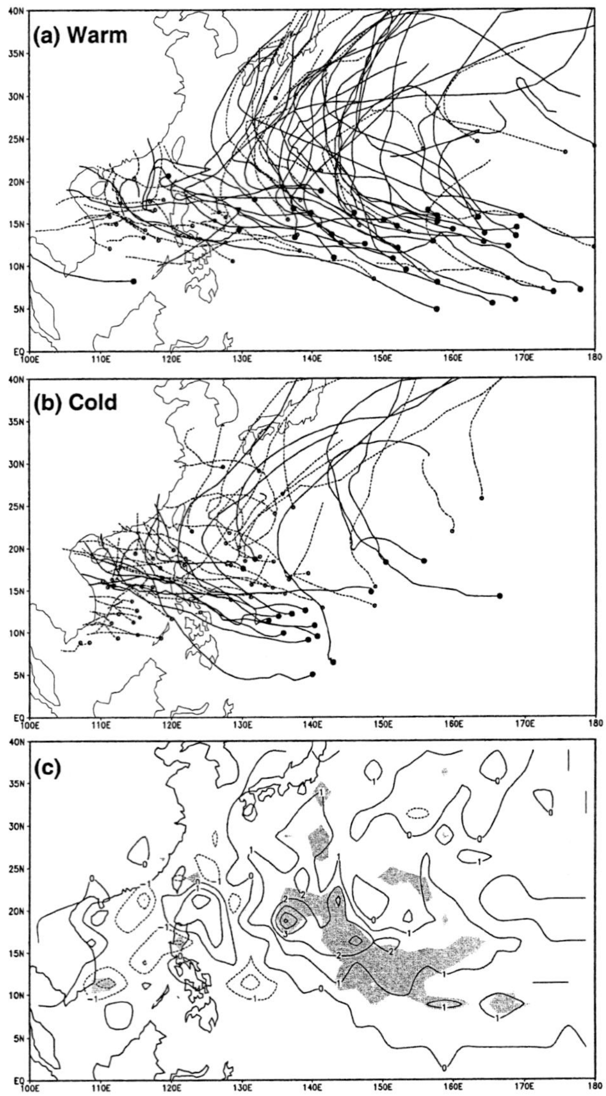

Figure 8. Typhoon tracks during September to November (SON) in six strong El Niño years (1965, 1972, 1982, 1987, 1991, and 1997) (a) and six strong La Niña years (1970, 1973, 1975, 1988, 1998, and 1999) (b). (c) The difference in the total number of typhoon occurrence in each 2.5° by 2.5°grid during SON between the strong warm and cold years. Shading indicates areas where the difference is statistically significant at the 1% confidence level. After Wang and Chan [45] with the permission of the American Meteorological Society

Rainfall associated with typhoon has a strong impact on the environment. Rodger et al. [48] estimated monthly typhoon rainfall using the Special Sensor Microwave Imager (SSM/I) satellite data, and found about 12% of the climatological rainfall comes from typhoon over the WNP from June to November. In particular, local typhoon rainfall exceeds 60% in Taiwan from July to October [49]. Typhoon rainfall shows interannual variability associated with the ENSO. In the Philippines, summer rainfall increases and autumn rainfall decreases during El Niño and the contrast of summer and autumn rainfall comes from the combined contribution of typhoon rainfall increase of 29.8% in summer and decrease of 27.9% in autumn [44, 50]. Kubota and Wang [49] analyzed the regression of rainfall to winter Niño 3.4 SST in Figure 9. They separated rainfall into TC rainfall and non-TC rainfall and showed the increase in summer rainfall over the WNP islands was primarily caused by typhoon rainfall during the developing phase of El Niño. They found little change in autumn rainfall in El Niño years over the WNP islands, which is due to the offset of increasing typhoon rainfall and decreasing non-typhoon rainfall. Although the Walker circulation shifts eastward away from the WNP islands during El Niño, major typhoon tracks travel over the WNP islands west-northwestward and produces typhoon rainfall.

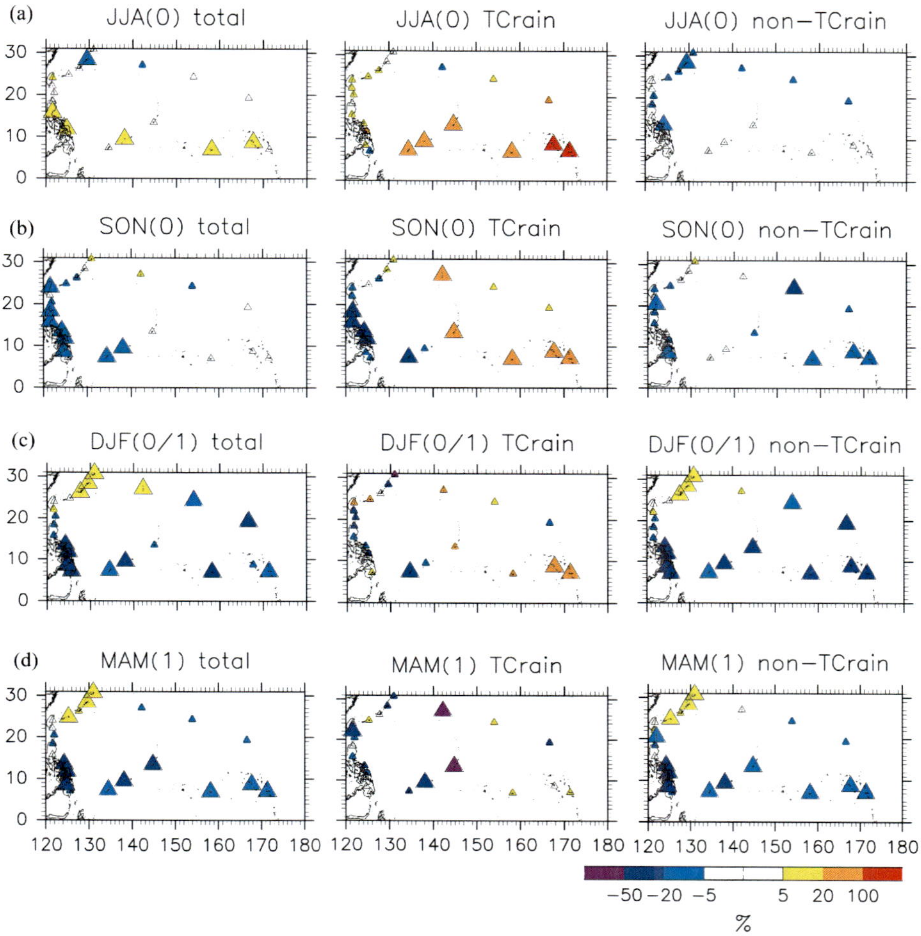

Figure 9. (a) (left) Regressed total rainfall, (middle) TC rainfall, and (right) non-TC rainfall anomalies during summer of El Niño/ La Niña developing year onJJA (0). The regression was made with respect to the Niño 3.4 SST in the mature phase of DJF(0/1), where 0 and 1 denote the El Niño/ La Niña developing and decaying years, respectively. The regressed rainfall has been divided by the corresponding seasonal means. (b)- (d) Same as (a), but for SON (0), DJF (0/1), and MAM (1), respectively. Larger symbols represent that the correlation coefficient is statistically significant at 95% confidence level. Color shadings indicate the regressions of rainfall anomalies. After Kubota and Wang [49] with the permission of the American Meteorological Society.

Recently a different type of El Niño was observed with a central Pacific SST warming near the dateline [51]. TC activity is also affected differently due to the different impact on atmospheric circulation [52].

Pacific-Japan Pattern

One of the dominant modes during Asian summer monsoon is the so-called Pacific-Japan (PJ) teleconnection pattern, defined by Nitta [53]. The PJ pattern is characterized by anomalous circulation between tropical convections over the Philippine Sea and anticyclones over the midlatitude. Anomalous convections over the Philippine Sea promote Rossby wave teleconnection poleward in intraseasonal and interannual time-scales [53-58]. Kawamura and Ogasawara [59] showed typhoon activity can produce anomalous convections as a heat source

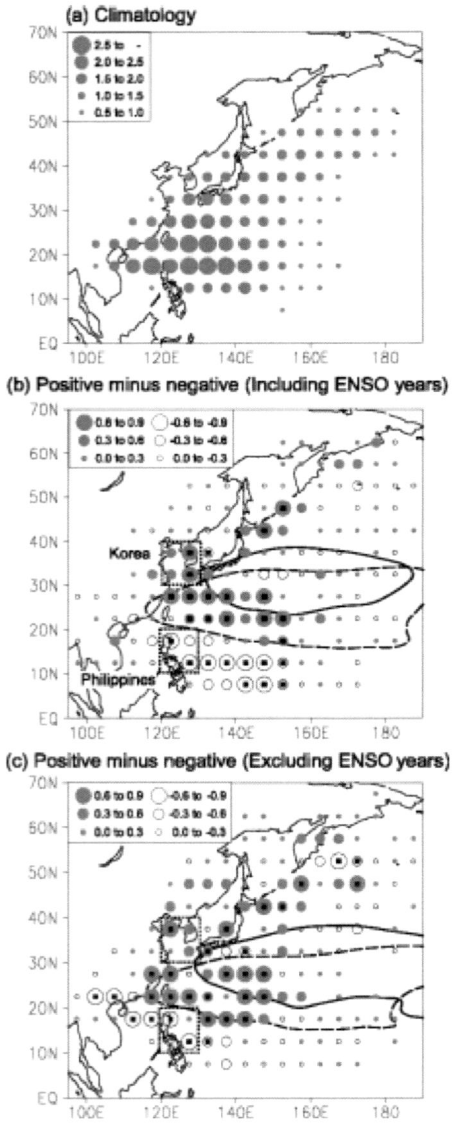

Figure 10. Summer JAS TC passage frequency for climatology, and difference between positive and negative PJ phases (b) with ENSO years and (c) selected from normal years. Solid and dashed contours indicate 5870 gpm for the positive and negative PJ phases, respectively. Small squares within circles are significant at the 95% confidence level. After Choi et al. [61] with the permission of the American Geophysical Union.

and, therefore, induces the PJ pattern. Yamada and Kawamura [60] mentioned the dynamical link between typhoon and PJ pattern in summer and autumn, that is, the PJ pattern remotely activates frontal system around Japan. From the view point of the Asian monsoon variability associated with the PJ pattern, typhoon genesis and track are strongly influenced by the PJ pattern [61]. Figure 10 shows typhoon forms in a larger number in northern locations during positive PJ pattern,which has more anomalous active convections over the Philippine Sea. Typhoon passes more frequently over northern regions toward Okinawa of Japan and Korea. The ENSO and the following summer PJ pattern are negatively correlated. Convections are active over the Philippine Seaduring the following summer of La Niña. The linkage between PJ pattern and typhoon variability becomes similar to that during the following summer of ENSO.

Indian Ocean Influence

El Niño teleconnection induces broad tropical Indian Ocean warming about three to six months after the peak SST of the equatorial eastern Pacific [62]. El Niño teleconnection also forces anomalous anticyclone over the WNP during the mature winter to next spring, which suppresses convections [63]. The delayed warming in the Indian Ocean acts as a capacitor, being charged up over the Indian Ocean for the following summer of El Niño year and influencing anomalous convective suppression over the Philippine Sea in summer [64]. The warm tropospheric Kelvin waves can trigger anomalous anticyclone over the WNP. Typhoon genesis number decreases over the WNP associated with the suppression of convective activity in the following summer after a strong El Niño[65]. Figure 11 shows the typhoon number, genesis potential [66], and zonal wind vertical shear (200 hPa minus 850 hPa) in the WNP (120°-180°E, 5°-25°N) during strong El Niño years. Typhoon number decreases in the following year of El Niño, due to unfavorable condition for typhoon genesis. The SST anomaly over the western and eastern equatorial Pacific is almost disappeared in the following summer after El Niño peak. Therefore, El Niño teleconnection becomes weak in the following summer. The capacitor effect of Indian Ocean warming may also suppress the convection and typhoon activity over the WNP.

Zhan et al. [67] discussed the influence of ENSO and eastern Indian Ocean SST anomaly (EIO SSTA) on typhoon activity over the WNP, respectively. The EIO SSTA affects the typhoon genesis number of weak to normal typhoons over the entire WNP; however the ENSO affects the number of intense typhoon activity. The mechanism for the warm EIO SSTA impact on the typhoon activity over the WNP is hypothesized by Xie et al. [64], as the reduction of land-sea thermal contrast in the maritime continent through the weakening of monsoon trough activity and warm equatorial Kelvin wave.

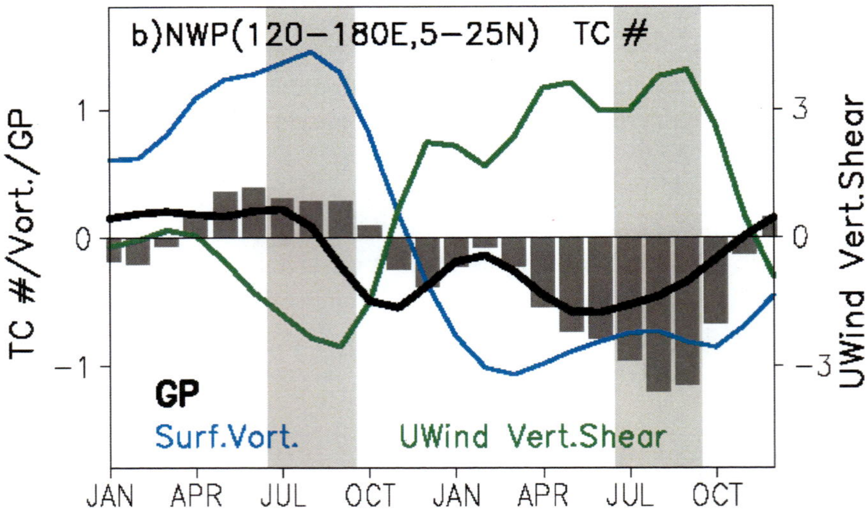

Figure 11. TC number (gray bars), genesis potential (GP) [66], zonal wind vertical shear (200hPa minus 850hPa; ms^{-1}), surface wind vorticity ($10^{-6}s^{-1}$) anomalies in the WNP (120°-180°E, 5°-25°N) during three strong El Niño events (1972/73, 1982/83, and 1997/98). After Du et al. [65] by the permission of the American Meteorological Society.

Quasi-Biennial Oscillation

The relation of typhoon activity to stratospheric Quasi-Biennial Oscillation (QBO) was the focus of Chan [68]. The QBO is an alternation of downward propagating westerly and easterly wind regimes in about 28 months over the equatorial stratosphere up to 50km [69]. The typhoon number over the WNP tends to increase during the westerly phase of the QBO. Gray et al. [70] hypothesized that upper-troposphere vertical shear decreases during the westerly wind regime, which creates favorable condition for convection and typhoon. However, the influence of the QBO is not large compared to that of the ENSO in terms of modulating typhoon activity over the WNP.

Long-Term Variability

When typhoon track data are accumulated for decades, the study for long-term variability of typhoon number becomes possible. Webster et al. [71] concluded the number of strong typhoon of Categories 4 and 5 had an increasing trend from 1970 to 2004 based on the JTWC typhoon track data. Emanuel [72] defined a power dissipation index as a potential destructiveness of typhoons and showed an increasing trend from the mid 1970s. Both studies showed a correlation of typhoon number with global warming of SST. However, Chan [73] claimed that this increasing trend of typhoon numbers is only part of the interdecadal variability. Kamahori et al. [74] showed an opposite trends of intense typhoon days in the JTWC and JMA typhoon track datasets. Chan and Liu [75] pointed out that the seasonal average of typhoon activity had no significant relationship with the SST in the WNP.

Chan and Shi [76] analyzed interdecadal variability of annual number of typhoon in the WNP from 1959 to 1994. High typhoon number appeared in the 1960s, low number in the late 1970s, and it increased again toward the mid 1990s. Yumoto and Matsuura [77] analyzed interdecadal variability of typhoon number from 1951 to 1999 over the WNP (Figure 12). There were enhanced periods of typhoon activity during 1961-72 and 1986-94 and weakened periods during 1951-60 and 1973-85. The authors suggested that favorable condition for deep convections was found during the period of high typhoon number. Anomalous high SST, anomalous low-level relative vorticity, and anomalous high-level divergence were seen east of 150°E in the WNP. Matsuura et al. [79] explained the mechanism of the interdecadal variability of typhoon activity using the NCEP/NCAR reanalysis data and a coupled general circulation model. However, for the explanation of long-term variability in typhoon number, interdecadal variability in the NCEP/NCAR reanalysis data has a discrepancy with the observation data [80, 81]. The explanation based on the reanalysis data for interdecadal variability needs to be verified.

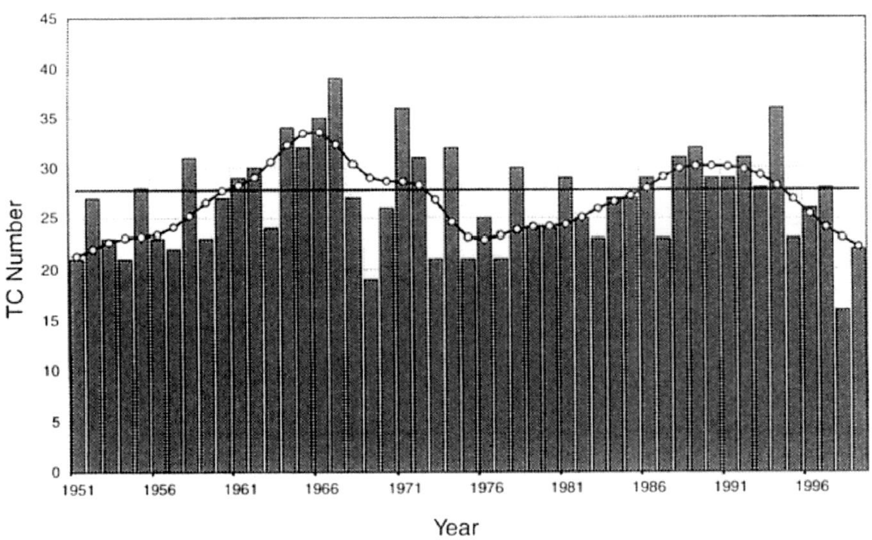

Figure 12. Annual typhoon numbers (bar chart) and smoothed annual number by 4(3RSR) running median smoothing [78] (line graph) during 1951-1999. After Yumoto and Matsuura [77] with the permission of Meteorological Society of Japan.

Interdecadal change of typhoon track was discussed by Ho et al. [82] and Wu et al. [83]. Both studies showed westward shift of typhoon tracks in the past 40 years. However, there exist discrepancies in typhoon track shift locations in the South China Sea and East China Sea, depending on typhoon track datasets the authors used. Both studies explained the typhoon track shift by the expansion of the WNP high using NCEP/NCAR reanalysis data.

Interdecadal variability of regional typhoon activity was carried out in Taiwan and Philippines. Tu et al. [84] showed a shift of typhoon number in terms of their passage in the vicinity of Taiwan before and after 2000 with the northward shift of typhoon tracks from the South China Sea and Philippine Sea region to Taiwan region. Kubota and Chan [16] recovered historical typhoon track data from the Philippine Weather Bureau from 1902 to 1940 and counted typhoon landfall number in the Philippines from 1902 to 2005 (Figure 13). Interdecadal variability of landfall number was studied but, no trend was found.

indicated a global decrease in global numbers of TCs in the future. After the IPCC AR4 [1], a number of high resolution models consistently showed a reduction of TC frequency due to global warming [2]. However, it has been noted that there are large uncertainties in regional TC frequency changes projected by various models, although the models consistently show a reduction of global TC frequency [2]. In this regard, we believe that the projections for the global change and regional change in the TC frequency should be considered separately, the former being more reliable than the latter. Moreover, a background behind the IPCC's conclusion that the projection of the TC frequency is less reliable than that of the TC intensity may be that the reduction of TC frequency is an unexpected projection. As a natural guess, we tend to think that we will have more intense TCs and more number of TCs in a warmer climate with increased moisture, because the energy source of TCs is the atmospheric moisture. For an unexpected projection to be accepted, a reasonable physical explanation is required. In this regard, understanding the mechanism of the reduction of global TC frequency in the future warmer climate is very important. The objective of this chapter is to review the GCM studies on the changes in TC frequency and intensity due to global warming, and present a coherent explanation for the mechanism of the projected changes, with a main focus on the change in the global TC frequency.

TC Frequency Changes Projected by GCMs

Results from medium and high resolution GCM experiments before the IPCC AR4 on the TC frequency changes are summarized in Table 1 [3-6]. The GCM experiments consistently show a reduction of global TC frequency, while there is substantial disagreement in the regional TC frequency changes (in Table 1, the regional changes are shown for North Western Pacific and North Atlantic). It should be noted that the physical processes of the JM8911 model and the Japan Meteorological Agency (JMA)/ Meteorological Research Institute (MRI) model are quite different, and the four models in Table 1 are independent each other. Although the resolution of the JMA/MRI model used by Oouchi et al. [6] is much higher than the other three models, the results of the four models are similar regarding the global TC frequency change. After the IPCC AR4, a number of medium and high resolution model projection experiments have been conducted [7-10], and summarized in Knutson et al. [2]. These medium and high resolution model projection experiments consistently also show a reduction of global TC frequency and substantial disagreement in the regional TC frequency changes.

Table 1. TC frequency changes as simulated by medium and high resolution GCMs. Changes are shown by the ratio of future TC frequency to present TC frequency

Reference	Model	Resolution	Experiment	Global	NWP	NA
Bengtsson et al. [3]	ECHAM4	120 km	5 years× 2	63 %	66 %	87 %
Sugi et al. [4]	JMA8911	120 km	10 years× 2	66 %	34 %	161 %
McDonald et al. [5]	HadAM3	100 km	15 years × 2	94 %	79 %	75 %
Oouchi et al. [6]	JMA/MRI	20 km	10 years × 2	70 %	62 %	134 %

upward mass flux, and the latter leads to the reduction of global TC frequency. From dynamical perspective, the reduced upward mass flux leads to a reduced low level vorticity convergence, which in turn leads to less TC genesis. On the other hand, from thermo-dynamical perspective, the enhanced precipitation associated with convections and reduced upward mass flux are favorable for development of TC warm core, which in turn leads to more intense TCs.

REFERENCES

[1] IPCC, Climate Change 2007: The Physical Science Basis, Cambridge Univ.Press, (2007).
[2] T. Knutson, J. McBride, J. Chan, K. Emanuel, G. Holland, C. Landsea, I. Held, J. P. Kossin, A. Srivastava and M. Sugi, Tropical cyclones and climate change. Nature Geoscience, 3,157 (2010).
[3] L. Bengtsson, M. Botzet and M. Esch, Tellus, 48A, 57 (1996).
[4] M. Sugi, A. Noda, and N. Sato, J. Meteorol. Soc. Japan. 80, 249 (2002).
[5] R. E. McDonald, D. G. Bleaken, D. R. Cresswell, V. D. Pope, and C. A. Senior, Clim. Dyn., 25, 19 (2005).
[6] Oouchi, K., J.Yoshimura, H. Yoshimura, R. Mizuta, S. Kusunoki, and A. Noda, J. Meteorol. Soc. Japan. 84, 259 (2006).
[7] L. Bengtsson, K. I. Hodges, M. Esch, N. Keenlyside, L. Kornblueh, J.-J. Luo, and T. Yamagata. Tellus. 59A, 539 (2007).
[8] S. Gualdi, Scoccimarro, E. and Navarra, J. Climate. 21, 5204 (2008)
[9] M. Zhao, I. Held, S.-J. Lin, and G. A. Vecchi, J. Climate. 22, 6653 (2009).
[10] M. Sugi, H. Murakami and J. Yoshimura, SOLA. 5, 164 (2009).
[11] I. Held, and B. J. Soden, J. Climate, 19, 5686 (2006).
[12] G. A. Vecchi, and B. J. Soden, J. Climate., 20. 4316 (2007).
[13] M. Allen and W. Ingram, Nature, 419, 224 (2002).
[14] M. Sugi, and J. Yoshimura, J. Climate, 17, 238 (2004).
[15] J. Yoshimura and M. Sugi, SOLA, 1, 133 (2005).
[16] T. Andrews, P. M. Forster and J.M. Gregory,J. Climate, 22, 2557 (2009)
[17] I. Held and M. Zhao, J. Climate. in press (2011).
[18] K. Emanuel, K., R. Sundararajan and J. Williams, Bull. Amer. Meteor. Soc., 89, 347 (2008).
[19] H. Murakami, H. and M. Sugi,. SOLA, 6, 73 (2010).
[20] T. Knutson, T. R., R. E. Tuleya, and Y. Kurihara, Science, 279, 1018 (1998).
[21] Walsh, K. J. E., K.-C. Nguyen, and J. L. McGregor, Clim. Dyn., 22, 47 (2004).
[22] W. M. Gray, Dept. of Atmos. Sci. Paper No. 234, Colorado State University, Fort Collins, CO, 121 pp (1975).
[23] Emanuel, K. and D. Nolan,: Preprints, 26th AMS Conference on Hurricanes and Tropical Meteorology, Miami, FL. (2004).
[24] S. Yokoi and Y. N. Takayabu, J. Meteor. Soc. Japan, 9, 525 (2009).
[25] H. Murakami, B. Wang and A. Kitoh, J. Climate, 23, 2699 (2010).
[26] H. Murakami, B. Wang and A. Kitoh, J. Climate, 24, 1154 (2011).

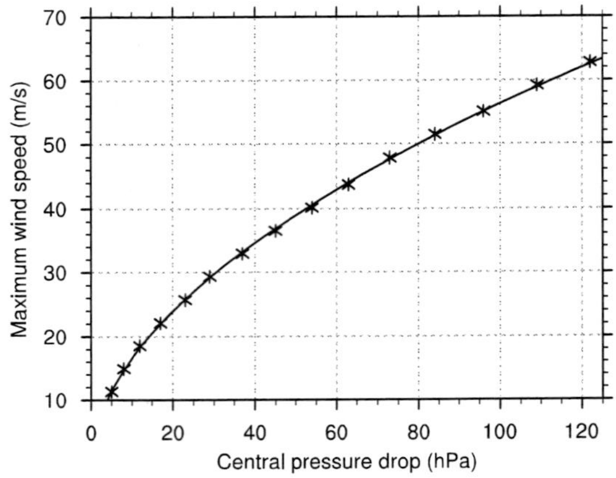

Figure 2. Relationship between the maximum surface (10-min) wind speeds and pressure drops at sea level of TCs in the western North Pacific. Asterisk markers represent values of the table used in the Japan Meteorological Agency [29], and the line is an approximate curve fitted to the values.

structure by integrating equation (3) with respect to r as

$$\delta T \sim \frac{UL}{R}\left(f + \frac{2U}{L}\right), \tag{4}$$

where U is the tangential velocity scale, which we regard as the maximum wind speed, and L is the horizontal scale of the warm-core structure. Assuming that the warmed climate does not change the horizontal scale of TCs, and considering an approximately proportional relationship between δT and δp_c in equation (1), we obtain

$$\frac{\Delta(\delta p_c)}{\delta p_c} = \left[1 + \frac{1}{1 + fL/(2U)}\right]\frac{\Delta U}{U}. \tag{5}$$

Since the Rossby number $U/(fL)$ is generally greater than unity for TCs, the term in the brackets [ ] in the previous equation is estimated to be slightly less than 2, and it turns out that the rate of change in wind speed is related to the rate of change in δp_c by a factor slightly greater than 0.5.

This factor can be determined from an empirical wind-pressure relation of TCs. According to Reference [28], one such empirical formula is given as

$$v_s = a(\delta p_c)^b, \tag{6}$$

where v_s is the surface maximum wind speed, and parameters a and b are determined, for example, as 4.9 and 0.53, respectively, by fitting this equation to the wind-pressure table used in the operational TC analysis at the Japan Meteorological Agency (JMA) [29], as shown in Fig. 2. v_s is defined as 10-min sustained wind at the reference (10 m) height, and the exponent 0.53 corresponds to the ratio of the rate of change in wind speed to the rate of change in the pressure drop.

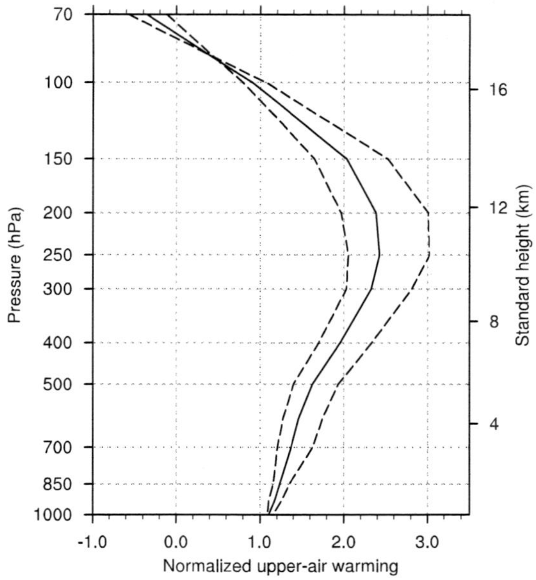

Figure 6. Normalized anomalies in August–September upper-air temperatures in the region of 29°N–31°N, 127°E–139°E. The solid line indicates ensemble means of all of the CMIP3 models, and each dashed line indicates those of subset models ranking as the six largest/smallest anomalies at the 200-hPa level. Source: Figure 2 of Reference [14].

3. Estimate the fractional change of the maximum wind speed near the TC center by using the ratio of the wind speed change to the δp_c change, which is typically 0.53.

4. Estimate the ω_e profile in the warmed climate by scaling that in the present climate in proportion to the maximum wind speed.

5. Compute the fractional change of the P_e from its difference between the present and warmed climate.

3.3. Representative changes

Table 1 summarizes results for the $\Delta\bar{T}_s$ of 1°C and 2°C. The rates of change in the δp_c and P_e for the 1°C warming are 6.5% and 9.3%, respectively, as the best estimates, which are nearly half that of the 2°C warming. The uncertainty of δp_c ranges from slightly less than the present level to almost twice that of the best estimate. This uncertainty range comes from the variations of the normalized temperature anomalies shown in Fig. 6, and the greater deviation of the normalized temperature anomalies for the larger anomaly models is responsible for the slight skewness of the uncertainty range.

The MPI change due to CMIP3 warming anomalies has been examined for broader applications to TC seasons and regions of development in the western North Pacific, where the average fractional change in the δp_c is 3.6%, 8.4%, and 19% in response to SST increases of 0.5, 1.0, and 2.0°C, respectively [13]. This relationship between TC intensification and SST increase is essentially comparable to representative studies based on observations [37]

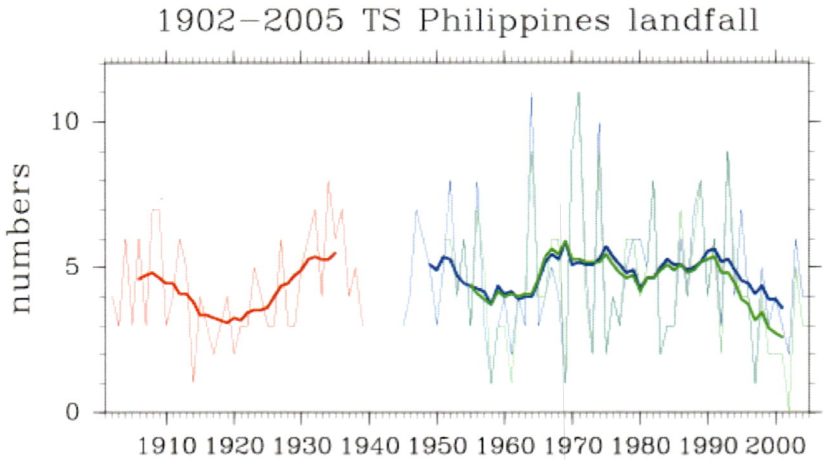

Figure 13. Annual typhoon landfall number in the Philippines using the Philippine Weather Bureau typhoon dataset (from 1902 to 1939; red curves), JTWC (from 1945 to 2005; blue curves), and JMA (from 1951 to 2005; green curves), where the thick curves are for 10-year-running mean. After Kubota and Chan [16] with the permission of the American Geophysical Union.

There are many discrepancies in various studies on the interdecadal variability of typhoon activity. One of the problems is that each typhoon track dataset over the WNP is made by different wind speed definition and used by different Dvorak technique to determine the typhoon intensity [85]. Discontinuity in these techniques used for creating typhoon track dataset due to the deactivation of the American aircraft reconnaissance for direct typhoon observation over the WNP in 1987 and the improvement of Dvorak technique also created problems. At present length of available typhoon track dataset is not sufficient for interdecadal studies. The trial to recover historical typhoon track data will be a great tool for long-term variability study.

Summary

The typhoon number and tracks in the WNP have intraseasonal to interdecadal time-scalesassociated with large-scale atmospheric and oceanic conditions. The conditions for deep convections and typhoon activity over the WNP have changed due to the phase change in intraseasonal and interannual time-scales, such as the MJO and ENSO. Typhoon number increases during the active phase of MJO. However, even in the absence of MJO typhoon number can increase in the active phase of intraseasonal oscillation. Typhoon activity was also related to shorter periods of 10 to 30 day westward propagating disturbances. The modulation of typhoon associated with ENSO is found in the WNP. During El Niño, typhoon genesis location shifts southeastward; landfall number increases in summer and decreases in autumn in Southeast Asia; typhoons become more intense and have longer life span; and typhoon rainfall increases in both summer and autumn but is offset in autumn by the reduction in non-typhoon rainfall over the WNP islands. The mechanism of southeastward shift of typhoon genesis locations is explained by the relocation of monsoon trough with the eastward shift of the Walker circulation. The Pacific-Japan pattern of Asian summer monsoon affects the typhoon number in northern regions of the WNP. Indian Ocean warming reduces typhoon number in the WNP by teleconnection. QBO also has an impact on the variability of typhoon number through the upper-troposphere vertical shear. Interdecadal variability of typhoon activity is found in the WNP. However, typhoon studies on interdecadal time-scale have many discrepancies due to the short period of typhoon track datasets. There are also differences in typhoon track datasets in terms of typhoon definition and the quality of datasets. Recently, climate change and global warming issues became a great concern, and their influences on typhoon are highlighted. The demand for long-term typhoon track data becomes higher. The trial to recover historical typhoon track data will be a great tool for long-term variability study. A unified typhoon track dataset over the WNP for a hundred years will help to answer this question.

Acknowledgments

The activity of collecting historical typhoon track data over the WNP was supported by "Global Environment Research Fund by the Ministry of the Environment Japan" B-061, "Data Integration & Analysis System" funded by the National Key Technology, Ministry of Education, Culture, Sports, Science and Technology (MEXT), Grant-in-Aid for Scientific Research 20240075 of the MEXT (Leader: Prof. Jun Matsumoto), and Grant-in-Aid for Young Scientific Research 21684028 of the MEXT. The author thanks Dr. Mong-Ming Lu ofthe Central Weather Bureau, Prof. Chih-Wen Hung of National Taiwan Normal University, Prof. Chung-Hsiung Sui of National Taiwan University, Mr. Edwin W. L. Ginn of the Hong Kong Observatory, Dr. Zhao Bingke of Shanghai Typhoon Institute, Dr. Esperanza O. Cayanan of the Philippine Atmospheric Geophysical and Astronomical Services Administration, University of Hawaii, and Japan Meteorological Agency for sharing historical typhoon books. The author thanks Prof. Jun Matsumoto, Dr. Koichi Masuda, Prof. Masumi Zaiki, Dr. Ikumi Akasaka, Dr. Kumi Kataoka, Dr. Mika Ichino, and Ms. Kazuko Fukuda for their help of collecting historical weather bulletins during the 19th and 20th century. GFD-DENNOU library was used for drawing the figures.

References

[1] P. K. MacKeown, *Early China coast meteorology,* Hong Kong University Press, (2010).

[2] J. Algué, J., *Baguiosó Tifones de 1894,* Philippine Weather Bureau, (1895).

[3] L. Froc, *Atlas of the tracks of 620 typhoons 1893-1918*, Zi-Ka-Wei Observatory, Chi China, (1920).

[4] S. S. Visher, *Mon. Wea. Rev.*, 50, 583, (1922)

[5] J. Bjerknes, *Mon. Wea. Rev.*, 97, 163, (1969).

[6] R. A. Madden, and P. R. Julian, *J. Atmos. Sci.,* 28, 702 (1971).

[7] R. A. Madden, and P. R. Julian, *J. Atmos. Sci.,* 29, 1109 (1972).

[8] A. Udías, *Bull. Amer. Meteor. Soc.*, 77, 2307 (1996).

[9] C. E. S. J., Deppermann, *Some characteristics of Philippine typhoons,* Weather Bureau, Manila, (1939).
[10] J. Algué, *The cyclones of the far east*, Philippine Weather Bureau, (1904).
[11] Central Meteorological Observatory Tokyo Japan, *Trajectories of tropical cyclones 1940-1950,* (1951).
[12] Y. Nyomura, *Typhoon stories,* Japan Weather Association, (1986).
[13] Y.-X. Gao, and Y.-S. Zeng, *Typhoon tracks and statistical analysis*, Chinese Science Publications, (1957).
[14] P. C. Chin, *Tropical cyclones in the Western Pacific and China Sea area*, Royal Observatory Hong Kong, (1958).
[15] S.-L. Shieh, S.-T. Wang, M.-D. Cheng, and T.-C. Yeh, *Tropical cyclone tracks over Taiwan and its vicinity 1897-1996*, Central Weather Bureau, Taiwan, (1998).
[16] H. Kubota, and J. C. L. Chan, *Geophys. Res. Lett.*, 36, L12802, doi:10.1029/2009GL038108, (2009).
[17] K. R. Knapp, M. C. Kruk, D. H. Levinson, H. J. Diamond and C. J. Neumann, *Bull. Amer. Meteor. Soc.*, 91, 364 (2010).
[18] Y. N. Takayabu, *J. Meteor. Soc. Japan,*72, 443 (1994).
[19] T. Matsuno, *J. Meteor. Soc. Japan*, 44, 25 (1966).
[20] A. E. Gill, *Q. J. R. Meteorol. Soc.*, 106, 447 (1980).
[21] R. A. Madden, R. A. and P. R. Julian, *Mon. Wea. Rev.,* 122, 814 (1994).
[22] R. A. Keen, *Mon. Wea. Rev.*, 110, 1405 (1982).
[23] T. Nitta, and T. Motoki, *J. Meteor. Soc. Japan,* 65, 497 (1987).
[24] H. Kubota, R. Shirooka, T. Ushiyama, J. Chen, T. Chuda, K. Takeuchi, K. Yoneyama and M. Katsumata, *J. Meteor. Soc. Japan*, 84, 115 (2006).
[25] B. Liebmann, H. H. Hendon, and J. D. Glick, *J. Meteor. Soc. Japan*, 72, 401 (1994).
[26] K.-M. Lau, and P. H. Chan, *Mon. Wea. Rev.*, 114, 1354 (1986).
[27] K.-M. Lau, G. J. Yang, and S. H. Shen, *Mon. Wea. Rev.*, 116, 18 (1988).
[28] D. L. Hartmann, M. L. Michelsen, and S. A. Klein, *J. Atmos. Sci.*, 49, 1277 (1992).
[29] Y. Fukutomi, and T. Yasunari, *J. Meteor. Soc. Japan*, 77, 753 (1999).
[30] K.-M. Lau, and D. E. Waliser, *Intraseasonal variability in the atmospheric-ocean climate system*, Springer, Heidelberg, Germany, (2005).
[31] E. M. Rasmusson, and T. H. Carpenter,*Mon. Wea. Rev.,* 110, 354 (1982).
[32] G.. D. Atkinson, *11th Tech. Conf. on Hurricanes and Tropical Meteorology,* Miami Beach, FL, Amer. Meteor. Soc., 645 (1977).
[33] J. C. L. Chan, *Mon. Wea. Rev.*, 113, 599 (1985).
[34] M. A. Lander, *Mon. Wea. Rev.*, 122, 636 (1994).
[35] M. A. Lander, *Mon. Wea. Rev.*, 121, 2137 (1993).
[36] J. C. L. Chan, *J. Climate*, 13, 2960 (2000).
[37] T.-C. Chen, S.-P. Weng, N. Yamazaki, and S. Kiehne, *Mon. Wea. Rev.*, 126, 1080 (1998).
[38] W. M. Gray, *Mon. Wea. Rev.*, 96, 669 (1968).
[39] C. S. Ramage, *Mon. Wea. Rev.,* 102, 745 (1974).
[40] H. H. Chia, and C. F. Ropelewski, *J. Climate*, 15, 2934 (2002).
[41] H. Fudeyasu, S. Iizuka, and T. Matsuura, *Geophys. Res. Lett.*, 33, L21815, doi:10.1029/2006GL027449, (2006).

[42] M. A.Saunders, R. E. Chandler, C. J. Merchant, and F. P. Roberts, *Geophys. Res. Lett.*, 27, 1147 (2000).
[43] M. C. Wu, W. L. Chang, and W. M. Leung, *J. Climate,* 17, 1419 (2004).
[44] B. Lyon, and S. J. Camargo, *Clim. Dyn.,* doi:10.1007/s00382-008-0380-z, (2008).
[45] B. Wang, and J. C. L. Chan, *J. Climate,* 15, 1643 (2002).
[46] S. J. Camargo, and A. H. Sobel, *J. Climate*, 18, 2996 (2005).
[47] S. J. Camargo, A. W. Robertson, S. J. Gaffney, P. Smyth, and M. Ghil, *J. Climate*, 20, 3654 (2007).
[48] E. B. Rodgers, R. F. Adler, and H. F. Pierce, *J. Appl. Meteor.,* 39, 1658 (2000).
[49] H. Kubota, and B. Wang, *J. Climate*, 22, 5495 (2009).
[50] B. Lyon, H. Cristi, E. R. Verceles, F. D. Hilario, and R. Abastillas, *Geophys. Res. Lett.*, 33, L24710, doi:10.1029/2006GL028182, (2006).
[51] K. Ashok, and T. Yamagata, *Nature*, 461, 481 (2009).
[52] H.-M. Kim, P. J. Webster, and J. A. Curry, *J. Climate*, 24, 1839 (2011).
[53] T. Nitta, *J. Meteor. Soc. Japan*, 65, 165 (1987).
[54] T. Nitta, *J. Meteor. Soc. Japan*, 64, 373 (1986).
[55] T. Nitta, *Meteo. Atmos. Phys.*, 41, 5 (1989).
[56] K. Kurihara, and T. Tsuyuki, *J. Meteor. Soc. Japan*, 65, 237 (1987).
[57] R. Kawamura, T. Murakami, and B. Wang, *J. Meteor. Soc. Japan*, 74, 867 (1996).
[58] Y. Kosaka, and H. Nakamura, *Q. J. R. Meteorol. Soc.*, 132, 2009 (2006).
[59] R. Kawamura, and T. Ogasawara, *SOLA*, 2, 37 (2006).
[60] K. Yamada, and R. Kawamura, *SOLA*, 3, 65 (2007).
[61] K.-S. Choi, C.-C. Wu, and E.-J. Cha, *J. Geophys. Res.*, 115, D19114, doi:10.1029/2010JD013866, (2010).
[62] S. A. Klein, B. J. Soden, and N.-C. Lau, *J. Climate*, 12, 917 (1999).
[63] B. R.Wang, Wu, and T. Li, *J. Climate*, 16, 1195 (2003).
[64] S.-P. Xie, K. Hu, J. Hafner, H. Tokinaga, Y. Du, G. Huang, and T. Sampe, *J. Climate,*22, 730 (2009).
[65] Y. Du, L. Yang, and S.-P. Xie, *J. Climate*, 24, 315 (2011).
[66] K. A. Emanuel, and D. S. Nolan, *Proc. 26th Conf. on Hurricane and Tropical Meteorology,* Amer. Meteor. Soc., 240 (2004).
[67] R. Zhan, Y. Wang, and X. Lei, *J. Climate*, 24, 509 (2011).
[68] J. C. L.Chan, *Mon. Wea. Rev.*, 113, 2567 (1995).
[69] M. P. Baldwin, L.J. Gray, T. J. Dunkerton, K. Hamilton, P. H. Haynes, W. J. Randel, J. R. Holton, M. J. Alexander, I. Hirota, T. Horinouchi, D. B. A. Jones, J. S. Kinnersley, C. Marquardt, K. Sato, and M. Takahashi, *Rev. of Geophys.*, 39, 179 (2001).
[70] W. M. Gray, J. D. Sheaffer, and J. A. Knaff, *J. Meteor. Soc. Japan*, 70, 975 (1992).
[71] P. J. Webster, G. J. Holland, J. A. Curry, and H.-R. Chang, *Science*, 311, 1713 (2005).
[72] K. A. Emanuel, *Nature*, 436, 686 (2005).
[73] J. C. L. Chan, *Science*, 311, 1713b, (2006).
[74] H. Kamahori, N. Yamazaki, N. Mannoji, and K. Takahashi, S*OLA*, 2, 104 (2006).
[75] J. C. L. Chan, and K. S. Liu, *J. Climate,* 17, 4590 (2004).
[76] J. C. L. Chan, and J. E. Shi, *Geophys. Res. Lett.*, 23, 2765 (1996).
[77] M. Yumoto, and T. Matsuura, *J. Meteor. Soc. Japan*, 79, 22 (2001).
[78] J. W. Tukey, *Exploratory Data Analysis*, Reading Massachusetts, Addison-Wesley, (1977).

[79] T. Matsuura, M. Yumoto, and S. Iizuka, *Clim. Dyn.,* 21, 105 (2003).
[80] R. Wu, and S.-P. Xie, *J. Climate,* 16, 167 (2003).
[81] R. Wu. J. L. Kinter III, B. P. Kirtman, *J. Climate,* 18, 3048 (2005).
[82] C.-H. Ho, J.-J. Biak, J.-H. Kim, and D.-Y. Gong, *J. Climate*, 17, 1767 (2004).
[83] L. Wu, B. Wang, and S. Geng, *Geophys. Res. Lett.*, 32, doi:10.1029/2005GL022937, (2005).
[84] J.-Y.Tu, C. Chou, and P.-S. Chu, *J. Climate*, 22, 3617 (2009).
[85] V. F.Dvorak, *Mon. Wea. Rev.*, 103, 420 (1975).

In: Cyclones: Formation, Triggers and Control
Editors: K. Oouchi and H. Fudeyasu
ISBN: 978-1-61942-976-5
©2012 Nova Science Publishers, Inc.

Chapter 5

REDUCTION OF GLOBAL TROPICAL CYCLONE FREQUENCY DUE TO GLOBAL WARMING

Masato Sugi*
Japan Agency for Marine-Earth Science and Technology,
Yokohama, Japan

ABSTRACT

In this chapter, we review recent medium and high resolution GCM studies on the changes in tropical cyclone (TC) frequency due to global warming, and present a coherent explanation for the mechanism of projected changes, with a main focus on the reduction of global TC frequency. The reduction of global TC frequency is explained by a weakening of tropical circulation, which is closely related to a large increase in atmospheric stability and a small increase in precipitation. An interesting point is that the overlap effect of CO_2 and water vapor absorption bands in long wave radiation is playing an important role in the small increase in precipitation.

INTRODUCTION

In the Summary for Policy Makers (SPM) of the Intergovernmental Panel for Climate Change Fourth Assessment Report (IPCC AR4) [1], it is concluded that: "Based on a range of models, it is likely that future tropical cyclones (typhoons and hurricanes) will become more intense, with larger peak wind speeds and more heavy precipitation associated with ongoing increases of tropical sea surface temperature (SST). There is less confidence in projections of a global decrease in numbers of tropical cyclones." This conclusion is based on the chapter ten of the IPCC AR4 [1], in which the discussion is based on the results of many coarse resolution (grid size of 100km or more) GCMs, one high resolution (grid size of 20km) GCM and several high resolution (grid size of 30km or less) regional models. The discussion overlooked that recent medium resolution (about 100km resolution) GCM results consistently

* E-mail address: msugi@jamstec.go.jp

Bengttson et al. [3] first showed that a significant reduction in the global TC frequency is likely in the future warm climate based on their GCM experiment (Table 1). They suggested that the reduction of global TC frequency is related to the changes in global scale fields such as Hadley circulation and subtropical jets. Sugi et al. [4] also found a significant reduction in the global TC frequency in their experiment (Table 1). They also suggested that the reduction in global TC frequency is closely related to a weakening of the tropical circulation. They further suggested that the weakening of the tropical circulation is explained by a large increase in the atmospheric dry stability and small increase in the tropical precipitation in the GCM experiment.

WEAKENING OF TROPICAL CIRCULATION

Held and Soden [11] and Vecchi and Soden [12] showed that a weakening of tropical circulation, a weakening of the upward mass flux associated with tropical convections, is a very robust feature in the climate change projected by Third Coupled Model Intercomparison Project (CMIP3) models. They attributed the weakening of upward mass flux to a small increase in precipitation and a large increase in moisture. The weakening of tropical circulation with small increase in precipitation indicates a weakening of hydrological cycle in terms of mass flux and a little intensification of hydrological cycle in terms of moisture flux. To understand the mechanism of the weakening of tropical circulation, Sugi et al. [4] considered a basic energy balance in the tropical circulation. The approximate energy balance equation in the tropics is expressed as,

$$\omega \frac{\partial \theta}{\partial p} \approx \frac{\theta}{T} \frac{Q}{C_p} \tag{1}$$

In the equation, θ is potential temperature, ω is vertical p-velocity, p is pressure, T is temperature, Q is diabatic heating and C_p is specific heat of dry air at constant pressure.

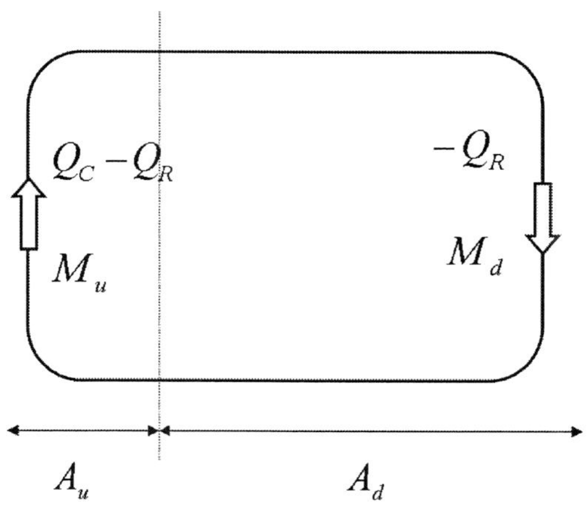

Figure 1. A schematic diagram showing the energy balance of tropical circulation.

As shown schematically in Figure 1, the basic tropical circulation consists of intense upward motions associated with convective updrafts in very small areas A_u and weak downward motions (compensating subsidence) in very large areas A_d, and upper level and lower level flows connecting these upward motions and downward motions. In the upward motion areas, the diabatic heating Q_C-Q_R (condensation heating in the convective clouds minus radiative cooling) is balanced by the adiabatic cooling associated with the upward motion, while in the downward motion areas, the radiative cooling Q_R is balanced by the adiabatic warming associated with the downward motion. Total upward mass flux $M_u=A_u \times \omega_u$ and downward mass flux $M_d=A_d \times \omega_d$ in the tropics should balance, neglecting the small net transport of mass between the tropics and mid-latitudes. Furthermore, the total diabatic heating in the upward motion areas in the tropics should be balanced by the total radiative cooling in the downward motion areas. Therefore, using the Eq.(1) and dry static stability S defined as,

$$S \equiv \frac{C_p T}{\theta} \left| \frac{\partial \theta}{\partial p} \right| \tag{2}$$

the energy balance at the mid-tropospheric level (e.g. 500hPa) for the upward motion area and downward motion area can be expressed as,

$$M_u S \approx (Q_C - Q_R) A_u \tag{3}$$

and

$$M_d S \approx Q_R A_d \tag{4}$$

The Eqs. (3) and (4) indicate that if the diabatic heating is increased and the stability does not change, then the upward and downward mass fluxes should increase. However, if the stability significantly increases and its increase is larger than the increase in diabatic heating, then the mass flux should decrease. Therefore, the weakening of tropical circulation can be explained by the significant increase in stability and little increase in precipitation (diabtic heating). The next question is why stability significantly increases, while precipitation increases only little in the global warming.

The reason for why the stability increases significantly is that the stability is closely related to the atmospheric moisture. We should note that the lapse rate of temperature in the tropics is close to the moist adiabat lapse rate, because the temperature profile in deep convective clouds is close to moist adiabat. The temperatures of deep convective clouds at level of free convection (LFC) near the cloud base and level of neutral buoyancy (LNB) near the cloud top are the same as the respective environment atmospheric temperatures. Therefore, if we define the mean tropospheric stability in the tropics as the difference in potential temperature at LNB and LFC, then it is the same as the potential temperature difference between LNB and LFC of a moist adiabat. This potential temperature difference $\Delta\theta=\theta_{LNB} - \theta_{LFC}$ is proportional to the condensation heating of the saturated air parcel rising from LFC to LNB along a moist adiabat:

$$\Delta\theta = \theta_{LNB} - \theta_{LFC} \approx \frac{L}{C_p} q_{LFC} \left(\frac{p_0}{p} \right)^{\frac{R}{C_p}} \tag{5}$$

where L is latent heat of condensation and R is gas constant for dry air. The potential temperature difference $\Delta\theta$ is nearly proportional to the atmospheric moisture near the cloud base. This is the reason for why stability of the tropical atmosphere is closely related to the atmospheric moisture and it significantly increases due to global warming.

PRECIPITATION CHANGE

The next question is why precipitation does not increase much, when the atmospheric moisture significantly increases due to global warming. To this question Allen and Ingram [13] answered as "changes in the overall intensity of the hydrological cycle are controlled not by the availability of moisture, but by the atmospheric energy". Figure 2 shows the relations between the changes in precipitation and atmospheric moisture and the change in surface temperature when CO_2 is doubled. The precipitation increases by about 3% as surface temperature increases 1°C, while atmospheric moisture increases about 7%. This is because the change in the precipitation is controlled by the change in radiative cooling in the atmosphere which is much less than the change in the moisture. It should be noted that this is consistent with the energy balance shown in the Figure 1. The total condensation heating in the upward motion areas is in balance with the total radiative cooling in the downward motion areas.

Another important point in Figure 2 is that the line showing the precipitation-temperature relation does not pass through the origin of the graph and there is a large negative intercept. This means that if the surface temperature does not increase at all when CO_2 is doubled, then the precipitation decreases as much as 5%. This precipitation decrease in response to CO_2 doubling can be explained by the reduction of radiative cooling due to the overlap effect of long wave radiation of CO_2 and water vapor long wave absorption bands [14]. Since CO_2 and water vapor have absorption bands of long wave radiation in the same wave length (overlap of absorption bands), the downward long wave flux emitted from CO_2 in the upper atmosphere is absorbed by the water vapor in the lower troposphere, and the radiative cooling there due to water vapor is significantly reduced (overlap effect). In general, when greenhouse gas in the atmosphere is increased, it increases downward long wave flux and increases atmospheric radiative cooling as well as heating of the earth's surface (greenhouse effect). On the other hand, when CO_2 is increased, the radiative cooling at the lower troposphere is significantly reduced by the overlap effect. Figure 2 indicate that the overlap effect is playing a significant role in the precipitation change due to CO_2 increase. However, in Figure 2 the overlap effect may be overestimated. Yoshimura and Sugi [15] and Andrews et al. [16] have shown that precipitation decreases by about 2 to 3% due to the overlap effect when CO_2 is increased without changing SST.

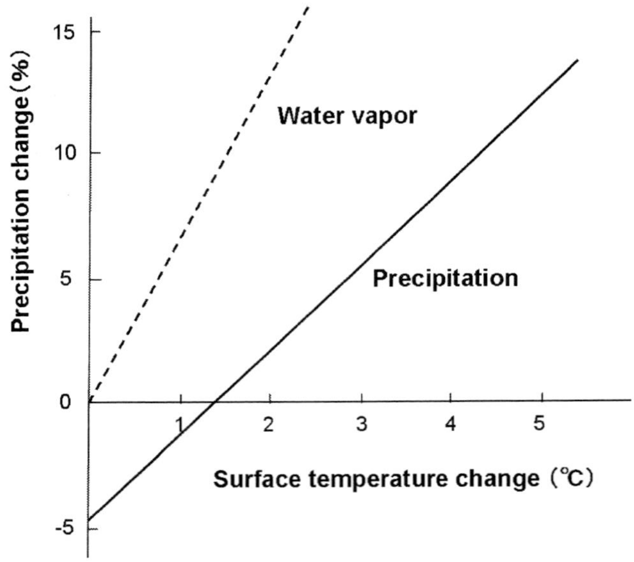

This figure is made based on the Figure 2 in Allen and Ingram [13].

Figure 2. Precipitation change (solid line) and water vapor change (dashed line) as a function of surface temperature change when CO2 is doubled.

MECHANISM OF THE REDUCTION OF GLOBAL TC FREQUENCY

We can summarize the mechanism of the reduction of global TC frequency as follows. The atmospheric moisture significantly increases due to global warming. This moisture increase leads to a large increase in the atmospheric stability, while precipitation (condensation heating) and radiative cooling do not increase as much. As a result, tropical circulation becomes weaker. The upward branch of the tropical circulation is the upward mass flux associated with convective updraft. The weakening of tropical circulation indicates a decrease in upward mass flux as a whole. And this decrease in the upward mass flux associated with convective updraft leads to the reduction of global TC frequency as simulated in the GCMs.

The last point in the above argument, the relationship between the decrease in total upward mass flux in the tropics and the decrease in the global TC frequency may not be straightforward. The mass flux associated with TCs is only a few percent of total mass flux in the tropics, because the precipitation associated with TCs is only a few percent of total precipitation in the tropics. It is not yet clear how a change in total tropical mass flux can constrain the number of TCs.

However, the above argument well explains the results of GCM experiments so far conducted. In particular, the experiments which examined separately the effects of SST increase and CO_2 increase on the TC frequency strongly support the above argument. In the experiment by Yoshimura and Sugi [15], TC frequency is reduced by 12% when SST is uniformly increased by 2K with fixed CO_2, although TC frequency does not change much when SST is uniformly decreased by 2K. On the other hand, global TC frequency decreases by 13% when CO_2 is doubled with fixed SST. Recently Held and Zhao [17] conducted a similar experiment. They found that the reduction of global TC frequency when CO_2 is doubled with fixed SST is about 10%, which is almost the same as the reduction of global TC frequency when SST is increased by 2K with fixed CO_2. These experiments indicate that both in the CO_2 increase experiment and SST increase experiment, the rate of reduction of global TC frequency is similar. They showed that the rate of reduction of the upward mass flux in the both experiments is also similar. It should be noted, however, that the reason for the reduction of the mass flux is different in the two experiments. The reason is a reduction of precipitation due to the overlap effect in the CO_2 increase experiment, while it is an increase in stability in the SST increase experiment, in which precipitation increases. An important point is that whatever the reason is, the reduction of mass flux leads to a reduction of TC frequency in both the CO_2 increase experiments and SST increase experiments. On the other hand, an increase in saturation-deficit, which was proposed by Emanuel et al. [18] as a mechanism for the TC frequency reduction, cannot explain the reduction of TC frequency in the CO2 increase experiment, in which SST is fixed and saturation deficit does not change.

Regional TC Frequency Change

Table 1 shows significant disagreement among the GCM experiments in the changes in the regional TC frequency, in contrast to the overall agreement in the global TC frequency change. The disagreement among the GCM experiments in the projected regional TC frequency is also found in recent medium and high resolution GCM experiment [2]. To explain this disagreement, we need to consider some other factors, which might affect the regional TC frequency, in addition to the argument discussed above with regard to the global TC frequency change. Sugi et al. [4] noted that the regional TC frequency change in their experiment is closely related to the pattern of the SST change. A significant decrease in the typhoons in the North Western Pacific and considerable increase in the Hurricanes in the North Atlantic are related to the relatively small increase and large increase in SST in the respective regions. A significant increase in the number of TC in the central Pacific in the McDonald et al. [5] seem to be related to a relatively large increase in SST over the region in their experiment. Recently, Sugi et al. [10] conducted a series of experiments with high resolution models, in which different SST changes projected by different climate models are prescribed. They found that the regional TC frequency changes projected by different models are similar to each other if the same SST change distributions are prescribed. The regional TC genesis frequency is closely related to the regional convective activity, and therefore, the changes in regional TC frequency are sensitive to relative SST change distribution which dominates the changes in regional convective activities. Thus, the difference among the models in regional TC frequency projection may be explained to a large extent by the difference among the distributions of SST change used in each experiment. The regional frequency of TCs clearly depends on the relative SST distribution.

Increase of Intense TC Frequency

Some high resolution models clearly show that the number of intense TCs may increase, even though the number of relatively weak TCs may decrease due to global warming [6, 7, 9, 19]. The increase in the number of intense TCs is consistent with regional model experiments by Knutson et al. [20] and Walsh et al. [21], while the decrease in the number of relatively weak TCs is consistent with the reduction in global total TC frequency. This suggests that these two changes are compatible and do not contradict with each other. Even though the total upward mass flux in the tropics decreases and the number of tropical cyclogenesis is reduced, once a TC forms, the TC may develops into a very intense storm by TC intensification mechanism which is not much affected by the general weakening of the tropical circulation.

Summary and Discussion

We have reviewed possible mechanisms of the changes in frequency and intensity of TCs as simulated by GCMs, with a focus on the change in global TC frequency. The reduction of global TC frequency is explained by a weakening of tropical circulation, that is, a reduction of total tropical mass flux associated with the convective updrafts. The reduction of total mass flux in the tropics is closely related to a large increase in the atmospheric stability caused by the significant increase in the atmospheric moisture due to global warming and a small increase in the precipitation (condensation heating) which drives the tropical circulations. A very interesting point is that the overlap effect of CO_2 and water vapor is playing an important role in the little increase in precipitation.

The mechanism of the TC frequency change due to global warming often discussed by using Yearly Genesis Parameter (YGP) by Gray [22] or Genesis Potential Index (GPI) by Emanuel and Nolan [23]. Sugi et al. [4] examined the changes in YGP parameter in their GCM experiment, and found that the dynamical factors such as vertical wind shear and low level relative vorticity play more important roles to determine the regional TC frequency than the thermo-dynamical factors such as SST and moist stability. In other words, a change in SST distribution can cause a change in tropical circulation and indirectly affect the tropical cyclogenesis in a remote region, rather than directly affect through a change in the thermo-dynamical structure over the region of the SST change. McDonald et al. [5] also noted the importance of dynamical factors in the changes in regional TC frequency. Yokoi and Takayabu [24] discussed the mechanism of changes in regional TC genesis frequency in western North Pacific by using GPI. Murakami et al. [25, 26] discuss the changes in regional TC genesis in North Atlantic and western North Pacific by using a modified GPI. These discussions on the mechanism of the future TC frequency changes using YGP or GPI must be viewed with caution. YGP or GPI is an empirical formula developed on the present day climate condition, and its applicability in the future climate condition is not guaranteed. Moreover, the several factors included in these formula are not independent each other, being most of them related to conditions favorable for development of deep convections.

The changes in TC frequency is no doubt related to the changes in the activity of deep convections in the tropics. The main point of our argument is that in the global warming the activity of deep convection is enhanced in terms of precipitation but it is reduced in terms of

In: Cyclones: Formation, Triggers and Control
Editors: K. Oouchi and H. Fudeyasu
ISBN: 978-1-61942-976-5
© 2012 Nova Science Publishers, Inc.

Chapter 6

ESTIMATION OF CHANGES IN TROPICAL CYCLONE INTENSITIES AND ASSOCIATED PRECIPITATION EXTREMES DUE TO ANTHROPOGENIC CLIMATE CHANGE

Junichi Tsutsui*
Central Research Institute of Electric Power Industry, Japan

Abstract

On the basis of theoretical models for the maximum potential intensity of tropical cyclones and general precipitation extremes, a simple scheme has been developed to estimate climatological changes in tropical cyclone intensities and associated precipitation extremes as a function of global surface temperature anomalies caused by anthropogenic climate changes. Since intensity changes strongly depend on the upper-air warming relative to the surface, this scheme incorporates the uncertainty of the variation in the upper-air temperature anomalies obtained from multiple-climate model experiments. A case study of Typhoon Flo in 1990 right before its landfall in Japan has revealed that its intensity, measured by a central pressure drop at sea level, and peak precipitation are projected to increase by 6.5% and 9.3%, respectively, under a globally 1-°C warmed environment relative to the present. These quantities and associated wind speed changes are directly calculated with uncertainties for arbitrary warming anomalies, which enables quantitative assessment for a wide range of greenhouse-gas emissions pathways in the future.

1. Introduction

The tropical cyclone (TC) is one of the most devastating atmospheric phenomena. To adequately prepare for their strong winds and torrential rains, societies affected by TCs require adequate preparation measures that include disaster prevention, appropriate infrastructure design, and water resource management. These measures can be adversely affected by socioeconomic circumstances in some regions and insufficient scientific understanding of

*E-mail address: tsutsui@criepi.denken.or.jp

TCs, in particular, the relationship between TC activity and natural climate variability and anthropogenic climate changes. This chapter considers the possible effects of long-term greenhouse gas-induced (GHG-induced) global warming on TC activity from a policymaker's perspective.

Recent modeling studies suggest an increase in intensity and a decrease in frequency of global TCs in warmed climates as summarized in the fourth assessment report (AR4) of the Intergovernmental Panel on Climate Change (IPCC) [1]. The intensity increase has a theoretical basis [2–4], and is believed to be relatively confident. Prospects for this trend have been discussed in a more recent review [5]. Global TC tendencies, however, are not necessarily the same as regional TC tendencies, which depend on large-scale circulations and spatial distributions of thermodynamic conditions in the tropics [6, 7]. Although different regional TC tendencies in numerical experiments appear to be explained by the modulation of natural climate variability [8], currently used atmosphere-ocean general circulation models (AOGCMs) cannot predict consistent trends in mean states and variability in the tropics [9, 10]. Moreover, fundamental challenges are encountered in deducing long-term regional tendencies from past observations; the duration of reliable and consistent TC records is too short to find any anthropogenic changes from relatively large fluctuations relevant to natural variations with different time scales [11]. Thus, at the current stage of scientific understanding, it is difficult to quantify the warming impact on all aspects of TC activity in a specific region.

From a policymaker's perspective, improvements to guard against possible TC intensification are rational, regardless of global warming impacts on TC activity [12]. Motivated by a growing need for appropriate adaptation to the warming environment, the author has applied a maximum potential intensity (MPI) theory to TCs in the western North Pacific, considering the uncertainty of thermodynamic conditions obtained through various AOGCMs [13]. Although the information from this study is limited to changes in the MPI, i.e., the maximum intensity of an idealized TC, this theory is a key aspect of risk management for avoiding complexities and uncertainties associated with the frequencies and motions of regional TCs. On the basis of this study and subsequent one [14], the present chapter provides a scheme to quantify changes in TC-caused strong winds and torrential rains prevalent in warmed climates.

As documented in Reference [15], TC-caused torrential rains involve complex physical processes relevant to water vapor supply and atmospheric instability in conjunction with embedded mesoscale circulation systems, orographic effects, and interactions with large-scale circulations including mid-latitude synoptic systems. Although these issues should be investigated from intensive observations and numerical experiments for better forecast and understanding of their mechanisms, a simple physically-based method has an advantage in quantifying GHG-induced background changes. Therefore, we employ a simple formula to assess changes in precipitation extremes from large-scale conditions resolved by AOGCMs [16].

The present scheme to estimate changes in the MPI and precipitation extremes can be formulated as a function of the global surface temperature anomaly by using appropriate pattern scaling to derive regional anomalies in thermodynamic conditions. This formulation is useful for risk assessment because it considers various types of uncertainties in a hierarchical structure. Moreover, planning adaptation to possible TC intensification depends on mitigation policies to reduce GHG emissions, which involves considerable uncertainties of climate sensitivity [17] and socio-economic development pathways. Since the global temperature anomaly is a common variable for both mitigation and adaptation studies, the formulation presented in this chapter is applicable to an integrated research framework for examination of mitigation and adaptation strategies, which leads to activation of rational plans and evaluation of priorities.

In a case study contained in this chapter, large-scale conditions used in calculating MPI and precipitation extremes in the present and future warmed climates were obtained from the JRA-25 long-term reanalysis of the global atmosphere [18] and recent multiple AOGCM experiments conducted in the framework of the Coupled Model Intercomparison Project phase 3 (CMIP3) [19]. The JRA-25 is one of the long-term reanalyses and has an advantage of better quality of precipitation analysis in the tropics [20]. The CMIP3 has formed the basis of the AR4 and is used here to derive regional anomalies in thermodynamic conditions. These datasets should be substituted by more advanced ones when available.

In the rest of this chapter, the methodology on the basis of the MPI theory is described in Section 2; a case study is presented and discussed in Section 3; and the summary and remaining issues are stated in Section 4.

2. Methodology

2.1. MPI theory

As distinct from an extratropical cyclone, the TC has a circularly symmetric warm-core structure. The warm core is formed by the release of latent heat in the upper air through convection originating from moist surface air. Upper air that is warmed directly reduces surface pressure to maintain the hydrostatic balance, and surface pressure drops over the warm sea surface to increase substantially the moist entropy, which accelerates upper air warming. This interaction between upper air warming and an increase in surface moist entropy is essential for TC development, where the surface temperature and water vapor amount are important factors in determining the upper boundary of the drop in surface pressure and the associated maximum wind speed, referred to as the MPI.

Two different MPI theories have been provided by Emanuel [2,3] and Holland [4]. The former considers the energy balance between the heat transfer from the ocean and the frictional dissipation at the surface to derive the MPI of a steady-state developed TC in terms of maximum surface wind. Available energy acquired from the ocean is formulated using an analogy with a Carnot cycle heat engine with the efficiency expressed by the temperature difference between the surface and the top of the storm. Meanwhile, Holland's theory considers thermodynamic processes in the inner core of a developing TC to calculate the upper-air warming associated with moist-adiabatic updraft in the eyewall and dry subsidence in the eye, leading to a reduction in the surface pressure. Although these theories have limitations in forecasting the intensity of individual TCs [21,22], they can provide basic information on climatological MPI to reflect global changes in thermodynamic environments including different upper-air warming anomalies [7,23–25].

A tropical atmosphere characteristic is the presence of warming anomalies amplified toward the upper troposphere, originating from the moist adiabatic ascent of convective air parcels [26]. TC intensification due to an increase in surface temperature is generally suppressed to some extent by greater upper-air warming in the environment, the degree of which is one of the major variables in climatological MPI in warmed climates. Here, we employ Holland's theory because it directly accounts for the warming structure and its variations in a simple manner. In this regard a care is needed because Holland's MPI appears to be more sensitive to global warming than Emanuel's MPI [5].

Essentially, Holland's MPI is expressed by the central pressure at sea level p_c and evaluated through the hydrostatic balance

$$\delta p_c = \frac{p_s}{T_v(p_s)} \int_{p_s}^{p_t} \delta T_v d \ln p, \tag{1}$$

where δp_c is the sea-level pressure drop, defined as the pressure difference between the environment and the TC center; p is the pressure; p_s is the surface pressure; p_t is the pressure at the top of the TC, and changes are negligible above it; T_v is the virtual temperature; and δT_v is its deviation from the environment.

The deviation of temperatures in the inner-core of a developing TC is determined for a given sea surface temperature (SST) and an upper-air temperature profile in the environment by calculating a moist-adiabatic ascent in the eyewall under isothermal conditions at the surface and dry subsidence in the eye. The surface relative humidity (RH) beneath the eyewall and the RH profile in the eye are assumed appropriately such that the maximum δp_c for the present climate becomes consistent with empirical findings. Computing implementation and sensitivity to key parameters are given in Reference [13].

The amount of water vapor providing latent heat increases with warmer SSTs, and the δT_v increases with cooler upper-air temperatures in the environment. The δp_c, therefore, increases with warmer SST and cooler upper-air temperatures in the environment. The calculation of the MPI does not consider SST cooling by TC-induced ocean upwelling and dynamic conditions in the atmosphere that suppress TC development. Although real TCs do not necessarily attain their MPI due to these restrictions, the climatological MPI constrains well the minimum central pressures of individual TCs, as illustrated in Fig. 1. Realized central pressures are more scattered at lower latitudes and more confined at higher latitudes, which is consistent with temporal and latitudinal MPI variations. This consistency ensures climatological use of the MPI.

It should be noted that some realized pressures exceed a suggested lower limit by the MPI, which is partly due to the incomplete spatial representation of the MPI for northward-moving TCs. Namely, it is possible that the intensity of such a TC was resulted from warmer conditions at lower latitudes that it had experienced previously. Conversely, realized pressures generally do not reach a suggested lower limit at lower latitudes from June to July, which implies that TC development in the early season is more constrained by the dynamic conditions of large-scale circulations.

2.2. Wind speed scaling

Wind velocities in a TC can be approximated by axisymmetric gradient winds. The gradient wind balance in cylindrical coordinates (r, θ, z) is given by

$$\frac{v_\theta^2}{r} + f v_\theta = \frac{1}{\rho}\frac{\partial p}{\partial r}, \tag{2}$$

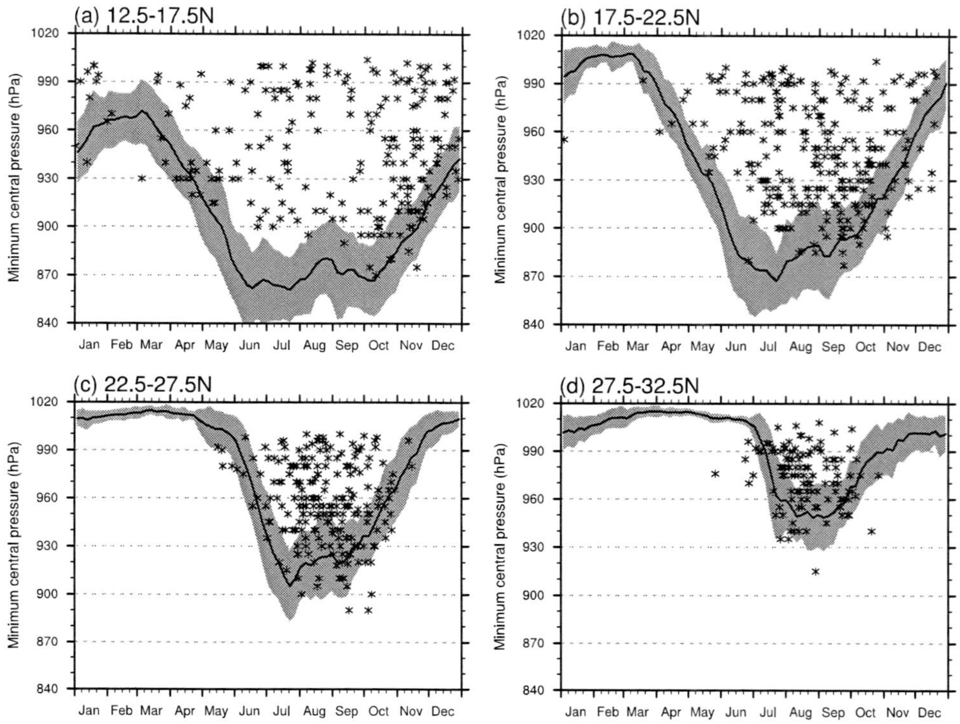

Figure 1. Climatological annual cycles of MPI in the western North Pacific region 12.5°N–32.5°N, 122.5°E–145°E, compared with minimum central pressures of individual TCs from the JMA best track during 1951–2009. Shading indicates plus/minus one standard deviation of interannual variability. The region is divided into four latitude ranges: (a) 12.5°N–17.5°N, (b) 17.5°N–22.5°N, (c) 22.5°N–27.5°N, (d) 27.5°N–32.5°N. Each individual TC is marked in one of the four panels, depending on the region including the latitude at which the TC attained minimum pressure. Updated from Fig. A1 of Reference [13].

where v_θ is the tangential velocity (positive for anticlockwise flow), f is the Coriolis parameter, and ρ is the density of air. According to Reference [27], this equation is transformed into a relationship between the vertical gradient of the absolute angular momentum and the radial temperature gradient considering the hydrostatic balance:

$$\frac{1}{r^3}\frac{\partial M_\theta^2}{\partial z^*} = \frac{R}{H}\frac{\partial T}{\partial r}, \tag{3}$$

where M_θ is the absolute angular momentum defined as $v_\theta r + fr^2/2$, z^* is the vertical independent variable in log-pressure coordinates defined as $-H\log(p/p_s)$, R is the gas constant for dry air, and H is a standard scale height defined as $R\bar{T}/g$; $\bar{T}$ is the global average temperature and g is the magnitude of gravity.

In Holland's MPI theory, the pressure drop is estimated from the vertical integration of temperature deviations with respect to the logarithm of pressure as expressed by equation (1). Then, the rate of change in wind speed can be estimated from that of the pressure drop through equation (3). Specifically, by considering a warm-core wind structure and assuming $\partial M/\partial z^* \sim v_\theta r/H$, we can approximate a temperature deviation in the warm-core

When surface pressures in a TC are specified, the gradient wind balance gives an idealized wind profile from the center to the outside environment [30]. One widely used empirical representation uses pressure as an inverse exponential function of radius [31]

$$p_s = p_c + \delta p_c \exp\left(-\frac{r_m}{r}\right), \tag{7}$$

where r_m is the radius of maximum wind. By using this equation and the relationship between the curvature radius of the streamlines R_s and the trajectories R_t [27]

$$R_s = R_t\left(1 - \frac{V_c \cos\gamma}{v_\theta}\right), \tag{8}$$

the gradient wind associated with a moving TC is obtained as

$$v_\theta = \frac{fr - V_c \cos\gamma}{2}\left[-1 + \sqrt{1 + \frac{4rG}{(fr - V_c\cos\gamma)^2}}\right], \tag{9}$$

$$G = \frac{1}{\rho}\delta p_c \frac{r_m}{r^2}\exp\left(-\frac{r_m}{r}\right), \tag{10}$$

where V_c is the speed of TC motion, γ is the angle between the streamlines (in this case, pressure contours) and the direction of TC motion, and G is the pressure gradient force. In this derivation, the r in the first term in the left-hand side of equation (2) (centrifugal force) is replaced with R_t. This wind profile is conventionally used for engineering applications [32].

Figure 3 shows an example of wind distribution, where the parameters are roughly adjusted for the specific TC described in the next section. The component of non-axisymmetric winds is relatively large near the center, and the maximum wind occurs at the right of the TC motion, as empirically known. This idealized wind distribution, however, differs from winds in an actual TC such that the winds in the outer circulation often vary independently of the vortex core [30]. Moreover, the winds associated with a TC moving into middle latitudes, such as that illustrated by this example, are modified by its transition to an extratropical system [33]. Nevertheless, the idealized distribution is useful for quantifying background changes created by warmed climates.

Figure 3 illustrates the rate of change in the wind speed relative to that in the δp_c. The value is 0.5–0.6 in the core region including the radius of maximum wind and increases toward the outer region. The factor of 0.53 is appropriate for the maximum wind. When scaling wind speed change by the δp_c change, however, this value should be larger for the winds in a remote region that may affect wind-induced waves and storm surges.

2.3. Precipitation extreme

Precipitation rates associated with severe events are expected to increase in warmed climates because atmospheric water vapor amounts increase in these conditions. Assuming that the water vapor is saturated when a severe event occurs, one can evaluate the rate of change in the precipitation rate with respect to warming according to a well-known relationship between temperature and saturation water vapor pressure. Here, we relate a change in

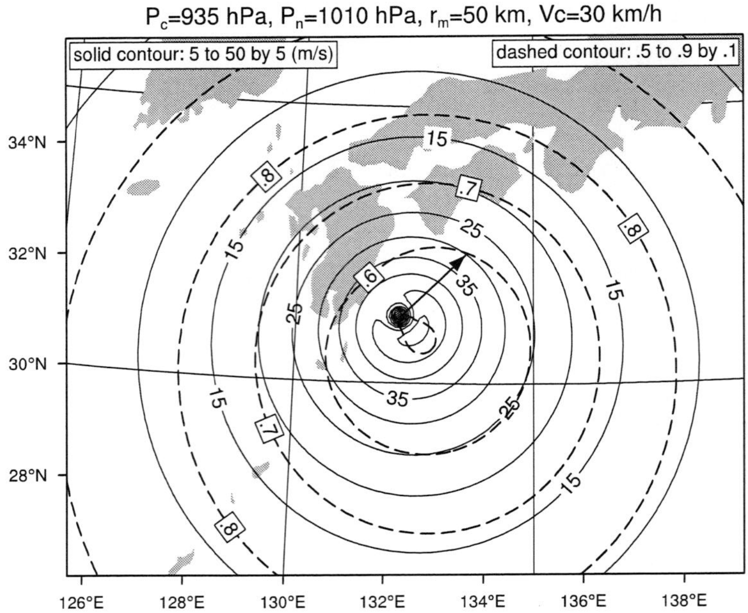

Figure 3. Example of the gradient wind distribution associated with a moving TC: p_c=935 hPa, δp_c=75 hPa, r_m=50 km, V_c=30 km h^{-1} (northeastward). Solid contours show the gradient wind speeds, and dashed contours show the ratios of the fractional change in the wind speed to that in the δp_c.

a peak precipitation rate associated with a TC to a change in the MPI by using the method provided in Reference [16]. This method is a more rational approach to warming impact because it considers changes in updraft profile in addition to the amount of water vapor.

When a severe precipitation event occurs, the rate of water vapor condensation with respect to time in the upper atmosphere is given by

$$c = -\omega \left. \frac{dq_s}{dp} \right|_{\theta^*}, \tag{11}$$

where ω is the pressure velocity, q_s is the saturation specific humidity, and the derivative of q_s with respect to p is evaluated along a constant equivalent potential temperature, denoted by θ^*. In this equation, the condensation rate is determined such that the rising air maintains the saturation amount of water vapor.

The precipitation rate is expressed as mass-weighted vertical integration of equation (12) and given by

$$P_e = -\left\{ \omega_e \left. \frac{dq_s}{dp} \right|_{\theta^*, T_e} \right\}, \tag{12}$$

where P_e is the precipitation extreme, defined as a high percentile of precipitation rate; ω_e is the corresponding vertical velocity; T_e is the corresponding temperature at which the θ^* is evaluated; and the braces { } denote a mass-weighted vertical integration. In our scheme, the precipitation extreme and MPI are calculated for given climate conditions obtained from datasets of the global atmospheric reanalysis and AOGCM experiments. Since these climate conditions are represented by a coarse horizontal resolution, typically 100–300 km,

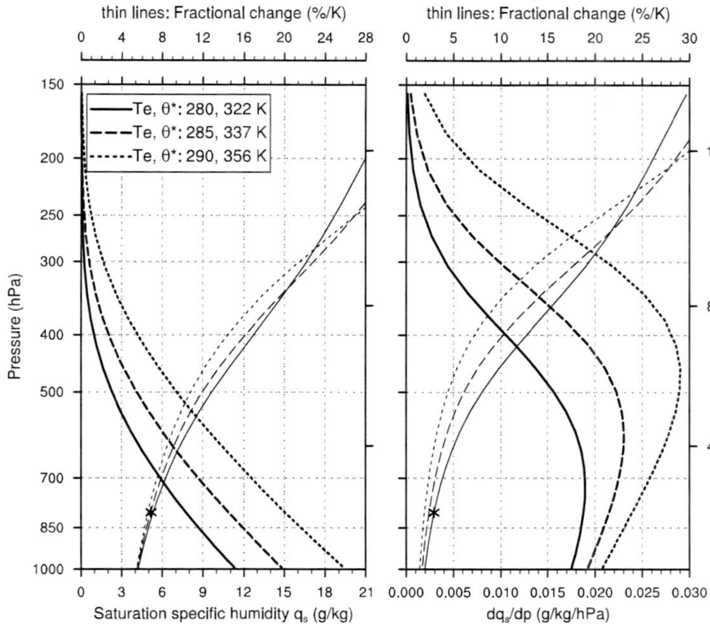

Figure 4. Profiles of saturated specific humidity q_s (left panel) and its vertical derivative $(dq_s)/(dp)|_{\theta*}$ (right panel) for three reference temperatures at the 800 hPa level. The rate of change in these quantities with respect to the reference temperature is indicated by thin lines with the upper auxiliary horizontal axis. The asterisk markers indicate the rate of change at the 800 hPa level for the reference temperature of 280 K, which are $6.9\%\,\mathrm{K}^{-1}$ and $2.9\%\,\mathrm{K}^{-1}$ for q_s and $(dq_s)/(dp)|_{\theta*}$, respectively.

we regard the variables in equation (12) as that of model-grid scale, whereby small-scale variations such as a convective downdraft and re-evaporation of condensated water vapor are implicitly included.

Here, we define the T_e as a temperature at the 800 hPa level. This level corresponds to about 2 km of standard atmosphere altitude, and is close to the level at which the amount of atmospheric water vapor is vertically divided into two equal amounts. As shown in Fig. 4, while the rate of change in q_s with respect to T_e is relatively large in the upper troposphere, that in the lower troposphere including a larger amount of water vapor is small, typically $7\%\,\mathrm{K}^{-1}$ at 800 hPa. With regard to precipitable water, defined as mass-weighted vertical integration of q_s, the rate of change is about $8\%\,\mathrm{K}^{-1}$. In contrast, the rate of change in dq_s/dp is significantly small in the lower troposphere, typically $3\%\,\mathrm{K}^{-1}$. An increase in precipitation extremes due to warming is, therefore, not as large as that inferred from the rate of change in specific humidity. In the case of a severe disturbance such as a TC reaching the tropopause, however, the rate of change becomes larger than $3\%\,\mathrm{K}^{-1}$, as determined from the rate of change in dq_s/dp, which is amplified at upper levels.

When estimating the P_e for a specific TC event, the ω_e and T_e can be specified according to the peak precipitation rate near the TC center. An increase in P_e due to climate warming is the result of two effects: a dynamic effect such that an intensified TC enhances water vapor transport, and a thermodynamic effect such that the amount of water vapor increases in a warmed environment. In equation (12), changes in ω_e represent the former relevant to TC intensity, and changes in dq_s/dp that depends on T_e represent the latter. The dynamic effect can be related to MPI change, and the thermodynamic effect is calculated with consideration of the T_e anomaly, which will be demonstrated in a case study.

Since the ω_e is related to the maximum wind speed on the basis of the continuity equation, the ω_e in warmed climate can be scaled to that of the maximum wind speed. For simplicity, the rate of change in ω_e can be substituted with that of the maximum wind speed at all of the pressure levels with the assumption that the vertical structure of ω_e is not affected by the warming.

3. Case study

3.1. Typhoon Flo (1990)

In this chapter, we examine Typhoon Flo (1990), which brought excessive rainfall to Japan, as a case study to evaluate the aforementioned scheme to estimate changes in the pressure drop δp_c and the peak precipitation P_e for a given warmed environment. Although not explicitly described, changes in wind speed are scaled with those of δp_c. International field experiments were conducted in August–September 1990, which includes the life cycle of Flo [34]. It is believed that intensive observations in the field experiments were incorporated into the best track [1], shown in Fig. 5 (left), and the JRA-25 reanalysis of the global atmosphere.

Flo intensified rapidly with northwestward movement. Its minimum central pressure of 890 hPa was recorded at 0000 UTC on September 17 near Okinawa. The typhoon then recurved northeastward and made landfall in Wakayama Prefecture at about 2000 JST September 19 with a central pressure of 945 hPa. The JMA reported that the total rainfall over Japan throughout the event (September 12–13, 15, and 17–20), including rainfall due to the front affected by Flo, exceeded 1100 mm in parts of Shikoku and Kinki districts. According to the JMA's Automated Meteorological Data Acquisition System (AMeDAS), which includes about 1300 surface observation sites throughout Japan, daily rainfall during Flo's landfall on September 19 exceeded 100 mm at 230 sites with a maximum of 638 mm.

It is found from six-hourly reanalysis at about 1.1°-grid resolution that the rainfall associated with Flo peaked at 0000 UTC on September 19 just prior to landfall. Here, we focus on the typhoon's maximum updraft at that time and apply its pressure velocity profile as the ω_e for this event. As shown in Fig. 5 (right), the maximum updraft is located at 32.0N, 132.8E, south of Japan's main islands, and the upward motion is analyzed throughout the troposphere with its maximum at about 500 hPa. The peak precipitation rate of the reanalysis is 241 mm d^{-1}, which coincides with the maximum updraft, and is regarded as the peak precipitation (P_e) over the ocean without the orographic effect.

An increase in precipitation rates caused by the orographic effect of Japan's main islands can be estimated by using a factor of 2.5–3.0 according to a case study on Typhoon Owen in 1979 [35], the characteristics of which are similar to that of Flo. The AMeDAS

[1]The track data is taken from the RSMC (Regional Specialized Meteorological Center) Tokyo-Typhoon Center `http://www.jma.go.jp/jma/jma-eng/jma-center/rsmc-hp-pub-eg/trackarchives.html`

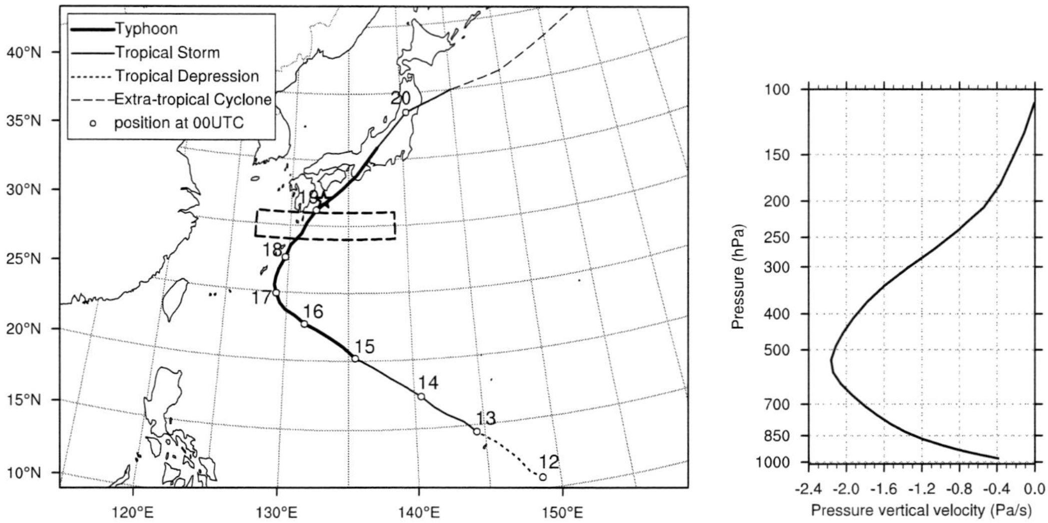

Figure 5. Track of Typhoon Flo and the vertical profile of the ω_e selected as the maximum updraft at 0000 UTC on September 19, 1990. The maximum updraft is located at 32.0°N, 132.8°E, indicated by the star marker on the map. The dashed rectangle on the map is the region of 29°N–31°N, 127°E–139°E, referred to for T_e and MPI calculation. Source: Figure 1 of Reference [14].

maximum of 638 mm is 2.6 times greater than the JRA-25 peak value of 241 mm, which is within the estimated range of the orographic effect.

The T_e in equation (12) is the temperature of the air parcel creating the updraft. Considering that a moist southerly caused Flo's torrential rains, we evaluate the T_e in the region of 29°N–31°N, 127°E–139°E (dashed rectangular in Fig. 5). The JRA-25 climatology (1979–2004 average) indicates an August–September mean temperature of 288.9 K in this region. By using this temperature as the T_e in addition to the ω_e profile, the P_e in equation (12) results in 260 mm d^{-1}. Despite ad hoc interpretation of T_e, this result is close to the peak value in the reanalysis, and the T_e value is acceptable as a base temperature in the present climate to assess changes due to warmed climate. Note that the rate of change in P_e in response to global warming, which will be shown subsequently, does not generally depend on the choice of T_e although the base value of P_e in the present climate is directly controlled by T_e.

For consistency, we use the same climatology for input data for MPI calculation (sea level pressure, SST, and upper-air temperatures). In this case, the SST is 28.3°C, which results in an MPI (δp_c) of 944 hPa (65 hPa) for the present climate. This MPI value is close to the observed central pressure at landfall, implying that Flo nearly attained the climatological MPI during the active TC season in the western North Pacific.

3.2. Pattern scaling of environmental changes

We need anomalies in sea surface and upper-air temperatures in the specified region to assess changes in the δp_c and P_e for a given anomaly in the global surface temperature relative to the present climate, denoted as $\Delta\bar{T}_s$. Here, we employ the ensemble of the CMIP3 multi-model climate experiments to derive these anomalies by linear pattern scaling, which proportionally adjusts a given basic warming pattern to a $\Delta\bar{T}_s$. Although the regional climate change is not necessarily proportional to the $\Delta\bar{T}_s$, it has been proved that the general spatial pattern does not significantly depend on the degree of warming [1].

The basic warming pattern of the anomalies has been obtained from the ensemble means of 23 AOGCMs and is defined as 2080–99 climatology from the SRES [36] A1B run relative to the 1980–99 climatology from the historical run. The SRES A1B is one of three future scenarios used in the CMIP3, providing a "medium-warming" baseline projection for the AR4. Specifically, the basic pattern has been created as follows:

1. For the individual AOGCMs, create anomalies of surface and upper-air temperatures in monthly climatology.
2. For the surface temperatures, compute the ensemble means over all the AOGCMs and divide them by the global annual mean for normalization.
3. For the upper-air temperatures, divide the individual anomalies by the surface temperature anomaly at the same grid point for normalization, and then, compute the ensemble means of the normalized vertical profiles over all of the AOGCMs and two subsets of AOGCMs.

Here, we set aside uncertainties of surface anomaly patterns and climate sensitivity to focus instead on uncertainties of upper-air warming amplification that controls the degree of TC intensification due to the surface warming. This amplification depends on the model's formulation; particularly, its physics parameterizations. To quantify this uncertainty range, the two subsets of AOGCMs mentioned in step 3 of the above procedure have been selected such that they rank as the six largest or smallest anomalies at the 200-hPa level for each month and grid point. The ensemble means over all of the models provide the best estimate while those over the larger and smaller anomaly models represent warmed environments corresponding to suppressed and enhanced TC intensification, respectively.

In the August–September climatology of the specified region, the normalized surface temperature anomaly is 0.86, and the normalized upper-air temperature anomalies are obtained as shown in Fig. 6. The normalized anomaly at the 200-hPa level is 2.4 for all of the CMIP3 models while the values are 3.0 and 2.0 for the selected larger and smaller anomaly models, respectively. For a given $\Delta\bar{T}_s$, the value multiplied by 0.86 gives the SST anomaly in the region, and scaling the three profiles in Fig. 6 by this SST anomaly yields corresponding upper-air temperature anomalies.

Changes in the δp_c and P_e for a given $\Delta\bar{T}_s$ are obtained through the following procedures:

1. Estimate the regional temperature anomalies from a given $\Delta\bar{T}_s$ by using the pattern scaling.
2. Compute the fractional change of the δp_c from the MPI difference between the present and warmed climate.

Table 1. Fractional changes in the δp_c and P_e for Typhoon Flo just prior to landfall. Numbers in brackets are uncertainty ranges reflecting the variations of upper-air temperature anomalies. Numbers in the column denoted by P_e(*) are the fractional changes due to the thermodynamic effect alone (δp_c unchanged)

$\Delta\bar{T}_s$ (°C)	δp_c (%)	P_e (%)	P_e(*) (%)
1.0	6.5 [−1.6, 12.2]	9.3 [4.7, 12.4]	5.6
2.0	14.2 [−2.7, 26.1]	19.7 [9.8, 26.8]	11.3

and numerical simulations [25, 38]. The δp_c change of 6.5% for the 1°C warming (corresponding to 0.86°C SST rise) in the present case study also agrees with this TC basin-wide relation.

The fractional change in the P_e is larger than that in the δp_c, and the thermodynamic effect accounts for more than half of this P_e change. For the 1°C warming, the thermodynamic effect is 5.6%. This value, although greater than the typical value of 3% for the dq_s/dp in the lower troposphere, is less than the typical value of 7–8% for the q_s in the lower troposphere and the precipitable water. The uncertainty range for the P_e change is relatively small because it reflects the variation of the dynamic effect alone in the present scheme through the ω_e scaling. The thermodynamic effect is essentially believed to be robust.

The present scheme can be applied to assess secular changes in the δp_c and P_e for future scenarios. Here, we consider the representative concentration pathways (RCPs) [39] used as the common forcing data for climate projections toward the fifth assessment report of the IPCC, planned for publication in 2013–14. The RCP consists of four different pathways of radiative forcings and related GHG concentrations and emissions. The four pathways are denoted as RCP 8.5, 6.0, 4.5, and 3-PD (2.6), where the first three numbers represent stabilized radiative forcing in W m^{-2} after the year 2100, and the last value stands for 3-W m^{-2} peak and decline to 2.6 W m^{-2}.

Figure 7 shows changes in the $\Delta\bar{T}_s$ along the four pathways, defined as the anomaly relative to the value in the year 1990, and corresponding fractional changes in the δp_c and P_e with their uncertainties for RCP 6.0. For simplicity, $\Delta\bar{T}_s$ has been calculated by a simple climate model [40] using the equilibrium climate sensitivity of 3°C, which is defined as the global equilibrium surface warming after a doubling of the atmospheric CO_2 concentration. This simple calculation can be adapted to a more sophisticated formula to reflect advanced climate modeling and relevant uncertainties.

The difference in $\Delta\bar{T}_s$ among the four pathways increases from a negligible level during the first half of the 21st century to a spread of about 3°C in the year 2100. The results for the 1°C warming included in Table 1 correspond to changes expected around the year 2040 in any pathways, while those for the 2°C warming correspond to 2055, 2080, and 2100 for RCP 8.5, 6.0, and 4.5, respectively. The values for RCP 3-PD (2.6) are almost constant during the second half of the 21st century. However, it should be considered that such dating estimates strongly depend on climate sensitivity.

The upper and lower bounds of the RCP have been determined such that they cover existing emissions scenarios. It is understood from the comparison between the differences

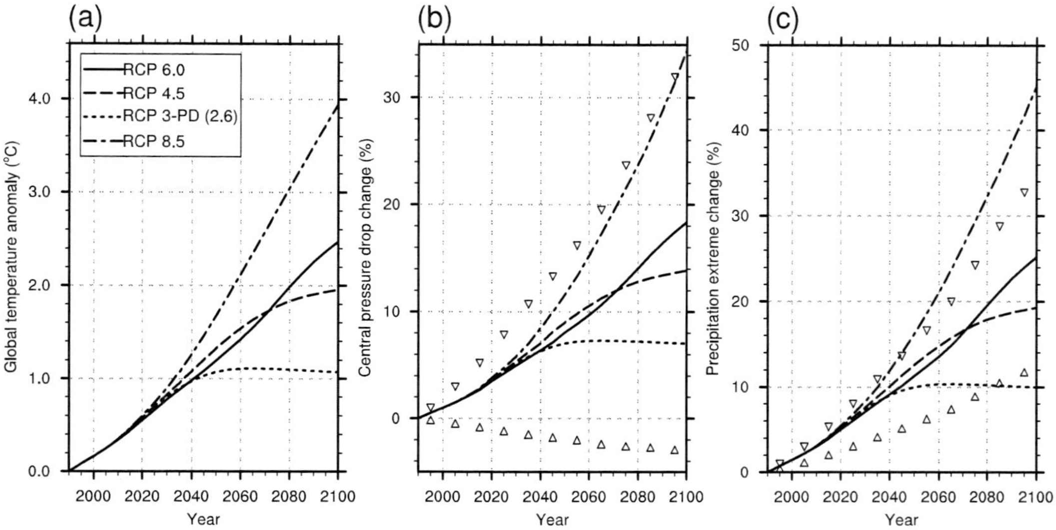

Figure 7. The global surface temperature anomaly $\Delta\bar{T}_s$ for the four pathways of the RCP (a), and corresponding fractional changes in the δp_c and P_e for Typhoon Flo just prior to landfall (b, c). The equilibrium climate sensitivity is assumed to be 3°C in computing $\Delta\bar{T}_s$. Markers in panels (b) and (c) indicate uncertainty ranges in the case of RCP 6.0. Source: Figure 3 of Reference [14].

among the pathways and the uncertainty ranges for a particular pathway (RCP 6.5 shown in the figure) that the uncertainty of the δp_c change is greater than the variation of possible future pathways. Although TC intensity is generally anticipated to be further intensified in warmed climates, the variation of the upper-air temperature anomalies in the current scientific understanding leads to a wide range of possibilities from virtually no change (or slightly negative) to double that of the best estimate. In contrast to the TC intensity measured by the δp_c, the uncertainty of the P_e change in 2100 is comparable to or smaller than the variation of the pathways. The increasing tendency of precipitation extremes is robust even if the large uncertainty of the TC intensity change is considered. The present scheme facilitates such a comparison of various uncertainties as in this example, which is particularly beneficial for determining priorities in adaptation planning.

4. Conclusion

This chapter describes a simple scheme for estimating changes in TC-caused strong winds and torrential rains resulting from GHG-induced warming on the basis of theoretical MPI and precipitation extremes. This scheme is designed for applications to risk assessment and policymaking associated with TC intensity in mitigation and adaptation strategies for warmed climates. For this purpose, changes in TC intensity measured by the central pressure drop δp_c at sea level and the precipitation extreme P_e representing a high percentile of precipitation rate are formulated as functions of the global surface temperature anomaly. Since the δp_c is related to the maximum wind speed and an idealized wind profile, changes in the δp_c quantify changes in TC-caused strong winds and the dynamic effect of precipitation increase caused by warmed climates.

The present scheme is suitable for assessing changes anticipated in the future for a high-impact TC. An extremely strong TC rarely forms in the case of large fluctuations of natural climate variability, regardless of the background warming. It is generally difficult to assess relatively small background changes in the intensity of such a rare event by observation statistics or numerical climate projections. The present scheme overcomes this difficulty by focusing on background changes in large-scale thermodynamic conditions alone with no consideration of the dynamic conditions that dominantly control TC frequencies. Since disaster prevention programs and infrastructure design guides are, in many cases, developed on the basis of past high-impact TCs, assessment results for such TCs are beneficial for examination of these issues. Moreover, this scheme is expected to provide a quantitative physical basis, as documented in Reference [16,41], for comparison of results from a large set of numerical experiments to improve understanding of TC activity in warmed climates.

As a case study in this chapter, Typhoon Flo in 1990 is considered a high-impact TC that produced excessive rainfall in Japan. When the global surface temperature rises by 1°C from the present, which is anticipated to occur around 2040, the δp_c and P_e observed just prior to Flo's landfall are projected to increase by 6.5% and 9.3%, respectively, and 5.6% in the case of P_e considering the thermodynamic effect alone. These values are the best estimates, based on the changes in the thermodynamic environment obtained from the CMIP3 AOGCMs. The variation in the AOGCMs of upper-air temperature anomalies has been used to quantify the uncertainty ranges of the intensity change and the dynamic effect of the precipitation change. Since the thermodynamic effect on increased precipitation due to climate warming is robust, the uncertainty range of the P_e change is relatively small. This assessment procedure is easily applied to arbitrary increases in global temperatures along various development pathways in the future, which facilitates comparison between various uncertainties relevant to development pathways and climate projections. With regard to the variation in the AOGCMs, the results presented in this chapter will be updated by the same procedure incorporating more advanced climate projections, which have been conducted for the next IPCC assessment report.

Although Flo is a typical high-impact event, increases in P_e should be further examined for many events because the degree of increase depends on the vertical structure of the updraft corresponding to the P_e. Moreover, the constraint on the vertical structure, which is assumed to be unaffected by the warming, should also be verified or relaxed to incorporate reliable tendencies, if available. This chapter has dealt with the maximum intensity and associated precipitation extreme alone. Additional aspects of TC activity should be addressed in the future in response to improved understanding of natural climate variations and continued advancement in climate modeling and projections.

Acknowledgment

The author acknowledges Dr. Masahiro Sugiyama of the Central Research Institute of Electric Power Industry for his valuable comments on precipitation extremes. The author also acknowledges the modeling groups, the Program for Climate Model Diagnosis and Intercomparison (PCMDI), and the World Climate Research Programme's (WCRP's) Working

Group on Coupled Modelling (WGCM) for their roles in providing the WCRP CMIP3 multi-model dataset. Support for this dataset is provided by the Office of Science, U.S. Department of Energy.

References

[1] IPCC WG1. *Climate change 2007: The physical science basis. Contribution of Working Group I to the fourth assessment report of the Intergovernmental Panel on Climate Change [S. Solomon, D. Qin, M. Manning, Z. Chen, M. Marquis, K. B. Averyt, M. Tignor and H. L. Miller (eds.)].* Cambridge University Press, Cambridge (2007).

[2] K. A. Emanuel. *J. Atmos. Sci.*, **43**, 585 (1986).

[3] K. A. Emanuel. *J. Atmos. Sci.*, **52**, 3969 (1995).

[4] G. J. Holland. *J. Atmos. Sci.*, **54**, 2519 (1997).

[5] T. R. Knutson, J. L. McBride, J. Chan, K. Emanuel, G. Holland, C. Landsea, I. Held, J. P. Kossin, A. K. Srivastava and M. Sugi. *Nature Geoscience*, **3**, 157 (2010).

[6] J. C. L. Chan and K. S. Liu. *J. Climate*, **17**, 4590 (2004).

[7] G. A. Vecchi and B. J. Soden. *Nature*, **450**, 1066 (2007).

[8] S. Yokoi and Y. N. Takayabu. *J. Meteor. Soc. Japan*, **87**, 525 (2009).

[9] W. J. Merryfield. *J. Climate*, **19**, 4009 (2006).

[10] G. A. Vecchi, A. Clement and B. J. Soden. *EOS*, **89**, 81, 83 (2008).

[11] WMO. "Statement on tropical cyclones and climate change". Prepared by participants of the WMO International Workshop on Tropical Cyclones, IWTC-6, San Jose, Costa Rica, November 2006, available from `http://www.wmo.ch/web/arep/press_releases/2006/iwtc_summary.pdf` (2006).

[12] I. Grossmann and M. G. Morgan. *Clim. Change*. doi:10.1007/s10584-011-0020-1 (2011).

[13] J. Tsutsui. *J. Meteor. Soc. Japan*, **88**, 263 (2010).

[14] J. Tsutsui. *Journal of Japan Society of Civil Engineers, Ser. G (Environmental Research)*, **67**, I17 (2011). In Japanese.

[15] L. Chen. In R. L. Elsberry, editor, *Global perspectives on tropical cyclones*, WMO/TD-No. 693, TCP-38, chapter 6. World Meteorological Organization (1995).

[16] P. A. O'Gorman and T. Schneider. *Proc. Natl. Acad. Sci., USA*, **106**, 14773 (2009).

[17] R. Knutti and G. C. Hegerl. *Nature Geoscience*, **1**, 735 (2008).

[18] K. Onogi, J. Tsutsui, H. Koide, M. Sakamoto, S. Kobayashi, H. Hatsushika, T. Matsumoto, N. Yamazaki, H. Kamahori, K. Takahashi, S. Kadokura, K. Wada, K. Kato, R. Oyama, T. Ose, N. Mannoji and R. Taira. *J. Meteor. Soc. Japan*, **85**, 369 (2007).

[19] G. A. Meehl, C. Covey, T. Delworth, M. Latif, B. McAvaney, J. F. B. Mitchell, R. J. Stouffer and K. E. Taylor. *Bull. Amer. Meteor. Soc.*, **88**, 1383 (2007).

[20] M. G. Bosilovich, J. Chen, F. R. Robertson and R. F. Adler. *J. Appl. Meteor. Climatol.*, **47**, 2279 (2008).

[21] J. P. Camp and M. T. Montgomery. *Mon. Wea. Rev.*, **129**, 1704 (2001).

[22] R. K. Smith, M. T. Montgomery and S. Vogl. *Quart. J. Roy. Meteor. Soc.*, **134**, 551 (2008).

[23] K. A. Emanuel. *Nature*, **326**, 483 (1987).

[24] A. Henderson-Sellers, H. Zhang, G. Berz, K. Emanuel, W. Gray, C. Landsea, G. Holland, J. Lighthill, S.-L. Shieh, P. Webster and K. McGuffie. *Bull. Amer. Meteor. Soc.*, **79**, 19 (1998).

[25] T. R. Knutson and R. E. Tuleya. *J. Climate*, **17**, 3477 (2004).

[26] D. L. Hartmann. *Global Physical Climatology*. Academic Press (1994).

[27] J. R. Holton. *An introduction to dynamic meteorology*. Academic Press, 3rd edition (1992).

[28] G. D. Atkinson and C. R. Holliday. *Mon. Wea. Rev.*, **105**, 421 (1977).

[29] H. Koba, T. Hagiwara, S. Osano and S. Akashi. *Geophysical Magazine*, **44**, 15 (1991).

[30] H. E. Willoughby. In R. L. Elsberry, editor, *Global perspectives on tropical cyclones*, WMO/TD-No. 693, TCP-38, chapter 2. World Meteorological Organization (1995).

[31] R. W. Schloemer. "Analysis and synthesis of hurricane and wind patterns over Lake Okechobee, Florida". Hydrometeorological Report 31, U. S. Government Printing Office, No. C30.70:31 (1954).

[32] T. Fujii and Y. Mitsuta. *Annuals, Disaster Prevention Research Institute, Kyoto University*, **29, B-1**, 229 (1986). In Japanese.

[33] F. Fujibe and N. Kitabatake. *J. Meteor. Soc. Japan*, **85**, 747 (2007).

[34] R. L. Elsberry. *Bull. Amer. Meteor. Soc.*, **71**, 1305 (1990).

[35] T. Muramatsu. "A study on the changes of the three-dimensional structure and the movement speed of the typhoon through its life time". Technical Reports of the Meteorological Research Institute 14, Typhoon Research Division, Meteorological Research Institute, Japan (1985). In Japanese.

[36] N. Nakicenovic and R. Swart, editors. *Emissions scenarios. A special report of working group III of the Intergovernmental Panel on Climate Change.* Cambridge University Press, UK (2000).

[37] J. B. Elsner, J. P. Kossin and T. H. Jagger. *Nature*, **455**, 92 (2008).

[38] K. Oouchi, J. Yoshimura, H. Yoshimura, R. Mizuta, S. Kusunoki and A. Noda. *J. Meteor. Soc. Japan*, **84**, 259 (2006).

[39] R. Moss, M. Babiker, S. Brinkman, E. Calvo, T. Carter, J. Edmonds, I. Elgizouli, S. Emori, L. Erda, K. Hibbard, R. Jones, M. Kainuma, J. Kelleher, J. F. Lamarque, M. Manning, B. Matthews, J. Meehl, L. Meyer, J. Mitchell, N. Nakicenovic, B. O'Neill, R. Pichs, K. Riahi, S. Rose, P. Runci, R. Stouffer, D. van Vuuren, J. Weyant, T. Wilbanks, J. P. van Ypersele and M. Zurek. *Towards new scenarios for analysis of emissions, climate change, impacts, and response strategies.* Intergovernmental Panel on Climate Change, Geneva (2008).

[40] J. Tsutsui. *Journal of Japan Society of Civil Engineers, Ser. G (Environmental Research)*, **67**, 134 (2011).

[41] M. Sugiyama, H. Shiogama and S. Emori. *Proc. Natl. Acad. Sci., USA*, **107**, 571 (2010).

In: Cyclones: Formation, Triggers and Control
Editors: K. Oouchi and H. Fudeyasu
ISBN: 978-1-61942-976-5
©2012 Nova Science Publishers, Inc.

Chapter 7

SENSITIVITY OF TROPICAL CYCLONES TO LARGE- SCALE ENVIRONMENTS IN A GLOBAL NON-HYDROSTATIC MODEL WITH EXPLICIT CLOUD MICROPHYSICS

Yohei Yamada[1, 2*], Kazuyoshi Oouchi[1], Masaki Satoh[1, 2], Akira T. Noda[1] and Hirofumi Tomita[1, 3]

[1]Japan Agency for Marine-Earth Science and Technology, Yokohama, Japan
[2]The University of Tokyo, Kashiwa, Japan
[3]RIKEN, Kobe, Japan

ABSTRACT

Using a global non-hydrostatic model with a horizontal 14 km mesh, we investigated intensities and spatial distributions of tropical cyclones (TCs) and their relationships to large-scale environments. We conducted three cases of boreal summer experiments (June-October) using explicit cloud microphysics schemes without cumulus parameterization. The differences among the three experiments are the implemented physics schemes (i.e., a radiation scheme, a cloud microphysics scheme, a land model, and treatment of the sea surface temperature). One of the experiments is regarded as a control case. Only the radiation scheme used in another case differs from that in the control case in terms of the number of absorption bands. The other case uses a more comprehensive cloud microphysics scheme compared to the control case. The land and ocean processes are also changed in the last case. The radiation scheme in the last case is the same as that in the second case.

We examined the sensitivity of the simulated TC frequencies, tracks, and intensities. The relationship between the maximum wind speed (MWS) and minimum sea level pressure (MSLP) is quite similar among the cases. However, TC frequencies and tracks depend on cases; in terms of the frequency, the last case best reproduces the global observed cyclogenesis number, while the second case reproduces the most realistic intensity histogram of TCs. Comparisons of the relationship between the tropical cyclogenesis and the large-scale environment using a genesis potential index (GPI) show that the spatial distribution of cyclogenesis is generally consistent with that of the GPI. Among the physical factors that contribute to the GPI, the absolute vorticity, relative humidity, and vertical wind shear are primarily relevant to the difference of the GPI. The change of these variables seems to be associated with convergence of the zonal wind at 850 hPa, and the zonal wind is affected by atmospheric circulation, such as the Walker circulation. Therefore, it is speculated that the changes in atmospheric circulation with the physical schemes play an important role for determining the spatial distribution of tropical cyclogenesis.

1. Introduction

Model projection of the change of tropical cyclone (TC) activity due to global warming is vigorously discussed [1] and remains a challenging problem and an interesting subject. A major problem in the approach with general circulation models (GCMs) comes from uncertainty in the cumulus parameterizations used in the hydrostatic framework [2]. GCMs were incapable of simulating TCs of realistic intensity under present-day climate conditions [3]. In addition, little consensus among models has been found in the projection of regional TC frequency under greenhouse-warmed conditions.

A reliable projection of regional TC change is sensitive to the pattern of relative sea surface temperature change which dominates tropical circulation and convective activity [4]. For a reliable model projection, it is absolutely imperative to improve the reproducibility of TC in present-day simulations. The sea surface temperature and associated atmospheric circulation are also important to understand TCs. Bengtsson et al. [5] emphasized that it would be misleading to focus only on the sea surface temperature. Vecchi and Soden [6] discussed a relationship between vertical wind shear and TCs over the tropical Atlantic and eastern Pacific under a global warming climate. They show that substantial increases in the tropical vertical wind shear are robust features of experiments that are operated with a suite of coupled ocean-atmosphere models forced by emissions Scenario A1B, and the substantial increases are connected to the model-projected decrease in the Pacific Walker circulation. Regarding to the Western Pacific, Yokoi et al. [7] examined a relationship between TC frequency and simulated behavior of monsoon troughs using atmospheric-ocean coupled GCMs from the third phase of the World Climate Research Programme Coupled Model Intercomparison Project. They showed that TC frequency is attributable to the trough migration. In this chapter, we examine whether or not atmospheric circulation affects TCs by replacing the physical scheme. For a more realistic representation of the climate system and more relevant predictions, much higher resolution is a fundamental prerequisite [8].

In this chapter, to avoid uncertainty in the cumulus parameterizations used in the hydrostatic framework [2], we use the Nonhydrostatic ICosahedral Atmospheric Model (NICAM) [9, 10] without any cumulus parameterization. NICAM, a prototype global cloud-resolving model, has exhibited great promise in reproducing cloud-associated disturbances in a series of case studies on Madden-Julian oscillation (MJO) [11] and TCs [12-16]. In addition, the model has successfully captured monsoon-related intraseasonal disturbances [17, 18]. The success of NICAM in simulating various tropical cloud systems forms the basis of this type of TC projection study.

Although computational resources have recently increased, high-resolution global experiments, such as those with a resolution higher than 20 km, are difficult to run frequently. Murakami et al. [19] examined the resolution dependency for the projected intensities, genesis number, and spatial distributions of the genesis frequency of TCs using the Japan Meteorological Agency/Meteorological Research Institute Atmospheric General Circulation Model, which is among the highest resolution models available. They suggest that the highest resolution model performs best at representing the observed intense TC. In addition, their model indicates that differences in the basin-scale annual mean TC genesis number relative to the observation and the spatial distribution of TC genesis frequency were not critically dependent on the resolution. Yamada et al. [16] used NICAM to show that the experiment with a horizontal 14 km mesh reproduced realistic TC tracks and intensities. In this chapter, following [16], we use NICAM with a horizontal 14 km mesh for examining the sensitivity of the spatial distribution, frequency of tropical cyclogenesis, and intensity of TCs to the atmospheric circulation depending on physical schemes.

The structure of this chapter is as follows: in "2. Methodology", the models and experimental setup and TC tracking methodology are described; in "3. TC genesis, track, and intensity", simulated TC features with observed TCs are shown. In addition, a comparison of TCs and the large-scale environment is presented in "4. Relation between TC activity and environment section"; "5 Summary" addresses material covered herein, and our future plans are outlined.

2. Methodology

Using NICAM, we conducted three cases of experiments that include different types of physical schemes. In the present experiments, the horizontal grid interval is approximately 14 km for all simulations. There are 40 vertical levels from the surface up to 38 km, and their vertical intervals increase from 160 m to 2.9 km in height. Hereafter, we refer to the three experiments as CASE1, CASE2, and CASE3. The major model physical schemes are listed in Table 1. CASE1 and CASE2 use an explicit cloud microphysics scheme [20], which includes 3 categories of prognostic variables for water substances (water vapor, air-borne hydrometer, and precipitating hydrometer), and liquid and ice phases are partitioned by temperature-dependent function for both air-borne and precipitating hydrometeors. CASE3 uses a more comprehensive cloud microphysics scheme [21] with 6 categories (water vapor, snow, cloud water, ice, rain, and graupel) as prognostic variables. The difference between CASE1 and CASE2 is only the radiation scheme [22]. The number of absorption bands in CASE1 is 22, and that of CASE2 is 26. CASE3 employs a more sophisticated land model, called MATSIRO [23], while a simpler BUKETS model was implemented in the model configurations of CASE1 and CASE2. In addition, we implement a single layer slab ocean model only in CASE3. In the model, the sea surface temperature is nudged toward the observation. The physical scheme of CASE1 is detailed in Oouchi et al. [13, 18] and Noda et al. [24]. CASE2 is the same as the present-day experiment in Yamada et al. [16]. The physical scheme in CASE3 is the same as that documented in Satoh et al. [17].

Table 1. List of physical schemes in each experiment

Physics / Case	CASE1	CASE2	CASE3
Cloud microphysics	Grabowski [20]	Grabowski [20]	Tomita [21]
Radiation	Sekiguchi and Nakajima [22] (number of absorption bands: 22)	Sekiguchi and Nakajima [22] (number of absorption bands: 26)	Sekiguchi and Nakajima [22] (number of absorption bands: 26)
Land	BUCKET	BUCKET	Takata et al. [23]
Ocean	Fixed (weekly updating)	Fixed (weekly updating)	Slab ocean

The three experiments are performed for the same time period. All the simulations started on 1 June 2004 and ran for about 5 months. The initial atmospheric conditions for the simulations are the National Centers for Environmental Prediction (NCEP) Global Tropospheric Analyses. The sea surface temperature was provided by a weekly interpolated National Organization of Atmospheric Agency (NOAA)-Objectively Interpolated (OI) sea surface temperature.

The tracking method follows that of Oouchi et al. [2]. The maximum wind speed threshold (traced at 10-m height) is 17.5 ms^{-1} for NICAM. Caution is required in determining the threshold, to which TC frequency is quite sensitive. When the horizontal resolution is finer than about 10 km, a threshold of 17.5 ms^{-1} for the 10-m wind speed is appropriate [25]. TCs in the NICAM simulations are compared with observation (OBS) from the International Best Track Archive for Climate Stewardship (IBTrACS v03r01) provided by the NOAA. To examine the climatology in the simulations, we use NCEP Global Tropospheric Analyses for June to October as OBS.

3. TC Genesis, Track and Intensity

Figure 1 shows the spatial distributions of precipitation averaged for five months in each case and an observation which is calculated from the Global Satellite Mapping of Precipitation project (GSMaP) [26-29]. Here, we regard the GSMaP as OBS. The distribution in CASE2 resembles that of CASE1. Elaborating only the radiation scheme does not significantly influence the horizontal distribution of precipitation. CASE1 and CASE2 roughly capture the features of the observed horizontal distribution. However, CASE3 differs from the others. With CASE3 in particular, the precipitation is small near the Philippine Islands. It is noteworthy that the differences in the physical schemes have a large influence on the simulations.

Table 2 lists the frequencies of cyclogenesis observed in 2004, CASE1, CASE2, and CASE3. We follow Oouchi et al. [2] regarding the definition of ocean basins. Based on OBS, cyclogenesis from June to October 2004 is active over the Western and Eastern Pacific and the Atlantic basins. In contrast, cyclogenesis is inactive over the Indian Ocean basin. NICAM simulations reproduce the active cyclogenesis over the Western and Eastern Pacific and the Atlantic basin. Although CASE1 and CASE2 overestimate cyclogenesis over the Indian Ocean basin, CASE3 yields an improvement of the overestimation. Although CASE1 and

CASE3 overestimate cyclogenesis over the Atlantic basin, CASE2 suppresses the overestimation.

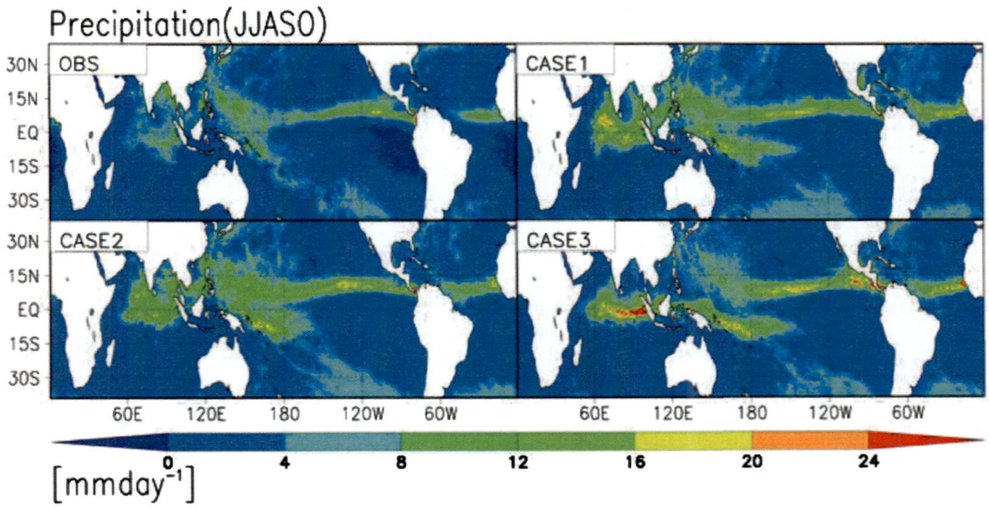

Figure 1. Averaged monthly precipitation [mm day^{-1}] from June to October 2004. The OBS is from GSMaP, and CASE1, CASE2, and CASE3 are from each simulation.

Table 2. Total number of cyclogeneses per experiment and ocean basin, where the Indian Ocean basin (IO) is defined from 30 to 100°E, the Western Pacific basin (WP) from 100 to 180°E, the Eastern Pacific basin (EP) from 180 to 90°W, and the Atlantic basin (AT) from 90 to 30°E. The OBS is from IBTrACS v03r01

Case / Basin	IO	WP	EP	AT	Globe
OBS	2	20	10	13	45
CASE1	8	21	10	21	60
CASE2	9	22	12	13	56
CASE3	3	14	12	20	49

Regarding the cyclogenesis distributions in OBS (Figure 2, upper left panel) over the Western Pacific, the cyclogenesis extends northward and southward from 15°N, and, over the Eastern Pacific, it concentrates at some distance from the western coast of Mexico. For the Atlantic, TCs occur over the western Gulf of Mexico, near the east coast of the Florida Peninsula, and from the Caribbean ocean to the western coast of Africa. The NICAM simulations generally reproduce these distributions. However, there are some discrepancies. For CASE1, a TC is generated over a more widespread area than that of the OBS over the Atlantic basin. In contrast, for CASE2, cyclogenesis concentrates in the sea around the western cost of Africa. In addition, more TCs than OBS are generated over the area southeast of the Hawaiian Islands and are not generated at all over the East China Sea in CASE2. Regarding CASE3, the distribution shifts eastward over the Western Pacific basin.

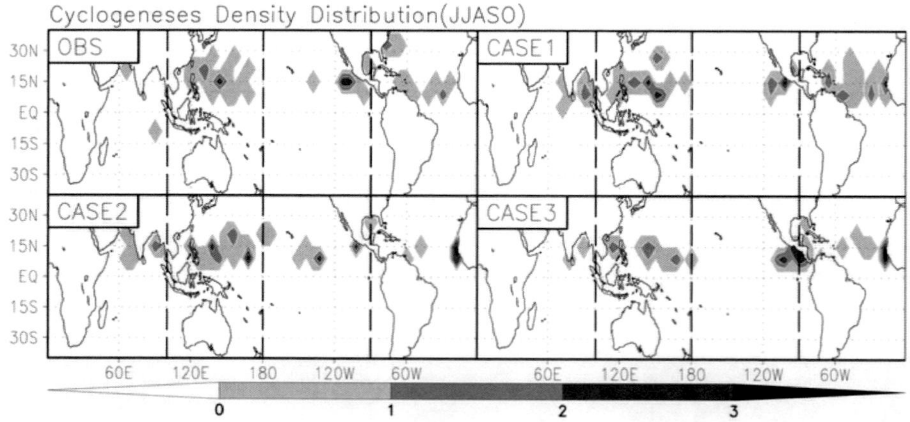

Figure 2. Cyclogenesis density (number per 5° lat.-long. area per 5 months) distribution from June to October. Long dashed line means partitions of ocean basin [2]. The OBS is from IBTrACS v03r01, and CASE1, CASE2, and CASE3 are from each simulation.

To compare the observed tracks of TCs with those in the NICAM simulations, the track densities in the OBS and each case are shown in Figure 3. Over the Western Pacific, the TC moves northwesterly toward the East China Sea and then veers away to the Japan Islands in the OBS. CASE3 reproduces the change in direction to some extent, while many TCs move to Eurasia without the veering that takes place in CASE1 and CASE2. When focusing on the Eastern Pacific, we find that TCs in the simulations, except in CASE1, originate over the central and eastern Pacific Ocean and travel across the dateline; however, no such TC is noted in the OBS. In addition, in CASE3, most TCs originate near the west coast of Mexico (Figure 2, bottom right panel). Nevertheless, TCs travel across the dateline, as shown in Figure 3. This means that the TCs last longer than those in the observations and other simulations. Figure 3 also indicates that, although many TCs are generated near the west coast of Africa and traverse longitudinally over the Atlantic Ocean in CASE2 and CASE3, CASE1 reproduces the track density observed.

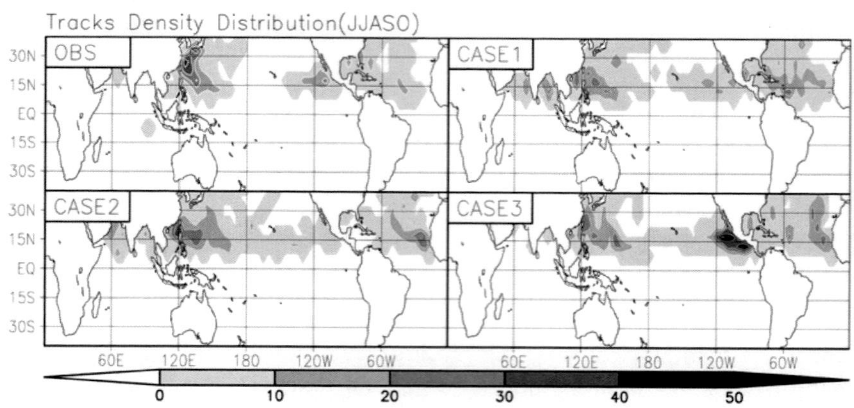

Figure 3. The same as Figure 2. However, these panels show Track density (number per 5° lat.-long. area per 5 months).

To illustrate the intensity of TCs, the histograms of the maximum 10-m wind speed (MWS) and the minimum sea level pressure (MSLP) in the life cycle of TCs are shown in Figures 4 and 5, respectively. In the OBS, the most intense TC develops to greater than 70 ms^{-1}, while about 35% of the TCs develop only to 30 ms^{-1} and less. In the NICAM simulations, however, the most intense TCs in each case do not develop to greater than 70 ms^{-1}. CASE1 and CASE3 reveal a local maximum of 40 to 50 ms^{-1}, while the OBS and CASE2 have a local maximum of 20 to 30 ms^{-1}. Weak TCs in CASE3, which do not develop to greater than 30 ms^{-1}, are smaller than those in the others.

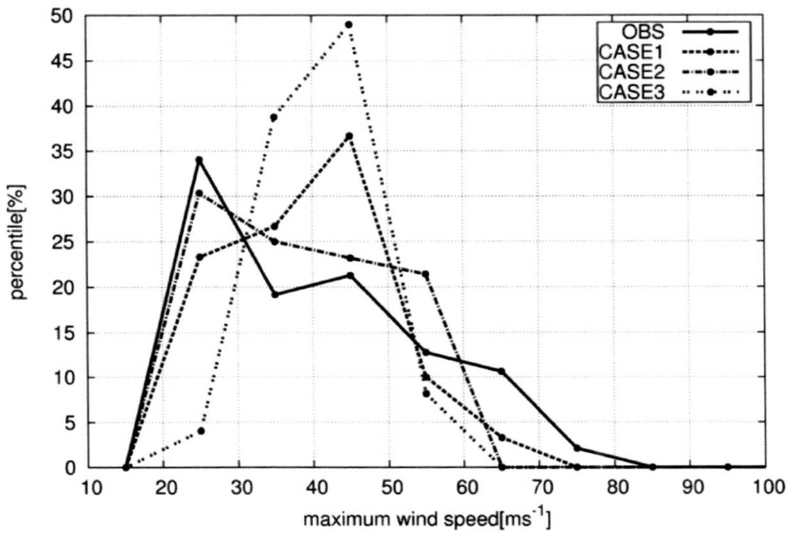

Figure 4. Rank histograms about the maximum attained 10-m wind speed [ms^{-1}]. The OBS is from IBTrACS v03r01, and CASE1, CASE2, and CASE3 are from each simulation.

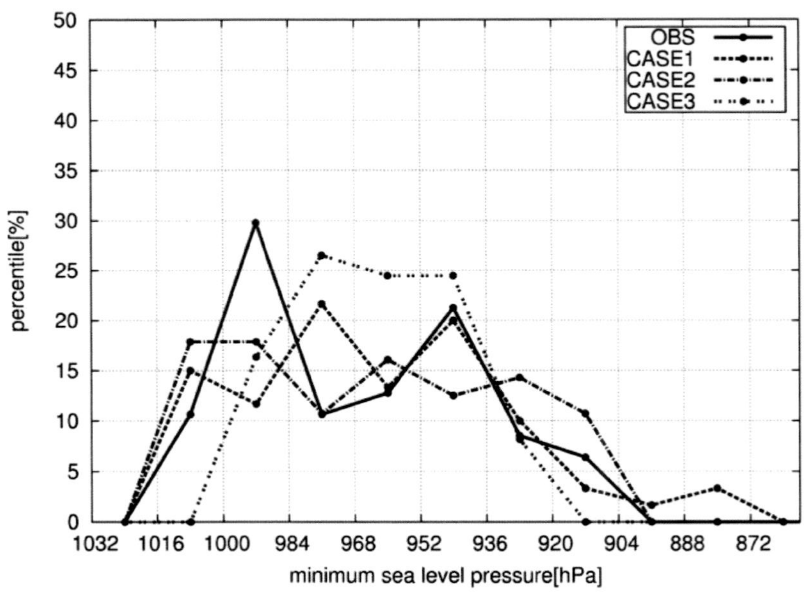

Figure 5. The same as Figure 4, except in the minimum sea level pressure [hPa].

In terms of MSLP, TCs in CASE1 and CASE2, which develop to less than 920 hPa, are more numerous than those in the OBS. In CASE3, there are no intense TCs that develop to less than 920 hPa. In contrast, there are also no weak TCs that maintain MSLP of greater than 1,000 hPa. Moreover, although the OBS show two peaks in 1,000 to 984 hPa and 952 to 936 hPa, the same peaks do not appear in these simulations.

A relationship between MSLP and MWS provides a way to assess the intensity and associated horizontal structure of the simulated TC. The scatter plots of both variables for all the TCs simulated in each case and observations are shown in Figure 6. The fitting line to the scatter plot in each case and the empirical relationship [30] are also shown in Figure 6. A comparison between the fitting lines in each simulation does not yield systematic differences. However, the scatter plots indicate that degrees of attainable TC intensity differ among the simulations. In a comparison of the simulations and the OBS, the maximum 10-m wind speed is less than that in the OBS when the MSLP is under 960 hPa.

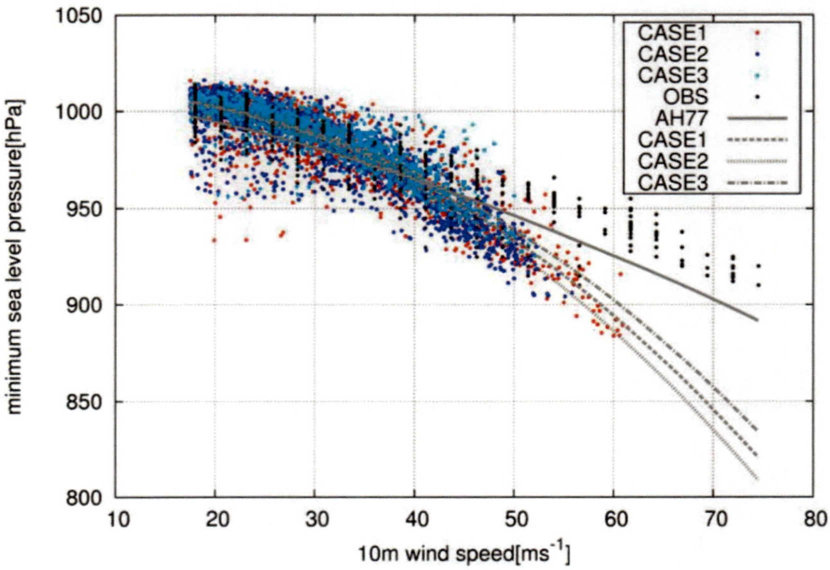

Figure 6. Relationships between the minimum sea level pressure (MSLP) [hPa] and the maximum 10-m wind speed (MWS) [ms^{-1}] when a tropical cyclone has MSLP in the observations and NICAM simulations; the lines indicate the fitting line to the scatter plot in each case and the empirical relationship between MSLP and MWS [30]. The OBS is from IBTrACS v03r01, and CASE1, CASE2, and CASE3 are from each simulation.

4. Relation between TC Activity and Environment

As discussed in the previous section, the intensity and genesis of a TC are sensitive to the model including physical schemes such as a cloud microphysics, a radiation scheme, a land model, and treatment of the sea surface temperature. Hereafter, we examine the factors that cause differences in intensity and genesis with the aid of a Genesis Potential Index (GPI) proposed by Emanuel and Nolan [31] and examined by Camargo et al. [32]. The normalized

GPI distributions in the NCEP Global Tropospheric Analyses and NICAM simulations are presented in Figure 7. Here, we regard NCEP Global Tropospheric Analyses as OBS. Comparing Figure 7 with Figure 2, the distributions of cyclogenesis are generally consistent with high GPI distributions in each simulation and the OBS. Therefore, comparing the change of the GPI and its components among cases is relevant. As CASE1 is a better simulation than the others to the extent of reproducing the distributions of cyclogenesis and the TC track, we regard CASE1 as a control case. The GPI is defined as:

$$GPI = \left|10^5 \eta\right|^{\frac{3}{2}} \left(\tfrac{H}{50}\right)^3 \left(\tfrac{V_{pot}}{70}\right)^3 \left(1 + 0.1V_{shear}\right)^{-2},$$

where η is the absolute vorticity [s^{-1}] at 850 hPa, H is the relative humidity [%] at 700 hPa, V_{pot} is the potential intensity [ms^{-1}] [32, 33], and V_{shear} is the vertical wind shear [ms^{-1}]. Here, to examine the GPI component that contributes most to the GPI change, we divide the GPI into four components, as shown below:

$$Vor = \left|10^5 \eta\right|^{\frac{3}{2}},\ RH = \left(\tfrac{H}{50}\right)^3,\ PI = \left(\tfrac{V_{pot}}{70}\right)^3,\ Vsh = \left(1 + 0.1V_{shear}\right)^{-2}.$$

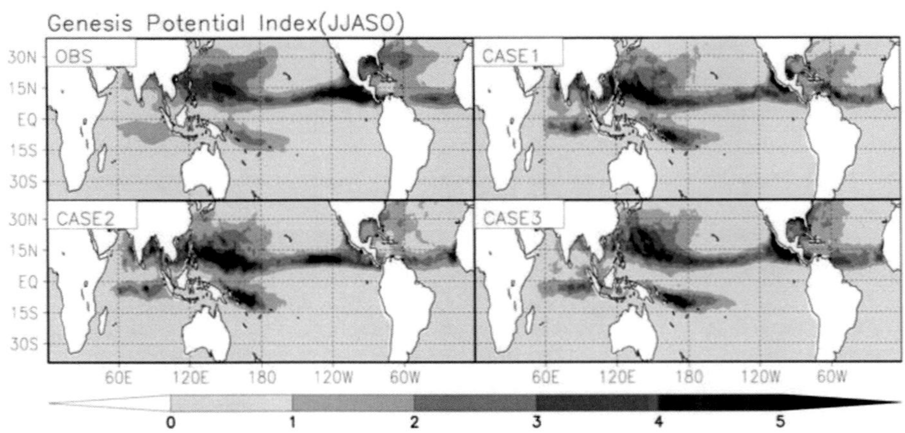

Figure 7. Averaged Genesis Potential Index from June to October 2004, which is normalized by 30°N-30°S area-averaged values. The OBS is from NCEP Global Tropospheric Analyses, and CASE1, CASE2, and CASE3 are from each simulation.

Figures 8a and 7.8f show the differences in GPI between CASE2 and CASE1 and between CASE3 and CASE1, respectively. Figures 8b-e and Figures 8g-j show that the contribution is assessed by taking the difference (CASE2 or CASE3 – CASE1) for only one of the contributing factors with the other factors set to CASE1. For example, when we examine the contribution of the Vor in CASE2, $\Delta GPI = (Vor_{CASE2} - Vor_{CASE1}) \times Vsh_{CASE1} \times RH_{CASE1} \times PI_{CASE1}$, where the subscript indicates a case of experiment.

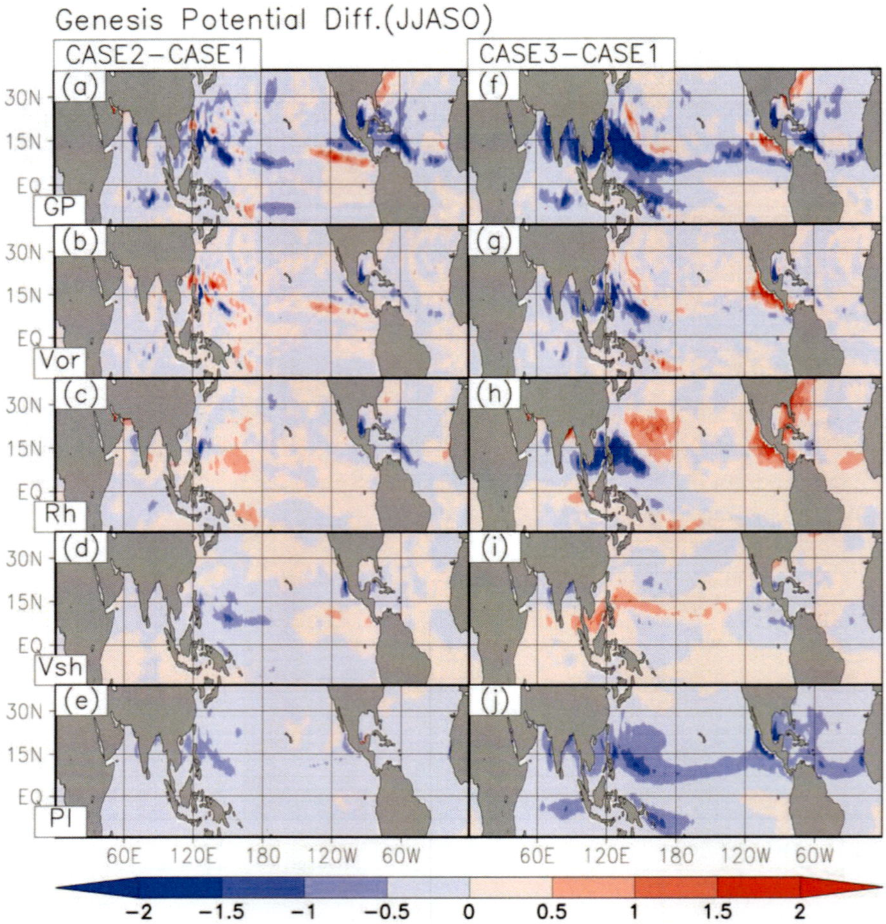

Figure 8. The difference in the GPI and its components by replacing the physics scheme, (a) CASE2 - CASE1 in the GPI, (b), (c), (d), and (e) show the contribution of factors in the GPI to the GPI change, which is calculated by taking the difference (CASE2 or CASE3 – CASE1) for only one of the contributing factors with the other factors set to CASE1 (i.e., $\Delta GPI=(Vor_{CASE2}-Vor_{CASE1})\times Vsh_{CASE1}\times RH_{CASE1}\times PI_{CASE1}$, where Vor indicates the term associated with the absolute vorticity at 850 hPa, Vsh indicates the term associated with the vertical wind shear (difference between 200 and 850 hPa), RH indicates the term associated with the relative humidity at 700 hPa, PI indicates the term associated with the Potential Intensity, and subscript indicates a case of experiment), (f) – (j) are the same as (a) – (e) but for CASE3 instead of CASE2. The blue shade is unfavorable for tropical cyclogenesis, and the red shade is favorable.

Over the Indian Ocean, the GPI components in CASE3 are less favorable than those in CASE1. Although it is difficult to determine what factor makes the largest contribution to the change of the GPI between CASE3 and CASE1, the reduction of the GPI is consistent with the reduction of cyclogenesis.

Figures 8a and 8f also show that the GPI in CASE2 and CASE3 is smaller than that of CASE1 over west of the Western Pacific. In CASE3, the cyclogenesis area shifts eastward (Figure 2), and the high-GPI area also shifts eastward over the Western Pacific. The translation in the cyclogenesis area is consistent with the high-GPI area shift. Focusing on the components of the GPI, we find that Vor and RH are unfavorable for cyclogenesis. Precipitation decreases near the Philippine Islands (Figure 1), so it seems that the convection weakens there. Comparing CASE2 with CASE1, although the GPI reduced west of the Western Pacific, the frequency of cyclogenesis does not largely change. Focusing on the difference in distribution of GPI components, although Vor (Figure 8b) and RH (Figure 8c) are increased, Vsh (Figure7. 8d) and PI (Figure 8e) are decreased. We find that the local distribution of reduction of the GPI west of the Western Pacific resembles that of PI in form. Therefore, we suppose that PI is not a significant contribution to the change of cyclogenesis over west of the Western Pacific, but it contributes to the change of the GPI in these simulations.

The distribution of cyclogenesis in CASE2 and CASE3 is reduced and increased, respectively, from that in CASE1 (Figure 2) near the west coast of Mexico. Focusing on the differences in the GPI (Figures 8a and 7.8b), a contrasting structure is shown; the distribution of the warm and cold colors lies north and south, respectively, in a contrasting manner. When we examine Figure 8, it seems that Vor plays a key role. In addition, Vsh in CASE3 is favorable for TCs over the central Pacific (Figure 8i). Therefore, TCs last longer in CASE3 than in other simulations.

In CASE2 and CASE3, many TCs are generated near the west coast of Africa (Figure 2) over the Atlantic basin. In the OBS, however, cyclogeneses occur not only over the ocean area but also over the western Gulf of Mexico, near the east coast of the Florida Peninsula and the Caribbean Sea. Regarding the change of the GPI components between CASE2 and CASE1 and between CASE3 and CASE1, in particular Vor and RH are elevated near the west coast of Africa (Figures 8b, c and Figures 8g, h). In addition, over the Atlantic Ocean, Vsh in CASE2 and CASE3 is relatively favorable, comparing with that in CASE1 (Figures 8d and 7.8i). Therefore, TCs are not suppressed in CASE2 or CASE3 over the Atlantic Ocean by the vertical wind shear.

With regard to the large-scale environments among the three experiments, the horizontal distribution of the absolute vorticity, the relative humidity, and the vertical wind shear differ from each other. These variables are associated with the vertical and horizontal circulation in the atmosphere. We estimate that the difference of the spatial distributions of the GPI and its components results from change of simulated atmospheric circulation depending on the physical scheme.

Focusing on horizontal circulation, Figure 9 shows the zonal wind at 850 hPa averaged for 5 months in the NCEP Global Tropospheric Analysis and simulations. In CASE1, CASE2, and OBS, we find a westerly wind region from the Indian Ocean to the Philippine Islands. However, the westerly wind does not exist over the Philippine Islands in CASE3. CASE3 differs from CASE1 and CASE2 in cloud microphysics, land model, and treatment of the sea surface temperature; this is such that we cannot determine which scheme is most responsible. Alternatively, it is possible for the change in the westerly wind to result from interaction among replaced physics schemes. At a minimum, replacing physics schemes changes the horizontal circulation, and then the convergence region moves away from the Philippine Islands. Consequently, an updraft region disappears from the area near the Philippine Islands. It is also identifiable in Figure 1. Compared with CASE1 and CASE3, a reduction of precipitation in CASE3 is observed in the area over the Philippine Islands (Figure 1). The differences in physics schemes weaken the convection which produces the precipitation. Basically, updraft disappears or weakens so that water vapor is not supplied from the lower troposphere. Hence, we suppose that RH in CASE3 becomes lower than that in CASE1 in the area over the Philippine Islands (Figure 8h). In addition, the distribution of the zonal wind changes (Figure 9) so that the distribution of absolute vorticity (Figures 8b and 7.8g) and vertical wind shear (Figures 8d and 7.8i) changes from CASE1 to CASE3. Therefore, we think that the cyclogenesis frequency distribution differs between CASE1 and CASE3 (Figure 2).

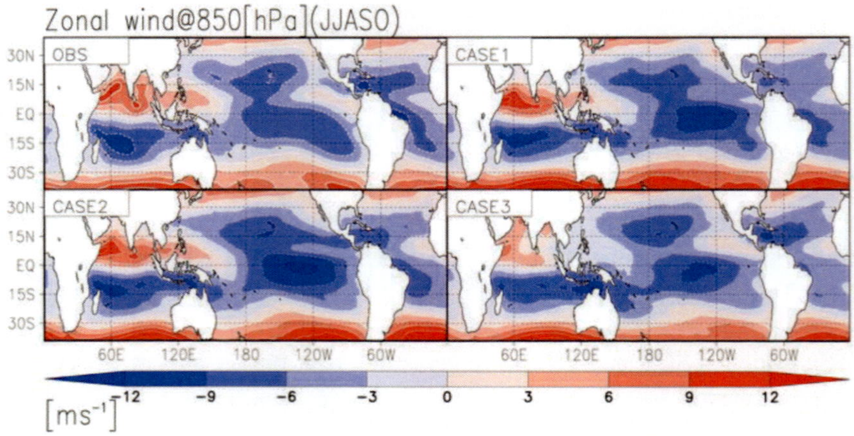

Figure 9. Averaged zonal wind speed at 850 hPa from June to October 2004. The blue shade is the easterly wind, and the red shade is the westerly wind. The OBS is from NCEP Global Tropospheric Analyses, and CASE1, CASE2, and CASE3 are from each simulation.

In terms of the intensity, Figures 8e and 7.8j show that PI in CASE2 and CASE3 are globally smaller than that in CASE1. In fact, the intensities of TCs simulated in CASE2 and CASE3 are weaker than those of the TCs simulated in CASE1 (Figures 4 and 7.5). In addition, the reduction in CASE3 is greater than that in CASE2. According to a comparison of the most intense TC in each simulation (Figure 4), that of CASE1 has the highest wind speed among all simulations, and that of CASE2 is stronger than that of CASE3. Therefore, we find that the PI reproduces actual intensity of the simulated TCs.

SUMMARY

In this chapter, we compare the frequencies, spatial distributions of the simulated cyclogeneses, and intensities of the simulated TCs in three experiments with those of OBS. The three experiments include different physical schemes such as a cloud microphysics, a radiation scheme, a land model, and treatment of the sea surface temperature. We thereby focus on differences in the physical schemes contributing to the spatial distribution of cyclogenesis and assess which large-scale environmental factors play an important role in those characteristics of TCs. The relationships between a TC and large-scale environments are discussed based on the GPI and its components. In the present experiments, the spatial distribution and frequency of tropical cyclogenesis are particularly sensitive to the physical scheme. In addition, the change of cyclogenesis is generally consistent with the change of the

GPI when CASE1 is regarded as a control case according to the comparisons of the spatial distribution of cyclogenesis and track densities.

A radiation scheme used in CASE2 is different from that in CASE1. Over the Atlantic basin, its effect is to cause tropical cyclogenesis in CASE2 to be smaller than that in CASE1, and its frequency of cyclogenesis to be closer to the OBS than that in CASE1. In contrast, the spatial distribution of CASE1 is more realistic than that in CASE2 over the Atlantic basin. Cyclogenesis in CASE3 is smaller than that in CASE1 and CASE2 over the Indian Ocean, and its area shifts eastward over the Western Pacific as compared to that in other cases and the OBS.

By comparing spatial distributions of the GPI and its components in CASE2 and CASE3 with those in CASE1 (Figure 8), those spatial distributions are different from those in CASE1. These components of the GPI are associated with the atmospheric circulation, such as convergence of zonal wind at the lower troposphere (Figure 9) and the location of updraft or subsidence in the atmosphere. Therefore, we assume that these changes would result from differences in the atmospheric circulation depending on the physical scheme used. Bengtsson et al. [5] shows that considering the distribution of the relative sea surface temperature and the associated atmospheric circulation is important for understanding a TC. The comparison of the present simulations also indicates that the atmospheric circulation depends on not only the sea surface temperature but also the physical schemes. Therefore, we ought to carefully select a physical scheme in order to reproduce realistic atmospheric circulation.

In terms of the intensity of TCs, MSLP and MWS differ among the cases. For MSLP, NICAM basically reproduces the intensity of a TC compared with the OBSs, while MWS for an intense TC is underestimated. The previous study shows that their experiment also underestimates MWS for intense TCs [5]. The differences in TC intensity are consistent with changes of calculated PI from each simulation. The most intense TC is generated in CASE1. The TC intensity in CASE2 is weaker than that in CASE1 as the radiation scheme is replaced. Moreover, the TC has the weakest maximum intensity in CASE3, while there are a few TCs with MSLP or MWS of more than 1,000 hPa or less than 30 ms^{-1}, respectively. The simulated relationships between MWS and MSLP are compared to those obtained from the observations [30]. Figure 6 indicates that MWS for a given MSLP is weaker than that in the OBS for intense TCs. In addition, we examined the relationship in the three simulations. There is a slight difference in the relationship (Fig 7.6). The 7 km mesh case [17] is much closer to the empirical relationship than the 14 km mesh cases. However, 14 km-mesh experiment used in this chapter reproduces more realistic intensity than the previous studies [2]. This is because GCRM using 14km has impressive strength at representing tropical convection and the convectively coupled disturbance compared with the conventional GCM [13]. Moreover such a propagating tropical wave influences on not only cyclogenesis but also rapid intensification of TC [34]. Thus, we assume that cyclogeneses and TCs intensity in the present simulations are more realistic.

In a future work, we will clarify how different physical schemes change the atmospheric circulation, such as the Walker circulation, that can influence characteristics of tropical cyclogenesis. The relationship between MWS and MSLP is associated with distance between the center of a TC and the location of MWS. Although we have a limitation on observed data of maximum wind radius, we hope that we can examine the horizontal structure and the maximum wind radius of the simulated TCs to verify whether the simulated TCs are realistic or not. We anticipate that findings in the future work will contribute to a reliable projection of TCs in a future global warming climate.

ACKNOWLEDGMENTS

The numerical experiments were performed on the Earth Simulator of JAMSTEC under the framework of the Innovative Program of Climate Change Projection for the 21st century (KAKUSHIN) project funded by the Ministry of Education, Culture, Sports, Science, and Technology (MEXT), Japan. The TC tracking tool was provided by Dr. Wataru Yanase and improved with the cooperation of Dr. Chihiro Kodama. The authors thank Drs. Tomoe Nasuno, Shin-ichi Iga, Hiroshi Taniguchi, Yoshiaki Miyamoto, Akira Noda, Masato Sugi, Taro Matsuno, and anonymous reviewers for their valuable comments. The GrADS and gnuplot were used for drawing the figures. GSMaP was provided by the Japan Aerospace Exploration Agency/Earth Observation Research Center and supported by "Production of a High-Precision, High-Resolution Global Precipitation Map Using Satellite Data" under the "Core Research for Evolutional Science and Technology" program of the Japan Science and Technology Agency.

REFERENCES

[1] T. R. Knutson, J. L. McBridge, J. Chan, K. Emanuel, G. Holland, C. Landsea, I. Held, J. P. Kossin, A. K. Srivastava, and M. Sugi, *Nature Geo.* 3, 157, doi:10.1038/NGEO779 (2010).

[2] K. Oouchi, J. Yoshimura, H. Yoshimura, R. Mizuta, S. Kusunoki, and A. Noda, *J. Meteor. Soc. Japan,* 84, 259, doi:10.2151/jmsj.84.259 (2006).

[3] Intergovernmental Panel on Climate Change (IPCC), *Climate models and their evaluation, in Climate Change 2007*, Cambridge Univ. Press (2007).

[4] M. Sugi, H. Murakami, and J. Yoshimura, *SOLA*, 5, 164, doi:10.2151/sola.2009-042 (2009).

[5] L. Bengtsson, K. I. Hodges, and M. Esch, *Tellus* 59A, 396 (2007)

[6] G. A. Vecchi and B. J. Soden, *Geophys. Res. Lett.* 34, L08702, doi:10.1029/2006GL028905 (2007).

[7] S. Yokoi, Y. N. Takayabu, and J. C. L. Chan, *Clim Dyn.* 33, 665 (2009).

[8] J. Shukla, R. Hagedorn, B. Hoskins, J. Kinter, J. Marotzke, M. Miller, T. Palmer, and J. Slingo, *Bull. Amer. Meteor. Soc.* 80, 2027, doi:10.1175/2008BAMS2759.1 (2009).

[9] H. Tomita and M. Satoh, *Fluid Dynamics Research,* 34, 357, DOI 10.1007/s00382-009-0593-9 (2004).

[10] M. Satoh, T. Matsuno, H. Tomita, H. Miura, T. Nasuno, and S. Iga, *J. Comput. Phys.* 227, 3486, doi:10.1016/j.jcp.2007.02.006 (2008).

[11] H. Miura, M. Satoh, T. Nasuno, A. T. Noda, and K. Oouchi, *Science,* 318, 1763 (2007).

[12] H. Fudeyasu, Y. Wang, M. Satoh, T. Nasuno, H. Miura, and W. Yanase, *Geophys. Res. Lett.* 35, L22808, doi:10.1029/2008GL036003 (2008).

[13] K. Oouchi, A. T. Noda, M. Satoh, H. Miura, H. Tomita, T. Nasuno, and S. Iga, *SOLA,* 5, 65, doi:10.2151/sola.2009-017 (2009).
[14] W. Yanase, H. Taniguchi, and M. Satoh, *J. Meteor. Soc. Japan,* 88, 497, DOI;10.2151/jmsj.2010-314 (2010).
[15] H. Taniguchi, W. Yanase, and M. Satoh, *J. Meteor. Soc. Japan,* 88, 571, DOI:10.2151/jmsj.2010-317 (2010).
[16] Y. Yamada, K. Oouchi, M. Satoh, H. Tomita, and W. Yanase, *Geophys. Res. Lett.* 37, L07709, doi:10.1029/2010GL042518 (2010).
[17] M. Satoh, K. Oouchi, T. Nasuno, H. Taniguchi, Y. Yamada, H. Tomita, C. Kodama, and J. Kinter, COLA & ECMWF people, *Clim Dyn.* submitted (2011).
[18] K. Oouchi, A. T. Noda, M. Satoh, B. Wang, S.-P. Xie, H. G. Takahashi, and T. Yasunari, *Geophys. Res. Lett.* 36, L11815, doi:10.1029/2009GL038271 (2009).
[19] H. Murakami and M. Sugi, *SOLA,* 6, 73, doi:10.2151/sola.2010-019 (2010).
[20] W. Grabowski, *J. Atmos. Sci.* 55, 3283 (1998).
[21] H. Tomita, *J. Meteor. Soc. Japan,* 86A, 121, doi:10.2151/jmsj.86A.121 (2008).
[22] M. Sekiguchi and T. Nakajima, *J. Quantitative Spectroscopy & Radiative Transfer,* 109, 2779, doi:10.1016/j.jqsrt2008.07.13 (2008).
[23] K. Takata, S. Emori, and T. Watanabe, *Global and Planetary Change,* 38, 209, doi:10.1016/S0921-8181(03)00030-4 (2003).
[24] A.T. Noda, K. Oouchi, M. Satoh, H. Tomita, S. Iga, and Y. Tsushima, *Atmos. Res.* 96(2-3), 208, doi:10.1016/j.atmosres.2009.05.007 (2010)
[25] K. J. E. Walsh, M. Fiorino, C. W. Landsea, and K. L. McInnes, *J. Clim.* 20, 2307, doi:10.1175/JCLI4074.1 (2007).
[26] K. Okamoto, T. Iguchi, N. Takahashi, K. Iwanami, and T. Ushio, "The Global Satellite Mapping of Precipitation (GSMaP) project," 25th IGARSS Proceedings, (Seoul, Korea) (2005).
[27] T. Kubota, S. Shige, H. Hashizume, K. Aonashi, N. Takahashi, S. Seto, M. Hirose, Y. N. Takayabu, K. Nakagawa, K. Iwanami, T. Ushio, M. Kachi, and K. Okamoto, *IEEE Trans. Geosci. Remote Sens.,* 45, 2259 (2007).
[28] K. Aonashi, J. Awaka, M. Hirose, T. Kozu, T. Kubota, G. Liu, S. Shige, S. Kida, S. Seto, N. Takahashi, and Y. N. Takayabu, *J. Meteor. Soc. Japan,* 87A, 119, doi:10.2151/jmsj.87A.119 (2009).
[29] T. Ushio, T. Kubota, S. Shige, K. Okamoto, K. Aonashi, T. Inoue, N. Takahashi, T. Iguchi, M. Kachi, R. Oki, T. Morimoto, and Z. Kawasaki, *J. Meteor. Soc. Japan,* 87A, 137, doi:10.2151/jmsj.87A.137 (2009).
[30] G. D. Atkinson and C. R. Holliday, *Mon. Wea. Rev.* 105, 421 (1977).
[31] K. Emanuel and D. S. Nolan, "Tropical cyclone activity and the global climate system," The 26th Conference on Hurricanes and Tropical Meteorology, (Miami, FL, USA)(2004) http://ams.confex.com/ams/26HURR/techprogram/paper_75463.htm
[32] S. J. Camargo, A. H. Sobel, A. G. Barnston, and K. A. Emanuel, *Tellus,* A59, 428 (2007).
[33] K. A. Emanuel, *J. Atmos. Sci.* 52, 3969 (1995).
[34] W. Yanase, M. Satoh, H. Yamada, K. Yasunaga, and Q. Moteki, *Geophys. Res. Lett.* 37, L08809, doi:10.1029/2010GL042516 (2010).

In: Cyclones: Formation, Triggers and Control
Editors: K. Oouchi and H. Fudeyasu
ISBN: 978-1-61942-976-5
©2012 Nova Science Publishers, Inc.

Chapter 8

VARIABILITY OF EXTRATROPICAL CYCLONIC ACTIVITY IN THE NORTHERN HEMISPHERE ASSOCIATED WITH GLOBAL PROCESSES IN THE OCEAN-ATMOSPHERE SYSTEM

Alexander Polonsky[1], Mikhail Bardin[2] and Elena Voskresenskaya[1]
[1]Marine Hydrophysical Institute, Sevastopol, Ukraine
[2]Institute of Global Climate and Ecology, Moscow, Russia

ABSTRACT

In this chapter we describe connections between the seasonal statistics of extratropical synoptic-scale atmospheric vortices in the Northern Hemisphere, and particularly in the North Atlantic and Pacific regions, and phases of principal interannual to multi-decadal climatic signals, or leading climate modes, such as El Ninio-Southern Oscillation (ENSO), North Atlantic Oscillation (NAO), Pacific Decadal Oscillation (PDO) and Atlantic Multi-decadal Oscillation (AMO). These modes account for the essential proportion of both SST anomalies and associated atmospheric signal in the Northern Hemisphere on the interannual to multi-decadal scale. The relationships are derived from empirical data. Archives of Northern Hemisphere extratropical cyclones and cyclone tracks from 1951 to 2010 were compiled based on 6-hrs the 1000 hPa geopotential heights from NCEP/NCAR reanalysis data. The monthly characteristics of the cyclonic activity (cyclone frequency, area and intensity) for the cells of a 5-degree grid and for selected regions were used to calculate climate statistics and evaluate relationships between cyclonic activity and climate modes. Analysis of the interannual to multi-decadal variability of cyclonic characteristics associated with the NAO, ENSO, PDO and AMO showed that their joint influence accounts for significant proportion of total variance of winter cyclone parameters in the neighborhood of principal stormtracks and in other regions of the Northern Hemisphere.

[1] E-mail: apolonsky5@mail.ru

1. INTRODUCTION

Extratropical atmospheric cyclones not only create the weather in the mid latitudes. They are one of the key elements of general atmosphere circulation (GAC). Synoptic eddies can be considered as a large - scale quasi geostrophic turbulence. In the idealized model of the atmospheric circulation, this synoptic – scale turbulence is generated by baroclinic instability within the middle and high latitude cells of the GAC. Cyclones and anticyclones are effectively generated in shear flows of mid-latitude tropospheric jets which are maintained through the mechanism generally known as "negative viscosity" (that is energy cascade from smaller to larger scales) [1]. However, while tropospheric eddies are mainly associated with "stormtracks" over the oceans, the traveling cyclones are found everywhere throughout extratropics and affect weather conditions virtually in all parts of both the oceans and continents. Over the oceans they provide the major part of exchange by momentum, sensible and latent heat between atmosphere and the ocean [2].

Cyclones generate the ocean response with the spectrum characterized by the red shift (as a result of integrated of by the highly inertial ocean relatively high-frequency atmospheric fluctuations with spectrum close to that of white noise [3,4]). On the other hand, the global and regional statistics of cyclones depends on the land/ocean surface conditions. Cyclones are most frequent over the areas with strong temperature gradients, i.e. at the ocean/land/sea-ice boundaries and strong oceanic fronts, in the vicinity of the mountain ranges and above the internal water pools (for example, in the Mediterranean - Black Sea region, where we find a combination of the latter two conditions, see e.g., [5]). Interannual-to-interdecadal sea surface temperature (SST) anomalies cause the atmospheric response, which can manifest itself through the anomalous intensity of Hadley, Ferrel and Walker cells, long planetary waves, cyclones'/anticyclones' statistics, etc., resulting in the certain teleconnection patterns (see [6-9] and others).

The extent to which generation of extra-tropical synoptic atmospheric disturbances by tropical and mid-latitude SST anomalies is effective was the matter of long-term and intense debates. The principal issue in this concern is a signal-to-noise ratio. Many publications argue that the interannual atmospheric fluctuations in mid-latitudes (including the synoptic-scale eddies) are rather inherent random atmospheric signals than a response induced by ocean variability. However, most of the recent studies performed within the scope of international programs and projects such as, TOGA, CLIVAR, ENSEMBLE, etc, have demonstrated that the atmospheric large/synoptic-scale response in the middle latitudes to large-scale SST anomalies is statistically significant. Moreover, this response is likely the only possible basis for seasonal to decadal predictability of atmospheric variability (see, e.g. [8,10,11] and others).

The main goal of the present chapter is to describe and discuss the connections between the seasonal statistics of synoptic-scale atmospheric vortices in high/mid-latitudes and subtropics of the Northern Hemisphere with strong focus on the North Atlantic and Pacific regions, and phases of principal interannual to multi-decadal climatic signals, or leading climate modes, such as NAO, ENSO, PDO and AMO. These modes combine atmospheric and oceanic processes and account for essential proportions of both SST anomalies and associated atmospheric signal in the Northern Hemisphere on the interannual to multi-decadal scale [12,13].

The general scheme of the chapter is as follows. First, we give a brief review of the four modes. After that, we describe datasets and methodology used in the study. Thereafter the general properties of the extra-tropical cyclones in the Northern Hemisphere in different seasons are discussed. In the main part of the chapter we present numerous pieces of evidence of significant change of seasonal cyclone statistics in response to large-scale climatic signals. Finally, the principal conclusions are presented.

2. Principal Interannual to Multi-Decadal SST Modes and their Atmospheric Manifestations

Empirical orthogonal functions (EOFs) of global, Pacific and Atlantic SST and sea level pressure (SLP) clearly demonstrate that the leading interannual to multi-decadal modes are ENSO, PDO, NAO and AMO. For example, four leading SST space complex EOFs extracted by Enfield and Mestas-Nunes [13] using global historical data prepared by Kaplan et al. [14] and reduced optimal interpolation technique, are ENSO, SST rise, PDO and AMO. The magnitude of the first mode is about two times as large as the magnitude of the other natural modes. Enfield and Mestas-Nunes [13] also clearly showed that low-frequency SST variability forces significant large-scale atmospheric response. They demonstrated the impact of low-frequency SST variability in the Atlantic and Pacific oceans, i.e. AMO and PDO, on global pressure fields in the middle troposphere. Numerous evidence of other SST-induced atmospheric patterns are given in [8, 10, 11, 15-20] and many other publications.

NAO is a leading interannual/sub-decadal mode of SLP variability and mid-troposphere pressure field in the Northern Hemisphere (see e.g., [12,19]). That is why it is described first. Then, we discuss the ENSO (interannual) SST/SLP signal. Decadal-scale modes (PDO and AMO) are discussed after the interannual ones. Taken into account the phenomenological focus of this chapter, the governing mechanisms of NAO, ENSO, PDO and AMO are not analyzed in details.

2.1. North Atlantic Oscillation

The North Atlantic Oscillation is quasi-periodic shifts in the relative strength of the quasi-permanent centers of high atmospheric pressure over the Azores and low atmospheric pressure between Greenland and Iceland (Azores High and Icelandic Low). It is a leading large-scale mode of the atmospheric circulation over the extratropical ocean in the Northern Hemisphere throughout the year except during a few summer months [12, 15, 19, 21-24]. NAO governs the large-scale variability of hydrometeorological fields over the North Atlantic, North America and northern Europe, as became clear after the pioneer paper published by Gilbert Walker and co-author in 1932 [25]. Recently, the NAO has been set in a broader spatial context of hemispheric climatic variations both at the surface and aloft (see generalized publications by Marshall et al. [17] and Polonsky et al. [19]). A special index is often used to measure NAO. This index, the North Atlantic Oscillation Index, is the normalized difference between SLP in the Azores High and Icelandic Low. Another measure of NAO is the time coefficient of leading EOF of atmospheric pressure over the Atlantic region [5].

Positive and negative NAO index values correspond to high and low NAO phases. The positive NAO phase is characterized by NAO index values exceeding the mean of the NAO index by one standard deviation or more. The negative phase is characterized by the opposite condition. Long-term series of the NAO index show that it varies at all time scales, from days and weeks, to decades and longer. This chapter is focused on interannual-to-sub-decadal scales of NAO variations which influence strongly the European climate. Spectral and wavelet analyses of fluctuations of NAO indices for more then 100 years between late 19th and early 21st centuries reveals dominant significant peak at the ~6 to 9 years period. The same scales of variability are typical for different meteorological fields over most of Europe. In winter, the significant portion of North Atlantic and European climate variability is due just to NAO. In fact, NAO-induced variations of zonal circulation intensity over the North Atlantic and displacement of atmospheric centers of action lead to significant climate changes over Europe. The pressure gradient between the Azores High and the Icelandic Low determines the strength of the mid-latitude Westerlies and, hence, the transport of relatively warm and humid air from the North Atlantic to Europe. During the 1990s, Hurrell and others [7, 21, 26] showed that, on the interannual-to-sub-decadal scale, NAO intensification (or positive phase of NAO) is accompanied by displacement of Azores High and Icelandic Low to the north-northeast. At the same time, the zonal atmospheric circulation over the North Atlantic and Europe, and zonal winds in the low troposphere within the band between 50 and 60 degrees N are strengthened. This accompanies displacement of pathways of most North Atlantic cyclones to the northern Europe and associated increases of air temperature and precipitation there. Meanwhile anticyclonic meteorological conditions prevail over central, southern Europe, and the Mediterranean region. The opposite tendency is observed during negative NAO phases, when the Azores High and Iceland Low shift to the south-southwest, and the Westerlies weakens. As well, the negative phase is accompanied by more frequent cyclonic conditions over European regions close to the Mediterranean Sea and Middle East, while more frequent anticyclones occur over northern Europe.

Climate change in late 20th century is accompanied by a tendency of a stronger NAO and reduced latitudinal distance between the Azores High and Icelandic Low. This and superimposed multi-decadal North Atlantic variability (see, paragraph devoted to AMO) lead to much more pronounced impacts of NAO on European climate after approximately 1960. Note however, that in the present chapter we consider the NAO as interannual-to-sub-decadal signal. Multi-decadal NAO mode is analyzed as inherent component of AMO.

Some ideas about the NAO focus on the North Atlantic basin itself [17, 19, 27]. At the same time, as shown by Eshel [28], NAO may be generated by processes in the North Pacific coupled system, which impacts the North Atlantic through the atmospheric bridge. It is worth to emphasize that the atmospheric waves are considered to play an important role not only for NAO generation, but also for North Pacific-North Atlantic interaction [18, 29, 30]. Teleconnection patterns associated with the NAO are described by Wallace and Gutzler [31], Barnston and Livezey [12] and others. There is also considerable evidence to suggest that NAO strongly impacts global and regional cyclone parameters [5, 32-34].

2.2. El Niño - Southern Oscillation

ENSO is the main climatic interannual signal of the global ocean-atmosphere system. It is a quasi-periodic climate pattern that occurs across the tropical Pacific Ocean every two to seven years. It is defined by prolonged positive/negative SST anomaly and trade-wind weakening/strengthening. This positive/negative SST anomaly in the tropical (mostly central and eastern) Pacific Ocean, the El Niño/La Niña events and associated variations of zonal SLP gradient in the tropical Pacific (the Southern Oscillation) are coupled, hence the term El Niño-Southern Oscillation. The positive phase of ENSO marked by positive SST anomalies in Central and Eastern Equatorial Pacific, is El Niño. During El Niño, SLP is lower than normal in the east and higher than normal in the west. The surface easterlies (trade winds) are relaxed. The negative phase of ENSO (La Niña) is characterized by colder than normal SST in the eastern and central Pacific and warmer than normal SST in the west is La Niña. SLP is higher than normal in the east and lower than normal in the west. The trade winds are stronger than normal. El Niño/La Niña cause meteorological anomalies in the equatorial Pacific Basin. The weather is wetter than normal in the east and drier than normal in the west during El Niño. The reverse anomalies occur during La Niña [35, 36].

The accepted definition of El Niño is a sea surface warming of at least 0.5°C (0.9°F) averaged over the east-central equatorial Pacific Ocean. The strength of El Niño events is measured by the El Niño Index (ENI). The ENI is computed as SST anomaly in the so-called NIÑO 3.4 region (5°S-5°N, 170°W-120°W). The strength of the Southern Oscillation is measured by the Southern Oscillation Index (SOI), which is computed as the standardized SLP difference between Tahiti and Darwin, Australia. El Niño episodes are associated with negative values of the SOI. La Niña events are characterized by SLP anomalies of the reverse sign. Note, the absolute magnitude of SST/SLP anomalies associated with La Niña is usually smaller than during El Niño [35, 37].

Global ENSO signal manifests in the interannual variability of the ocean-atmosphere-land system not only in tropics and subtropics, but also in the middle and high latitudes [20, 38-40]. It often causes extreme weather such as strong cyclones, floods and droughts in many regions of the world and associated economical and social loses there. For instance, the strongest 20th century El Niño (1997-1998) caused 25 billion USD in losses in the USA [41].

ENSO manifestations in the regions close to the Pacific Ocean have been described in detail and successfully simulated (e.g., [39, 43-45]*)*. However, it has long been debated whether and how El Niño influences the interannual climate variability in the North Atlantic and European region. Ropelewsky and Halpern [38], Fraederich and Muller [46], Voskresenskaya and Polonsky [47], Polonsky [48], Mariotty et al. [49], and other authors analyzed the relations between Equatorial Pacific anomalies and some associated manifestations over the Atlantic and European/Mediterranean region. In particular, it was shown that in winter-spring, ENSO events have more significant impact on European climate anomalies with ~1 to 5 months delay.

Study of a large number of events shows differences among El Niño events. These differences include the time of event beginning, duration, intensity, pattern of ocean and atmosphere interaction, localization, reasons of origin. Each of the indicated signs was used in different publications as basis for classification [50-58]. For instance, three El Niño types were classified by Voskresenskaya and Mikhailova [58], namely: spring, summer and summer/autumn. They differ from one another by season of its beginning, its intensity and duration. The first of them (spring) is close to the classic (canonical) El Niño event described by Rasmusson and Carpenter [59]. In principle, existence of "uncanonical" El Niño events was found earlier by Trenberth and Stepaniak [51], Larkin, Harrison [54], Ashok et al. [60] and others. These "uncanonical" El Niño events were called in different works as "Trans-Niño", "Dateline El Niño", "Modoki" (pseudo EN), "Central Pacific El Niño", etc. Different El Niño types manifest themselves in remote regions in various manners [18]. In spite of differences between the El Niño events we use all events for the composite analysis of cyclonic characteristics because there are not enough data for the reliable evaluation of the regional manifestations for each type of the El Niño events. In fact, presented results mostly refer to the most numerous and intense spring (or canonical) events.

Likely, the differences between El Niño events are due to plurality of governing mechanisms. Numerous authors invoke different explanations of ENSO origin, such as sporadic instability of the coupled ocean-atmosphere system in the Pacific Ocean, theory of nonlinear stochastic oscillator, monsoon-trade wind and cross-basins interaction, etc. ([9, 18, 20, 35, 61, 62], and very many others). Several mechanisms have been proposed through which the disturbances, induced by warmth/cold SST Tropical Pacific anomalies, spread over the globe. One of them is an atmospheric bridge connecting the North Pacific and Atlantic oceans through the Northern Pole.

2.3. Pacific Decadal Oscillation

Studies of the last 10-15 years have shown that the PDO can be briefly described as quasi-periodic variations of the ocean–atmosphere system in the North Pacific Ocean on inter-decadal time scale. Hare and Francis first coined the term "Pacific Decadal Oscillation" in 1995, when they studied the connections between Alaska salmon production cycles and Pacific climate [63]. PDO was described for the first time in climatological terms at the end of the 1990s [64-66] as quasi-periodical alternation of warm and cool surface waters in the Pacific Ocean, north of 20°N. During a "warm" (or "positive"), phase, the west Pacific becomes cool and part of the eastern ocean warms; during a "cool" (or "negative") phase, the opposite pattern occurs. Just two full PDO cycles in the past century were found and described by different authors, namely: the "cold" PDO regime of 1890-1924 and 1947-1976, and the "warm" PDO regimes which prevailed from 1925 to 1946 and from 1977 to the end of the 1990's ([13, 67] and others). Minobe [67] has shown that 20th century PDO fluctuations concentrated in two general periodicity bands, one from 15-to-25 years, and the other from 50-to-70 years. PDO phase change has significant importance for global climate while it impacts hurricane activity in the Pacific and Atlantic oceans, droughts and floods, marine ecosystems state, and global air temperature.

As a measure of its variability, it is customary to use the PDO index defined as the average anomaly of the SST in the North Pacific between 20 and 60°N. The time series of this index clearly illustrate the typical scales of PDO variability [65, 66]. They can be considered as quasi-20-yr periodical positive and negative SST anomalies within the general 50–60-yr periodicity. This is close to the Minobe's conclusion [67]. In fact, however, these periodicities are accompanied by higher-frequency oscillations (with periods of 8–10 yr) in the ocean–atmosphere system. The other option for the assessment of PDO index is to calculate the leading EOF of SST in the North Pacific. The associated time coefficient can be used as a measure of PDO. A distinctive feature of the corresponding atmospheric pattern is a basin-scale area of negative/positive SLP anomalies between 30-60ºN and positive/negative SLP anomalies centered at 50ºN over the Rockies. It is characterized by any of the several indices calculated from the SLP field over the North Pacific region. One of them is North Pacific (NP) index (mean SLP anomaly over the central North Pacific, [68]) and it is used in the present chapter. It should be emphasized that the North Pacific Oscillation described in the early papers [25, 69] is associated with the second Pacific SST EOF, not with the leading one (that is, PDO, [70]).

There are different explanations of the PDO origin. The prevailing hypotheses are that the PDO is caused by a "reddening" of ENSO-induced forcing combined with stochastic atmospheric forcing [71] and advective processes in the North Pacific (mostly, in the Pacific subtropical gyre). At the same time, PDO has modulating influence on ENSO [65]. The most recent publications (e.g., [70]) show that different mechanisms can be more or less important in different periods of 20th and 21st centuries. A few publications prove that PDO impacts the properties of the Pacific cyclones [16, 72].

2.4. Atlantic Multi-decadal Oscillation

The quasi-periodic thermal variations in the North Atlantic with typical time scales of 50 – 100 years are called the AMO. The natural variations of the ocean–atmosphere system on these time scales are described in numerous works (see, e.g., [13, 18, 73-78]). Likely, the first time this phenomenon was analyzed by Kushnir [73] using century-long records of instrumental observations of temperature and atmospheric pressure for the North Atlantic region. Almost simultaneously, Schlesinger and Ramankutty [74], by using one-dimensional spectral analysis of global data on SST and SLP averaged over several large regions, showed that a ~65 yr oscillation is mainly a North-Atlantic mode because it manifests itself just in that region. Polonsky and Voskresenskaya [77] described in details the low-frequency variability of meteorological fields, SST and turbulent heat fluxes in the North Atlantic with the typical time scale of ~65 yrs. Later, the following important fact was established [13] namely, the inter-decadal mode of the global SST is one of the main signals in the temperature field of the World Ocean on scales from interannual to multi-decadal, which are not connected with the ENSO and that it induces an atmospheric response of the pressure field. The amplitude of this signal is at a maximum in the North Atlantic. In other words, this is in fact, the North Atlantic mode. Its amplitude is at a maximum to the southeast of Greenland. Therefore, in some works (see, e.g., [18, 77]), it was called the inter-decadal mode of the NAO. The term AMO has become generally accepted for this mode only in recent years. It emphasizes not only the Atlantic origin of this oscillation but also its low-frequency character. Somewhat later, the AMO was detected in modeling the global climatic system by using the coupled models of the ocean – atmosphere system [75, 76, 78].

As a quantitative characteristic of the AMO, one uses an index reflecting the annual mean SST anomaly in the North Atlantic normally averaged between the equator and 60°N. As an additional characteristic of the AMO, one can use the NAO index smoothed by a low-frequency filter. The analysis of the long-term series of the NAO index reconstructed from paleodata dating back to 1675 shows that the 54–68-yr variations of the NAO index are significant at a 1% confidence level [79]. Somewhat later, Eden and Jung [80] showed that there is a significant inter-decadal variability of the heat fluxes on the ocean–atmosphere boundary connected with the low-frequency mode of the NAO. So, the AMO and low-frequency NAO are, in fact, connected modes [18]. Quite recent analysis of reconstructed temperature in the Alpine region confirms the existence of ~65 yrs robust oscillations [81].

The quasi-periodic oscillations of the meridional heat transport (MHT) in the North Atlantic are usually regarded as the main governing mechanism of the low-frequency variations of the SST and heat fluxes on the ocean–atmosphere boundary in the North Atlantic (i.e., of the AMO) [78]. However, there are different viewpoints concerning the main cause of the low-frequency MHT variations (and, hence, AMO). Thus, some researchers believe that the quasi-periodic AMO appears as a consequence of the phase shift between the thermal and haline modes in the North Atlantic [82]. Other researchers insist on a predominantly thermal origin of the AMO. In their opinion, the AMO is supported by a phase shift between the variations of SST and heat fluxes on the ocean – atmosphere boundary. The importance of the last mechanism is confirmed by the early authors' results [77]. Regional manifestations of AMO in cyclones' statistics are described and discussed below.

3. Method and Data

3.1. Basic Definitions

We expect to investigate statistical properties of regional cyclonic activity, that is, frequency of cyclone occurrence and certain characteristics of individual cyclone (intensity and size). To determine these characteristics, we give more or less formalized definition of what we will consider as *"cyclone"*, and what are its "intensity characteristics". We start with the most common notion of a cyclone as a baric depression that can be seen on the (instantaneous) map of air pressure or absolute topography as an area enclosed within some closed isobar (isohypse). This definition lies at the base of many studies concerning climatology of cyclones ([83, 84], to name only a few*)*. Although Sinclair [85] argues in favor of using relative vorticity rather than pressure to determine cyclone centers, we prefer the more traditional approach, since we consider a cyclone as a vortex, which is characterized by closed streamlines. Technical considerations prevent us from developing a method based directly on finding closed contours. Instead we accept that a depression is simply a pit on the isobaric surface, and its deepest point (local minimum) is a *baric center.* This leads to the following definition (where ***Z*** denotes absolute topography of specified isobaric surface 1000 hPa in this chapter):

A (single-centered) baric depression is a connected domain consisting of points ***x(f, l)*** where ***Z(x)<Zo*** for some ***Zo***, which contains one local minimum of a field ***Z*** (*baric center*) and which is maximum in the sense that with ***Z>Zo*** a domain {***x: Z(x)<Z***} contains more than one center. We will denote a center as *c*, and the value at the center ***Z(c)*** = Z_c. For a sufficiently smooth field on a sphere a contour ***Z*** = ***Zo*** will be a simple closed contour on the isobaric surface.

The following characteristics of a single depression are used in the subsequent analysis (illustrated by the Figure 1; ***So*** denotes the depression domain, ***So*** = {***x: Z(x)<Zo***}, hatched in the Figure):

(f_c, l_c) *(deg)* - coordinates of a center ***c***
Z_c *(gpm)* geopotential height at the center
$d = |Zc - Zo|$ *(gpm)* intensity
$a = \int_{x \in So} ds$ *(sq.km)* area of the depression domain ***So***.

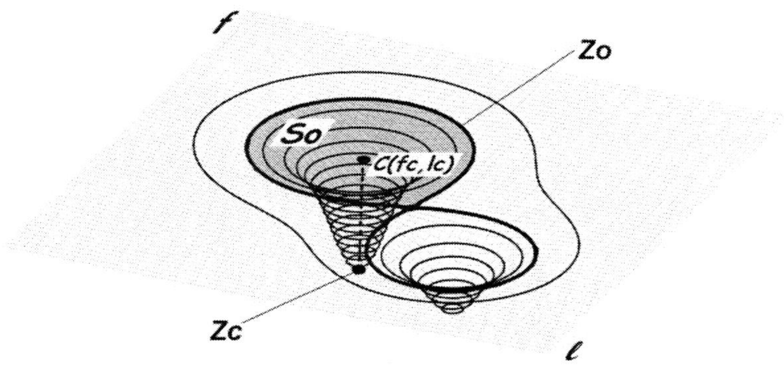

Figure 1. Scheme illustrating the basic definitions for a baric depression (for details, see text).

We intentionally avoid using the term "cyclone" because any analysis can produce relatively small noise that may manifest itself as spurious "centers". These spurious centers normally will occur in a single map and will not be found in later maps. They also will have small intensity. Therefore, we can filter them applying certain restrictions to lifetime and/or intensity. But we should take special precautions against filtering real cyclones, say, in the phase of cyclogenesis, when they have small intensity. So we speak of "cyclones" only when appropriate filtering is done. In those cases we provide filter thresholds. An identification algorithm first finds all local minima, and for each minimum determines a depression domain to which it belongs. The mentioned characteristics of baric depressions are then calculated and archived.

3.2. Data and Processing

For the calculation of a basic archive of individual cyclones we used gridded 1000 hPa (and optionally 500 hPa) geopotential height data from the NCEP/NCAR Reanalysis 1 from 1951-2010, 4 times daily, on a 2.5° regular grid. The NCEP/NCAR Reanalysis is an output from the NCEP data assimilation system, based on a global spectral model and a three-dimensional variational analysis scheme. Assimilated data include rawinsonde, land and marine observations, and some satellite observations from 1979 (mainly over the ocean). A reanalysis project utilized specially compiled historic database from 1948 to the mid-1990s, and operational data later. Details concerning the assimilation scheme and data used may be found in [86, 87]. This dataset was used in numerous climate studies, and some bugs both in the assimilation system and data were revealed and, where possible, corrected. Details may be found on the sites http://www.cpc.ncep.noaa.gov/products/wesley/ reanalysis.html#problem and (http://www.esrl.noaa.gov/psd/data/reanalysis/problems.shtml).

Some remarks concerning reliability of these data are necessary. Output reanalysis fields are divided into categories, and geopotential height is considered as “A”, i.e., most reliable. However, it is obvious that reliability of data over land and oceans is different. Further, we should keep in mind inevitable inhomogeneity of time series when considering long-term variations over several decades. The authors of the reanalysis mention two “milestones”: 1958, when the “modern rawinsonde network” was established, and 1979 (the beginning of satellite era). It also should be noted that in the period, started from June 1957, analysis was done at synoptic times 0000, 0600, 1200, and 1800 UTC, while in the earlier period it was done 3 hours earlier. Unfortunately, we cannot eliminate these inhomogeneities, since terabytes of observational data, preprocessing, assimilation and postprocessing are only available in NCEP. Moreover, all efforts in this direction cannot solve the problem of inhomogeneity in the cyclone time series in the regions with sparse data. Cyclones omitted due to sparse observations cannot be restored by any homogenization technique. Taking this fact in account, we use for quantitative assessment of cyclonic activity only the part of the time series from 1958. The restricted use of satellite data over the ocean at 100 hPa and higher from 1979 causes pronounced effect in stratospheric and upper tropospheric temperatures and only minor effect at the surface, where pressure observations dominate. Therefore, we decided to save the 21-years fraction of the series between 1958 and 1978. However, precautions are necessary when estimating trends of that period, which is not the matter of the present paper.

Our basic dataset of baric depressions (derived from the reanalysis AT1000 fields) contains the data from 1951 on all depressions in the Northern Hemisphere extratropics (20-80ºN) 4 times daily. These data include coordinates of the center of each depression, its intensity, area and geopotential height at the center (see paragraph ***Basic definitions***).

A derived archive of trajectories was calculated using a simple algorithm that takes into account the nearest centers on successive maps if the distance between them does not exceed a chosen threshold (600 km in this study; this figure is based on the analysis of distributions of cyclone speed, obtained by the tracking algorithm with different thresholds, which demonstrates that the number of episodes with cyclone speed exceeding 100 km per hour is close to zero). Initial and last points of each trajectory are regarded as cyclogenesis and cyclolysis, respectively. The list of cyclone characteristics is therefore completed by the date and coordinates of cyclogenesis and cyclolysis, cyclone lifetime, zonal and meridional components of its displacement (which characterizes the mean cyclone travel velocity vector between successive observation times). To avoid including spurious centers in the statistics, we then restrict the latter archive to the cyclones, observed at least on 6 successive maps, i.e., whose lifetime is greater than 30 hours.

For a qualitative description of long-term variations of cyclonic activity we also used a cyclone archive starting from 1901, derived from the so-called “20th Century reanalysis”, or R20C [88]. The R20C uses the same data assimilation scheme as Reanalysis 1. However, it assimilates only the data of surface stations and ship pressure observations from various sources, compiled in the most complete data set of pressure observations ever assembled, called the International Surface Pressure Databank. This reanalysis provides the most accurate gridded data of near-surface baric fields (SLP and AT1000) today. Of course, the development of the observation network in the early decades of the 20th century along with increasing number and wider geography of ship observations result in serious inhomogeneities in these datasets, which in mid-latitudes of the Northern Hemisphere are *a priori* most pronounced over the oceans and Asian region. Besides, the gradual increasing of observational activity broke twice in the periods of World Wars, when marine observations were decreasing to the level of the early 1900s [89-91]. Therefore we use R20C output only for the restricted regions.

3.3. Methodology Used in the Analysis of Interannual-to-Multi-Decadal Signals in the Seasonal Cyclone Statistics

We analyze some features of the cyclonic activity associated with certain "global-scale" modes of climate variability (NAO, ENSO, PDO and AMO) using the same technique for each mode. It is the technique of "composites" of characteristics of the cyclonic activity with regard to the phase of the climatic modes. More specifically, we first prepare two "catalogues" of winter (DJF) months (for each climate mode and for the period 1958-2010), when extreme ("positive" and "negative" or "high" and "low") phases of this mode were observed. Each catalogue includes ~15% of winter months (i.e., ~25 months) on each side of the ranged sample of some mode-specific index. The cyclone characteristics for each 5-degree grid cell were then averaged over all months separately for the years in the two catalogues. The averages obtained for two samples (that is, composites) were compared to identify patterns of the cyclonic activity associated with a signal produced by the mode.

The significance of the composite differences was estimated using different statistical tests for different characteristics. The significance of differences in the composites of intensity and size (area) of synoptic vortices in each grid cell was estimated using the Student's t-test [92]. Note that the number of degrees of freedom cannot be considered equal to the number of occurrences of a cyclone center in a given square, because one center may fall into a cell several times in a row (especially in regions of cyclolysis, where cyclones move slowly) and, thus, consecutive intensities (areas) cannot be considered as independent. Therefore, the number of degrees of freedom was defined as ($n^+ + n^- - 2$), where n^+ and n^- are the numbers of occurrences of *different* cyclones in the "positive" and "negative" phases, which centers fall into a cell. The significance was estimated for each 5-degree grid cell if it contained more than five cyclone centers for each sample. If the frequency of occurrence of eddies in some cell was less, the significance of the composite was not estimated.

The significance of the composite differences of cyclone frequency was estimated from the following considerations. To identify a region where frequencies are "significantly" different in the opposite phases, a probabilistic model is required. If the occurrence of a cyclone center in a grid square is regarded as a "success" (1) and the absence as a "failure" (0) in the series of independent trials, these experiments can be treated as Bernoulli trials. The number of successes (failures) has a binomial distribution with parameters f, $1 - f$, and n, where n is the sample size (the total number of observations is greater than $3000 = 25$ *sample months* $\cdot$ 30 *days* $\cdot$ 4 *obs.times* for each sample), $f = n_c / n$ is the probability of success (the frequency): n_c is the number of occurrences of a center in a grid cell, and $(1 - f)$ is the probability of failure. Since the sample size is large it is possible to approximate the probability by the normal distribution $N(f, [f(1 - f)/n]^{1/2})$ (with mean f and variance $f(1-f)/n$) and to test whether the frequency in the opposite phases falls into a proper confidence interval. Otherwise, the composites can be considered as significantly different in two phases of the mode. Further details of significance tests can be found in [5]. Below we consider only significant results (at least at 1% confidence level for the frequency of synoptic eddies and at 5% level for their intensity and area).

4. Results and Discussion

4.1 General Features of Cyclonic Activity

In this section we describe general features of cyclonic activity in the Northern extratropics. As mentioned above, to calculate general cyclones' properties we chose the period from 1958 taking into account the data inhomogeneity problem (see previous section).

4.1.1. Average Properties of Cyclones in the Northern Hemisphere

Most general features of geographic distribution of yearly cyclone frequency, shown in Figure 2, are well-known from numerous studies (e.g., [83, 84, 93, 94]). There are two principal elongated areas, where the cyclone centers occur most frequently. They are located over the North Atlantic and North Pacific and are known as the Atlantic and Pacific stormtracks. Origin of their axes is at the west coasts of the oceans near 35ºN. Then, the axes are directed northeastward. The Pacific stormtracks end at the northeastern coast of the Ocean (in the Gulf of Alaska with a spot of very high frequency). The same spot in the Atlantic is found between Greenland and Iceland, but the stormtracks, although less intense, extend further to the Scandinavian Peninsula. There are also a few additional areas of enhanced cyclone frequency. They are located in the Mediterranean region (where the local cyclogenesis occurs) and in the Eastern Europe from the Baltic Sea to the Urals. That is why in this chapter we focus on the variability of cyclonic activity in the North Atlantic, European, Mediterranean and North Pacific regions. There are restricted secondary areas of enhanced cyclonic activity in Siberia, over Hudson and Baffin Bays, and the North American cyclogenesis region east of the Rockies.

Specific seasonal features of the distribution of cyclone centers are less known. Cyclones over the oceans in winter are apparently more frequent, than in summer (on average, by ~20%) due to more strong temperature contrasts between the equator and pole, the ocean and land, and to the southward extension of the sea-ice shield. In summer, the Atlantic stormtrack becomes more zonal. The same is not true for the Pacific stormtrack (Figure 2). The cyclone frequency over the Mediterranean region in summer strongly decreases. Expected summer weakening of the cyclonic activity over the North Atlantic, North Pacific and Mediterranean region is accompanied by the growing frequency of cyclones over the land, especially pronounced in the Eastern Asia and the East of the Rockies.

Two large areas of the most frequent cyclogenesis are located in the western part of Atlantic and Pacific Oceans. Other cyclogenetic regions are located between Greenland and Iceland, around the Apennines Peninsula and east of the Rockies. Areas of the most frequent cyclolysis over the oceans are located at the north margins of the stormtracks. In the North Pacific, this maximum looks like a narrow meandering band between 50 and 60ºN (Figure 3).

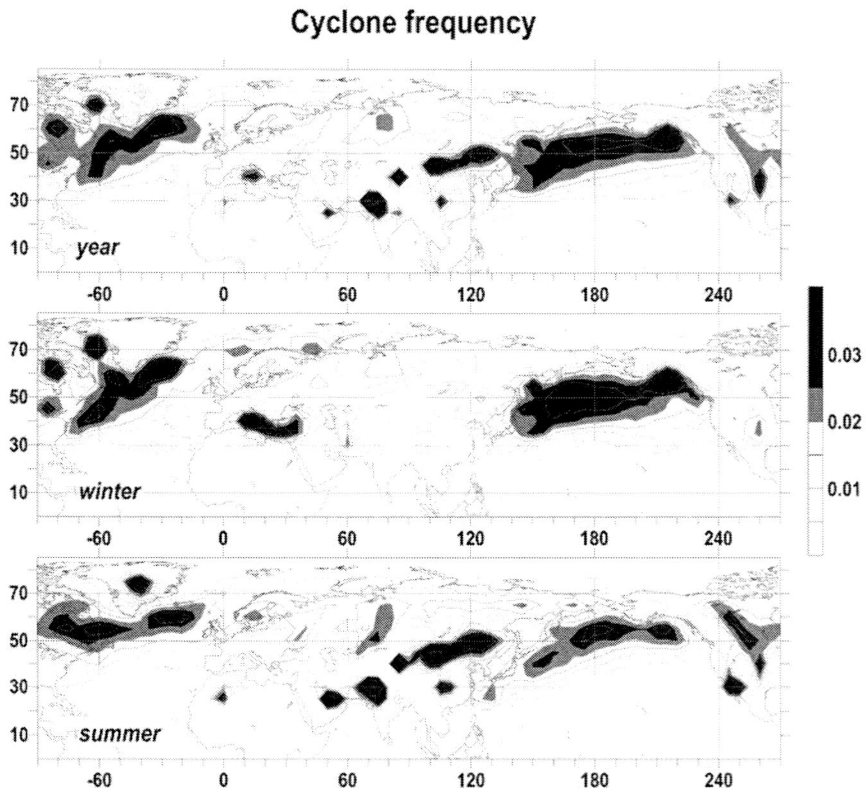

Figure 2. Relative frequency (fractions, 0.01=1%) of cyclones' centers in 5-degree grid cells for the entire year (top), winter (DJF, middle) and summer (JJA, bottom). Reference period: 1958-2010.

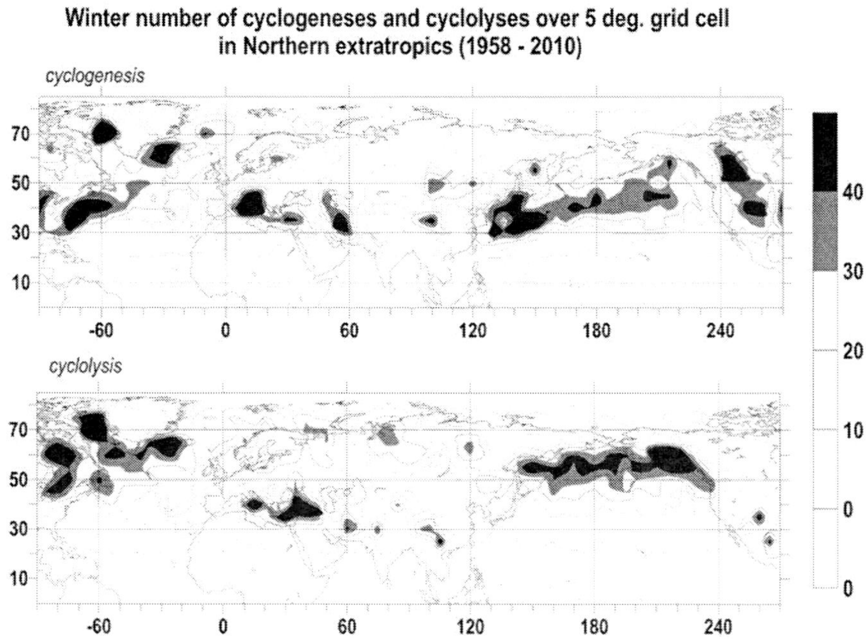

Figure 3. Winter cyclogenesis and cyclolysis: total number of episodes over 5-degree grid cell for 1958-2010.

Seasonal differences in cyclone intensity and area are even more pronounced. The greatest differences are found in the vicinity of the Pacific stormtrack, where winter cyclone intensification exceeds 50 gpm. Note that intensities of cyclones in the vicinity of the Pacific and Atlantic stormtracks are generally similar. However, the Pacific cyclones in winter reach much larger horizontal sizes (Figure 4).

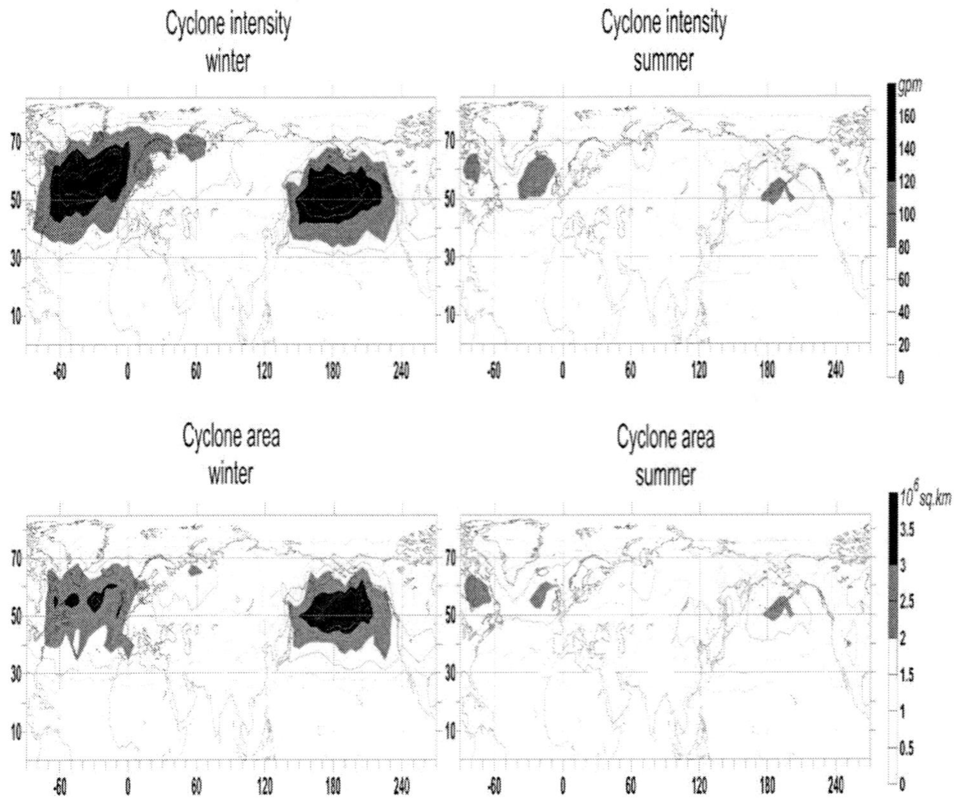

Figure 4. Winter (left) and summer (right) characteristics of cyclones' parameters: intensity (*gpm*; upper panel) and area (*10^6sq.km*; lower panel). Means are for 1958-2010.

An analysis of cyclone trajectories and speed reveal the following features (Figure 5). Cyclones mostly move from the cyclogenesis regions north-eastward. Generally, the meridional cyclone velocity component is smaller than the zonal one. However, in the North Atlantic and North Pacific there is large northward component of the cyclone displacement from the areas of cyclogenesis. Typical speed of the cyclones varies from ~400 km per day (100 km per 6 hours) to >1200 km per day (>300 km per 6 hours). The maximum cyclone speed is observed in the western parts of the North Atlantic and Pacific Oceans and over the cyclogenesis North American region in winter. In summer, the speed of the cyclone movement does not exceed 1000 km per day (250 km per 6 hours).

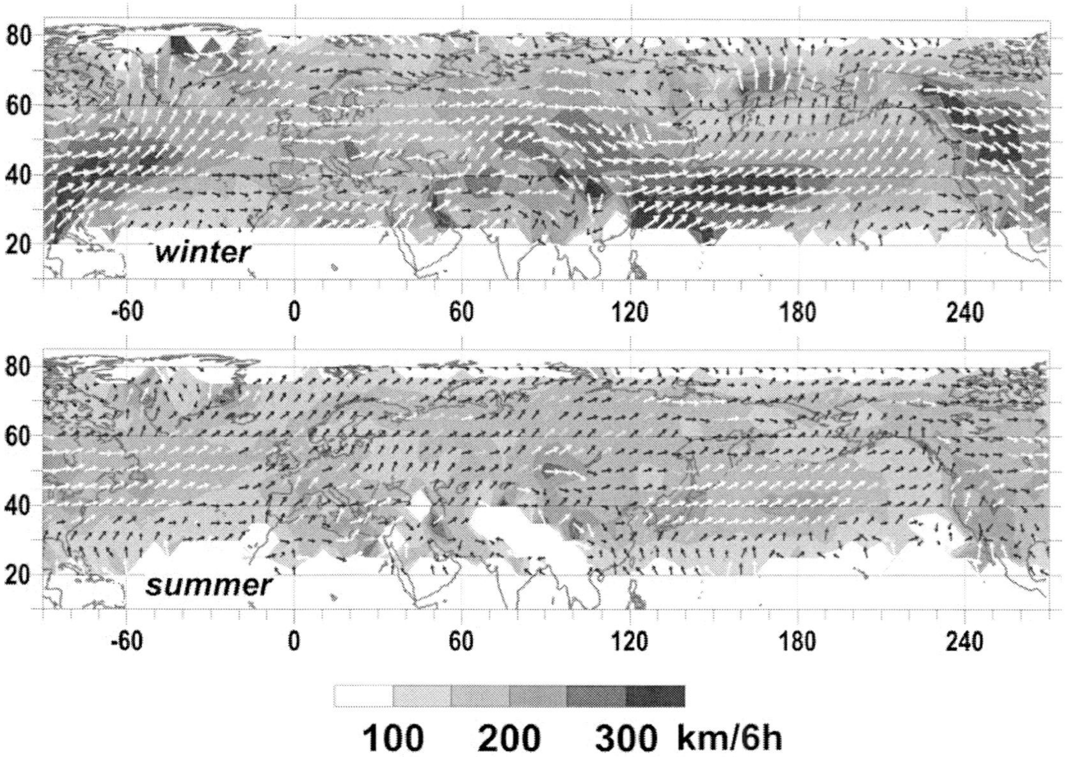

Figure 5. Mean cyclone velocity for the period 1958-2010 (center displacement over a 6-hour time interval). Contours demonstrate the velocity magnitude, and arrows are mean vectors (with lengths proportional to the magnitude: maximum 350 km/6 hours, and color range). Winter (top) and summer (bottom).

4.1.2. Statistical Distributions of Cyclones Properties in the Vicinityof Stormtracks

Comparative analysis of distributions of selected regional cyclone characteristics (lifetime and maximum intensity that a cyclone reaches during its lifetime) in the North Atlantic and North Pacific sectors reveals some interesting features (Figure 6). Note that in both regions total number of cyclones in winter is approximately 10% greater than in summer.

As expected, more shallow cyclones are observed in summer. Less than 1% of all summer cyclones exceed intensity of 300 gpm, while in winter this figure is about 20%. The winter cyclone intensity distributions at Figure 6 are bimodal, especially for the Atlantic. At the same time, the greater number of winter cyclones has shorter lifetimes, while more long-living cyclones are observed in summer. In the Atlantic sector, there is an apparent boundary of lifetime at 6.5 days. The number of longer living summer cyclones is much more numerous than number of winter ones. The summer distribution of cyclone lifetime is apparently bimodal. There are two maxima of lifetime. First of them is at 1.5 day, while the second one is at 6.5 days. The similar feature is true for the Pacific sector; however, the second mode is at 8 days here. Hence, we can conclude that cyclones observed over the North Atlantic and North Pacific are mix of eddies of different types.

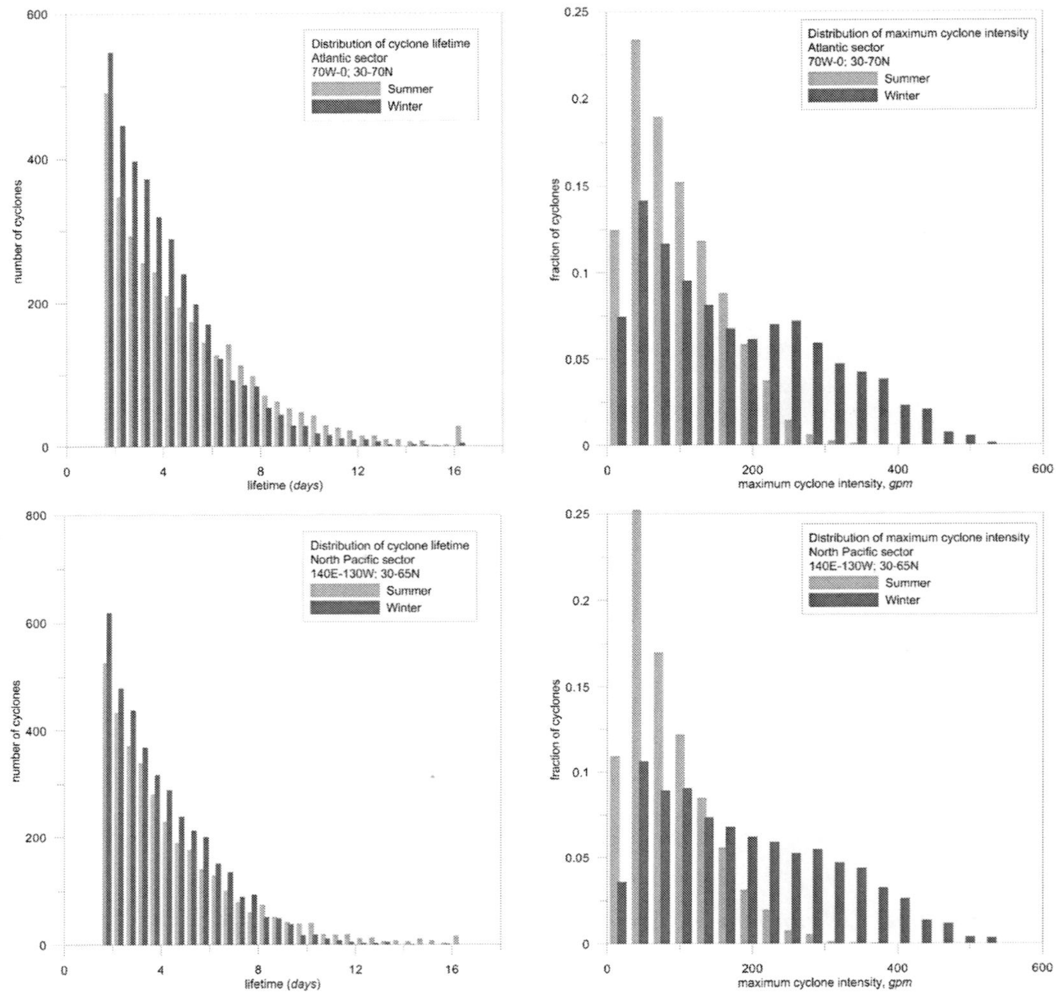

Figure 6. Left: Distribution of cyclone lifetime (number of cyclones in each half-day lifetime category). Right: Distribution of maximum cyclone intensity during lifetime. Distributions are for cyclones occurring in the North Atlantic sector (70°W-0, 30-70°N, upper panel) and North Pacific sector (140°E-130°W, 30-65°N, low panel), for summer (JJA, light) and winter (DJF, dark). Reference period: 1958-2010.

4.2. Cyclonic Activity at Extreme Phases of Principal Circulation Modes

In this section we present results of the phenomenological studies of the cyclonic activity associated with extremes of NAO, ENSO, PDO and AMO. In all cases we use the technique of composites of characteristics of the cyclonic activity in winter (see section **Method and data**). Winter was chosen because the cyclones in that time are more frequent and intense. So, the analyzed signals are much more significant (see, above paragraph). Because the NAO is the leading mode of interannual atmospheric variability in the Atlantic sector (see, paragraph ***North Atlantic Oscillation***), we are going at first to analyze just this signal.

4.2.1. NAO

As a NAO measure, we chose a leading principal component of SLP in the North Atlantic sector (20-80°N, 90°W-40°E), calculated in NCAR and available at the site (http://www.cgd.ucar.edu/cas/jhurrell/indices. data.html#naopcmon). Table 1 contains catalogue of years and winter months for the positive and negative NAO phases. It differs from that used by the authors in one of their previous works [5] (hereafter BP), because BP used the data from much narrower Atlantic sector (20-60°W) and for the other time interval (1952 – 2000's). Coincident (with BP) dates are marked by asterisk in the Table 1.

The composites of cyclone frequency (Figure 7) clearly show that a pattern of cyclonic activity crucially depends on the NAO phase over the ocean and some other regions, specifically Western Europe, Polar Russia, Baltic and Mediterranean regions. Over the North Atlantic, this pattern looks like dipole structure. In the positive NAO phase, numerous cyclones occur over the northern North Atlantic Ocean, while their frequency is lower in the subtropical and mid-latitudes. The negative NAO phase is characterized by opposite anomalies of cyclone frequency over the North Atlantic. This pattern resembles a leading mode of variability of extratropical synoptic disturbances in troposphere [32]. However, tropospheric synoptic disturbances over the North Atlantic are characterized by a single clearly distinguished maximum in the vicinity of stormtrack origin (near Newfoundland), while the positive NAO phase coincides with the most frequent cyclones between Greenland and Iceland. This pattern (Figure 7) corresponds to the first EOF of the winter cyclones frequency, described by Bardin [23]. He explained the existence of cyclone frequency maximum near Iceland during the positive NAO phase by higher frequency of cyclolysis there. However, Figure 8 demonstrates that this is only partly true. Obviously, not only cyclolysis constitutes the maximum, but cyclogenesis also has a maximum of similar magnitude in the same area in the positive NAO phase.

The NAO-related dipole structure of cyclone frequency is supported by the correlation analysis of interannual variations of NAO index and regional cyclonic activity. Thus, a seasonal regression of winter (DJF) frequency of cyclones over the north-eastern Atlantic region on NAO index explains almost 50% of its total variance (coefficient of determination R^2=0.49). For the Mediterranean region, this figure is not so large (R^2=0.34), but also highly significant. It should be emphasized, that regression coefficient in the former case is positive, while the later one is negative. This confirms the well-known relation between NAO phases and cyclonic activity in analyzed two regions (see, also paragraph ***Manifestation of climate modes in the key regions*** below).

In the positive/negative NAO phase, the northeastern branch of the North Atlantic stormtracks is more/less pronounced than its southwestern part. Both cyclogenesis and cyclolysis in the negative NAO phase are much more widespread over extratropics in the Atlantic – European sector. In the positive phase, they are concentrated within the narrower band in the northern part of the region and in the western Mediterranean – the Near East (Figures 7 and 8).

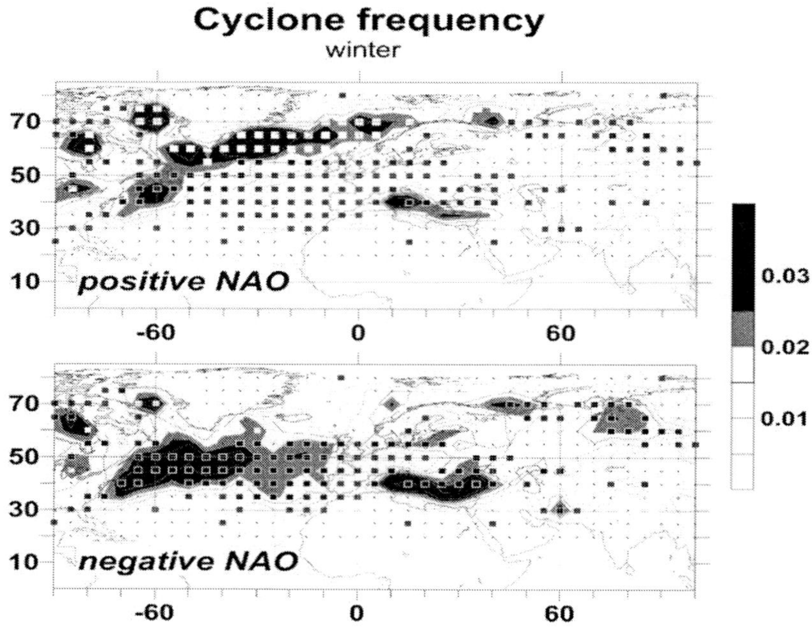

Figure 7. Relative frequency (fractions, 0.01=1%) of cyclones' centers in 5-degree grid cells in positive (top) and negative (bottom) NAO phases (composites); the contour interval is 0.01 (1%). The cells where the composite difference is significant at the 1% level are shown by different squares: white means that the frequency is higher in the positive NAO phase, and black means that the frequency is higher in the negative phase.

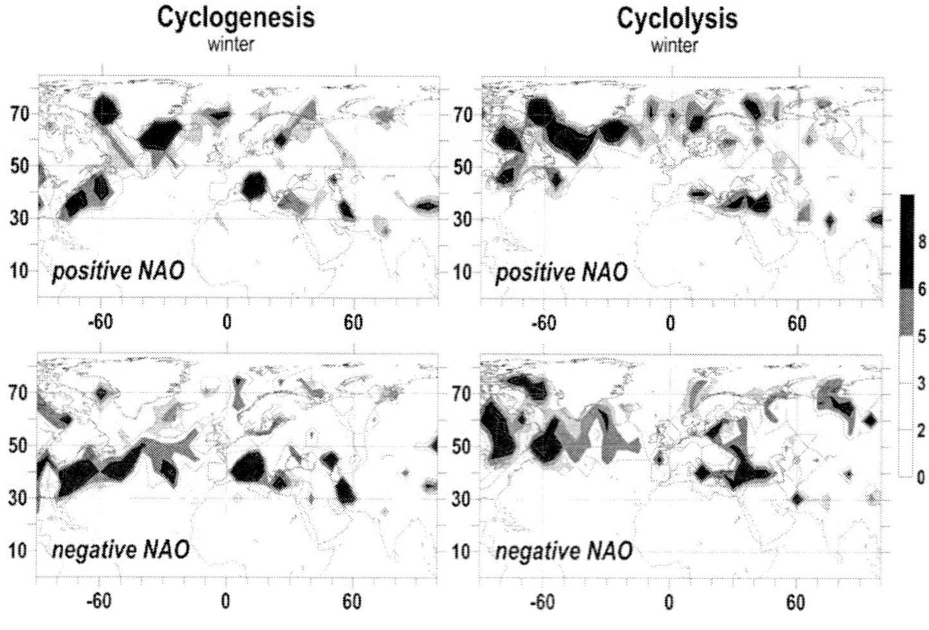

Figure 8. Number of cyclogeneses (left) and cyclolyses (right) over 5-degree grid cells in 25 winter months with extreme positive (top) and negative (bottom) NAO index. Catalogue of months is given in Table 1.

In the positive NAO phase, the intensity and area of the cyclones increase in the northeastern North Atlantic, while over the majority of other regions they decrease (Figure 9). It should be noted that there is the quite extended region in the Far East and China, where the significant NAO-induced signal in intensity and area of the cyclones occurs. The positive NAO phase is also characterized by faster displacement of the North Atlantic cyclones to the northeast than the negative one (Figure 10). This result confirms the quite well-known features of NAO discussed above.

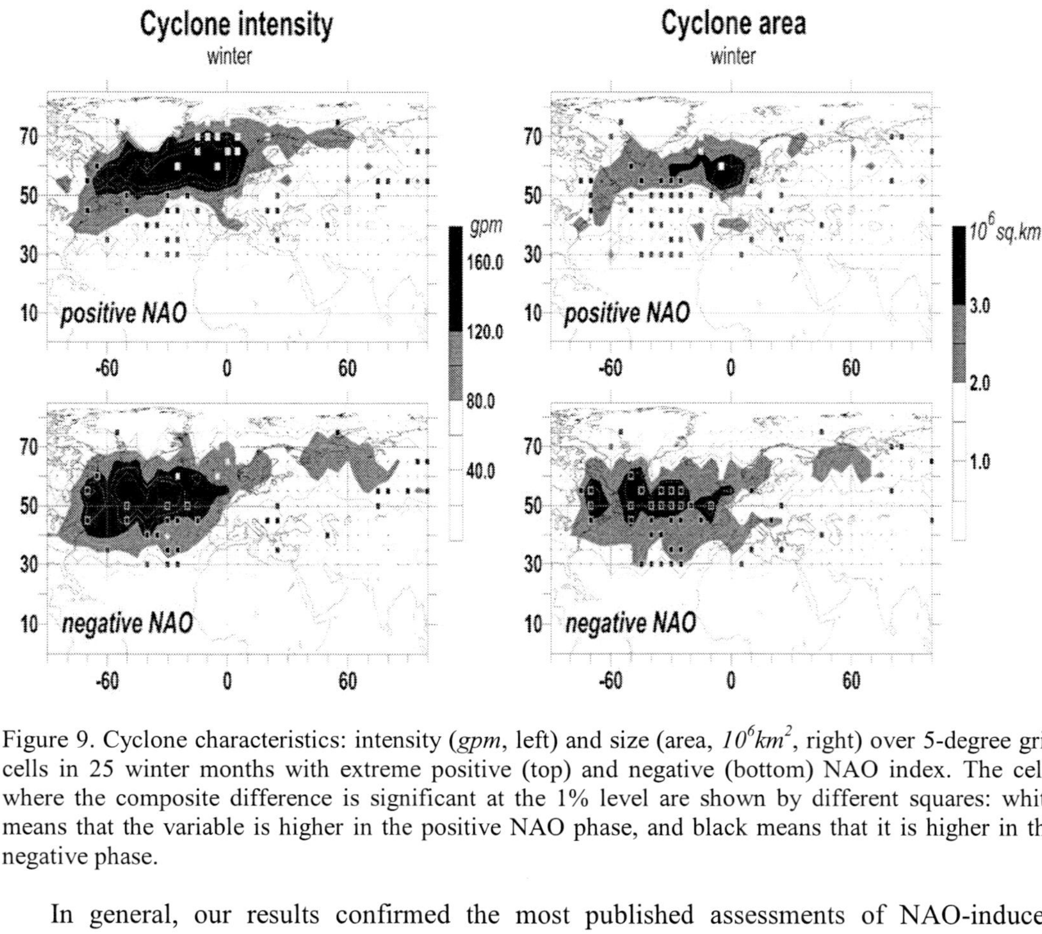

Figure 9. Cyclone characteristics: intensity (*gpm*, left) and size (area, $10^6 km^2$, right) over 5-degree grid cells in 25 winter months with extreme positive (top) and negative (bottom) NAO index. The cells where the composite difference is significant at the 1% level are shown by different squares: white means that the variable is higher in the positive NAO phase, and black means that it is higher in the negative phase.

In general, our results confirmed the most published assessments of NAO-induced features of the extratropical and Mediterranean cyclones in winter, when the NAO magnitude is at a maximum. At the same time, they give the generalized view on the global response of the eddy atmospheric field on the NAO.

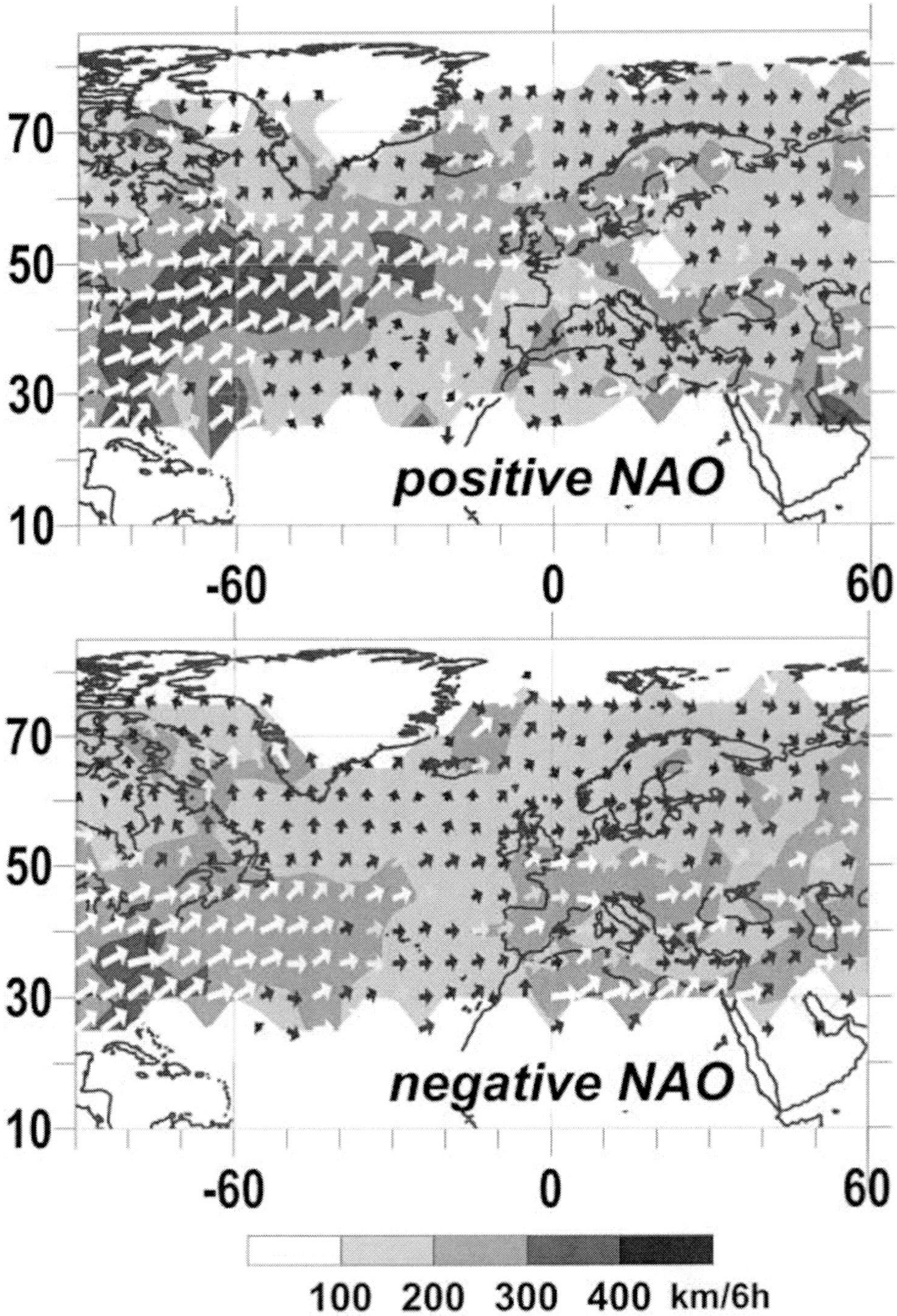

Figure 10. Winter cyclone velocity (as in Figure 5) in the positive (upper panel) and negative (lower panel) NAO phases.

Table 1. Extreme winter NAO episodes

Positive NAO		Negative NAO	
Year	**Month**	**Year**	**Month**
*1959	2	*1958	2
1959	12	*1959	1
1960	12	1960	1
*1962	1	*1960	2
1973	2	*1961	12
*1974	1	*1963	1
*1976	2	*1965	2
*1982	12	*1966	1
1983	1	1966	2
*1984	1	*1969	1
*1986	1	*1969	2
*1989	1	*1970	1
*1989	2	*1977	1
*1990	1	*1978	2
*1990	2	1978	12
*1992	2	*1979	1
*1993	1	*1981	12
1993	12	*1985	1
1994	1	*1986	2
1995	2	1987	1
1997	2	1995	12
1999	2	*1996	12
2000	2	2009	12
2004	12	2010	2
2006	12	2010	12

4.2.2. ENSO

The catalogues of winter El Niño/La Niña months were based on the ranged DJF monthly data of average SST anomaly in the Niño 3.4 region (5°N-5°S, 170-120°W) [95, site http://www.cpc.ncep.noaa.gov/ data/ indices/]. The catalogues thus compiled (Table 2) include not only classic El Niño and La Niña events (see paragraph ***El Niño - Southern Oscillation***), but also a few months from events of the other kinds.

The composites of the cyclone frequency show that the ENSO influence is traced over the extratropical Pacific, North America, and in the Atlantic - European sector (Figure 11). The pattern over the eastern North Pacific – western North America is reminiscent of leading NP mode (see below, paragraph ***PDO***). The El Niño episodes roughly correspond to the positive (strong) NP phase. However, the El Niño-induced response of the central North Pacific, which is the main feature of the strong NP phase, is much less pronounced. Atlantic manifestations of ENSO signal are more complicated. Although ENSO affects key Atlantic regions of maximum cyclonic activity and its signature is reminiscent of NAO-induced pattern, the ENSO signal is much weaker here than in the Pacific region.

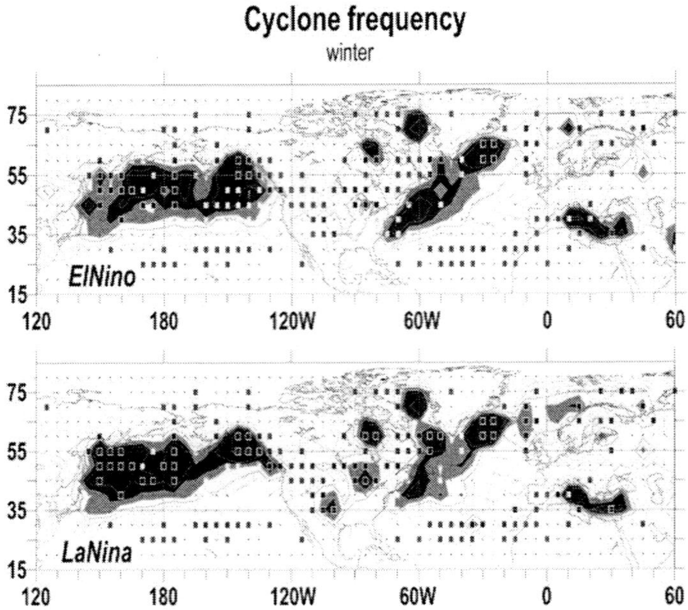

Figure 11. Relative frequency (fraction; 0.01=1%) of cyclone centers in 5-degree grid cells in positive (top) and negative (bottom) ENSO phases (composites); the legend see in Figure 7.

Table 2. El Niño / La Niña episodes

El Niño		La Niña	
Year	**Month**	**Year**	**Month**
1957	12	1964	12
1958	1	1965	1
1958	2	1965	2
1965	12	1970	12
1966	1	1971	1
1966	2	1971	2
1972	12	1973	12
1973	1	1974	1
1973	2	1974	2
1982	12	1975	12
1983	1	1976	1
1983	2	1976	2
1987	12	1988	12
1991	12	1989	1
1992	1	1989	2
1992	2	1998	12
1994	12	1999	1
1995	1	1999	2
1997	12	1999	12
1998	1	2000	1
1998	2	2000	2
2002	12	2007	12
2009	12	2008	1
2010	1	2008	2
2010	2	2010	12

In general the discussed pattern may be described as intermittent bands of enhanced cyclone frequency either during El Niño, or La Niña, almost around the entire hemisphere. Some of above results confirm the published finding. For example, this is rather well-known that during ENSO event, the cyclone frequency over the southeastern states of the USA and Mediterranean region changes. The other manifestations (e.g., in the eastern North Atlantic and Caspian regions) are not so wide-spread among the ENSO community. At least partly, this is due to differences in regional manifestations of El Niño events of various types (see paragraph ***El Niño - Southern Oscillation***).

4.2.3. PDO

To analyze this leading inter-decadal climate mode in the Northern Hemisphere we chose the NP index (see paragraph ***Pacific Decadal Oscillation***) and prepare a catalogue of winter (DJF) months when extreme phases of the Pacific mode were observed. Note that the NP index and SST-based PDO index are highly correlated at the decadal scale. However, the first index is more closely related to the atmospheric variables. That is why we used this index (Table 3) as more suitable characteristics of the cyclonic activity. It should be emphasized, that a term "strong NP" is use for large negative SLP anomalies in the central North Pacific. This NP phase associates with strong regional circulation (both atmospheric and ocean) and roughly corresponds to the “warm” PDO phase. The opposite phase, consequently, is referred as “weak NP”. It roughly corresponds to the “cold” PDO phase. The composites are presented in Figures 12, 13 and 15.

The most pronounced feature of the cyclonic activity in the strong NP phase is enhanced frequency of cyclones in the central part of the North Pacific Ocean and especially in the vicinity of Aleutian Islands. Over the other North Pacific regions, the cyclone frequency is significantly larger in the weak NP phase. Maxima of cyclone frequency locate in the western part of the Ocean as a dipole structure with centers east and west of the Kamchatka peninsula (in Okhotsk and the western part of the Bering Sea), and along the eastern coast of the Gulf of Alaska. Large continental areas of higher cyclone frequency in the weak NP phase are found over Asia and North America (north of 40ºN and 30ºN, respectively, see Figure 12). Figure 13 clearly shows that the North Pacific region in general is characterized by more numerous cyclogeneses and cyclolyses during strong NP phase. Decadal-scale pattern of cyclolysis variations is reminiscent of the pattern for the cyclone frequency. The pattern for cyclogenesis a little bit differs.

A strong connection between the cyclone frequency in the North Pacific region and NP-index is supported by the analysis of time series of winter cyclone characteristics in the Aleutian Low vicinity (170ºE-150ºW; 42-57ºN) and seasonal (DJF) NP index (Figure 14). NP mode explains almost 50% of the total variance of the cyclone frequency here (coefficient of determination in regression of the cyclone frequency on NP mode R^2=0.49). So, inter-decadal variations of the cyclone frequency in the North Pacific and the NP index are similar. Note the entire analyzed period from 1951 to 2010 is characterized by positive trend of the cyclone frequency.

Generally, the speed of movement of the NP cyclones is higher in the strong NP phase (Figure 15). This resembles the NAO case (see above) and may be a consequence of similarity of one of the generating mechanisms of NAO and PDO (namely, gyre one). The time scale difference between these two oscillations is due to the various sizes of the basins.

The difference between principal direction of the Atlantic and Pacific stormtracks may be explained by physical-geographical features of these basins (configuration of the coastlines, properties of the surrounding continents, etc).

Let us compare the NP and PDO-related patterns of the cyclone frequency. We defined the phases of PDO as near-decadal periods when positive or negative values of SST-based PDO index (leading SST principal component) prevailed - respectively 1977-1988 and 1962-1976. Composites of the cyclone frequency for these phases are presented in Figure 16. The general pattern is reminiscent of combination of the eddy field responses to the ENSO and NP signals, especially between 180 and 160°W. However, the PDO-related response demonstrates less pronounced maximum around Kamchatka in the negative PDO phase, than in case of La Niña, and less pronounced maximum in the central North Pacific in the positive PDO phase. These features might be treated in favor of the hypothetical PDO generation by the ENSO signal modulated by the NP Oscillation. Thus, the NP and PDO signals are not equal, in spite of their quite high correlation.

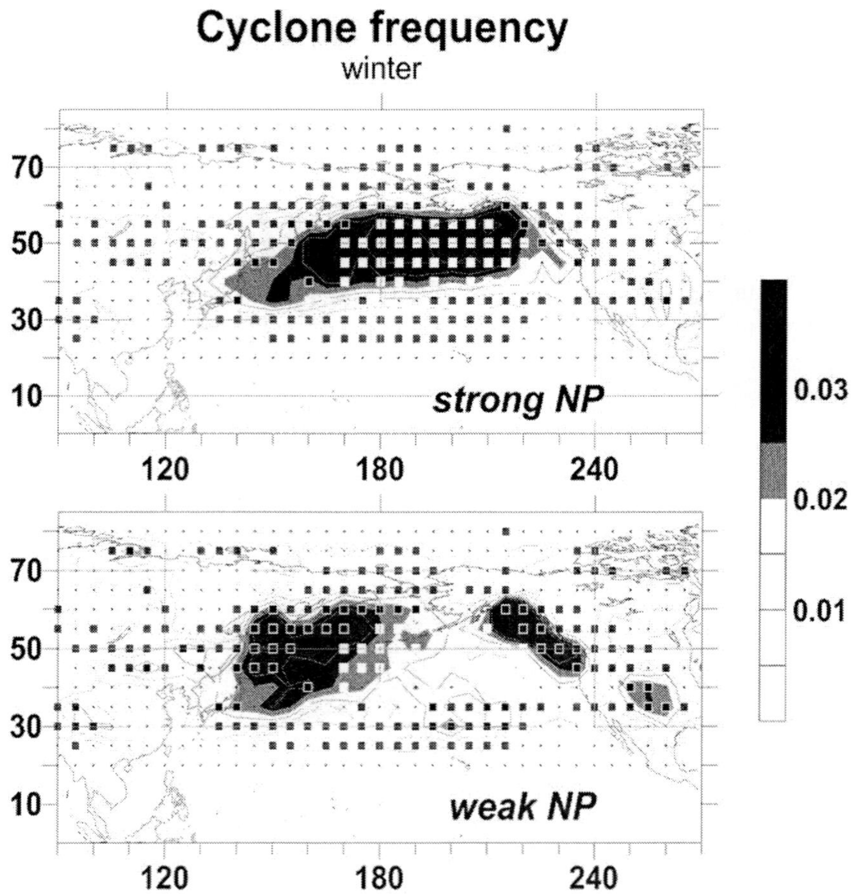

Figure 12. Same as in Figure 7, but for the strong (upper panel) and weak (lower panel) phases of the NP mode.

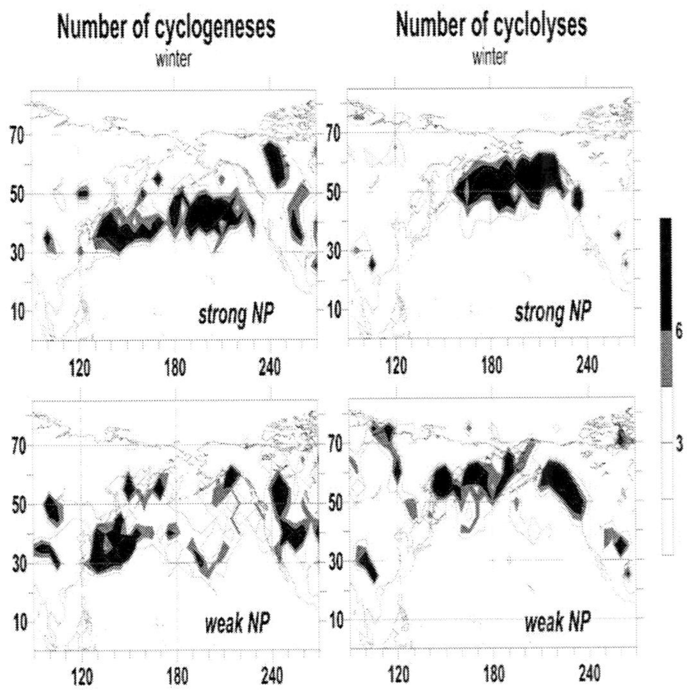

Figure 13. Number of cyclogeneses (left panel) and cyclolyses (right panel) in extreme NP phases.

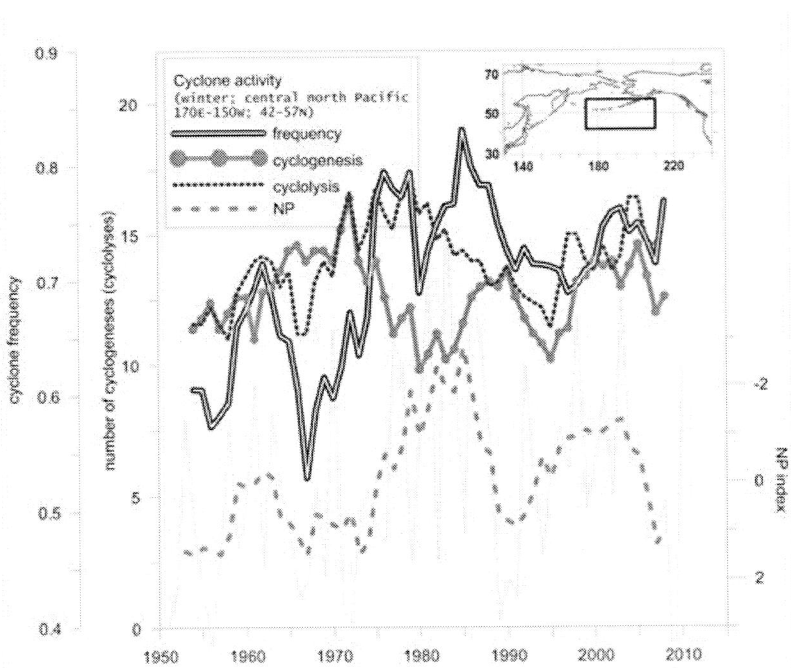

Figure 14. Time series of winter (DJF) characteristics of the cyclonic activity over the "key" region in the central North Pacific (170°E-150°W; 42-57°N): relative frequency and seasonal numbers of cyclogeneses and cyclolyses. 5-year running means for the entire analyzed region are shown. Data for the second part of the 20th century derived from NCEP Reanalysis 1. The seasonal NP index (seasonal values and 5-year running means) is also shown.

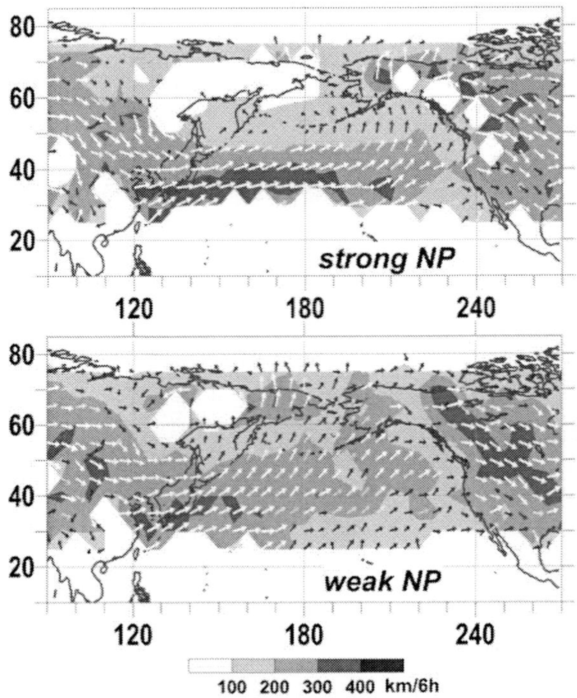

Figure 15. Winter cyclone velocity (as in Figure 5) in the strong (top) and weak (bottom) NP phases.

Table 3. NP index extremes

Strong NP		Weak NP	
1961	1	1959	2
1963	2	1962	1
1968	2	1965	12
1969	12	1966	2
1977	1	1968	12
1977	2	1969	1
1978	1	1971	1
1978	2	1971	12
1980	2	1972	1
1981	1	1975	1
1983	2	1976	2
1985	1	1978	12
1985	12	1979	2
1986	2	1982	2
1986	12	1985	2
1992	1	1989	1
1993	2	1989	2
1995	12	1990	2
1998	1	1992	12
1998	2	1993	1
2000	2	2000	1
2003	1	2006	2
2004	2	2008	12
2005	12	2009	2
2010	1	2010	12

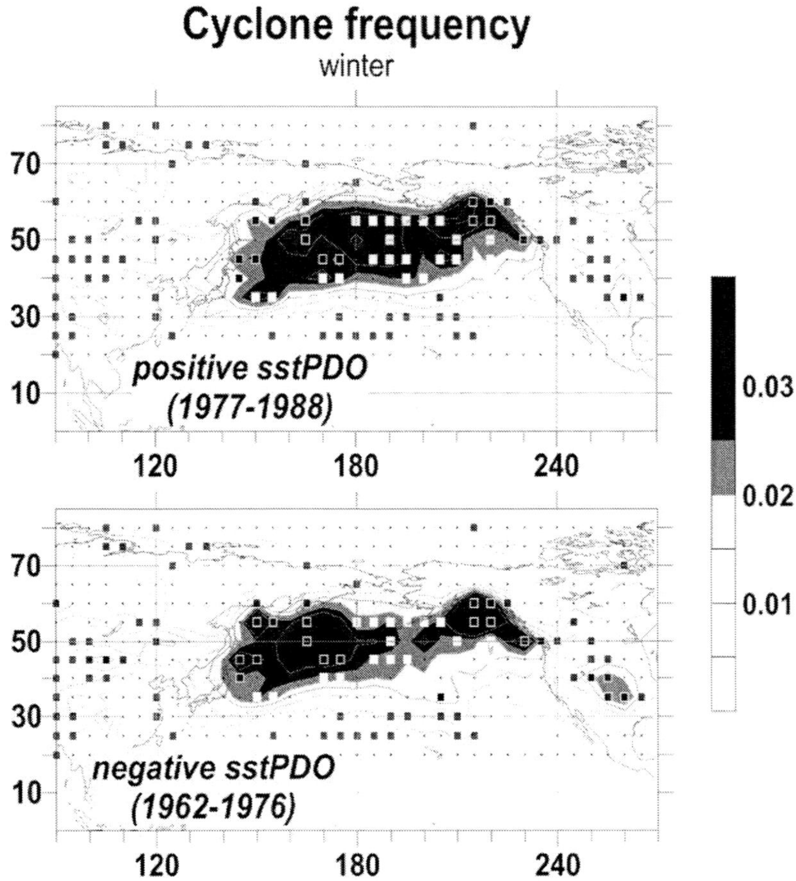

Figure 16. Same as in Figure 7, but for the positive and negative phases of PDO.

4.2.4. AMO

The AMO primarily manifests itself in SST variations, but also in low-frequency NAO variability (on timescales of several decades, see paragraph ***Atlantic Multi-decadal Oscillation***). That is why, we chose two 8 yr periods of low-frequency NAO extremes to analyze the behavior of cyclones namely: 1963-70 (negative phase) and 1988-95 (positive phase). The composites of the cyclone frequency in these phases (Figure 17) resemble those for the extreme phases of NAO (cf. Figures 17 and 7). However, more close inspection reveals a few important differences. First of all, AMO is not accompanied by strong difference in the shape of the Atlantic stormtracks. In both phases, the maximum between Iceland and Greenland (although stronger in the positive phase) is pronounced, and there is only a hint of the latitudinal elongation of the stormtracks at 45-50°N in the weak phase. This confirms the early authors' results [7, 96]. Secondly, the AMO pattern is more pronounced in the eastern Mediterranean region. There are more numerous cyclones here in the negative AMO phase. Besides, the maximum of cyclone frequency shifts to the eastern Mediterranean basin in the positive AMO phase and into Asia Minor in the negative phase.

Lifetime of the North Atlantic cyclones for the positive AMO phase exceeds their lifetime for the negative AMO phase (Figure 18). This is the apparent result of additional anticlokwise vortex generating over the warm ocean during the positive AMO phase.

This should be emphasized, the AMO-induced pattern of variability of the cyclonic activity was extracted using just one period of oscillation. Let us therefore consider more extended time series of cyclone characteristics in "key" regions, where major AMO-related differences of cyclonic activity are observed and more or less reliable data since 1900 are available. Such data for the first half of the century are derived from the R20C (see, section **Method and data**). It is possible to consider these data with certain precautions for the qualitative characteristics of long-term variations, since there were always quite large amount of marine observations in the Atlantic during the 20th century (except the periods of the World Wars), and meteorological station coverage over Europe was adequate for the identification of larger cyclones living at least 1.5 days. We consider only the total frequency of cyclone centers in the region, since other characteristics may be too sensitive to the current resolution of the observational network.

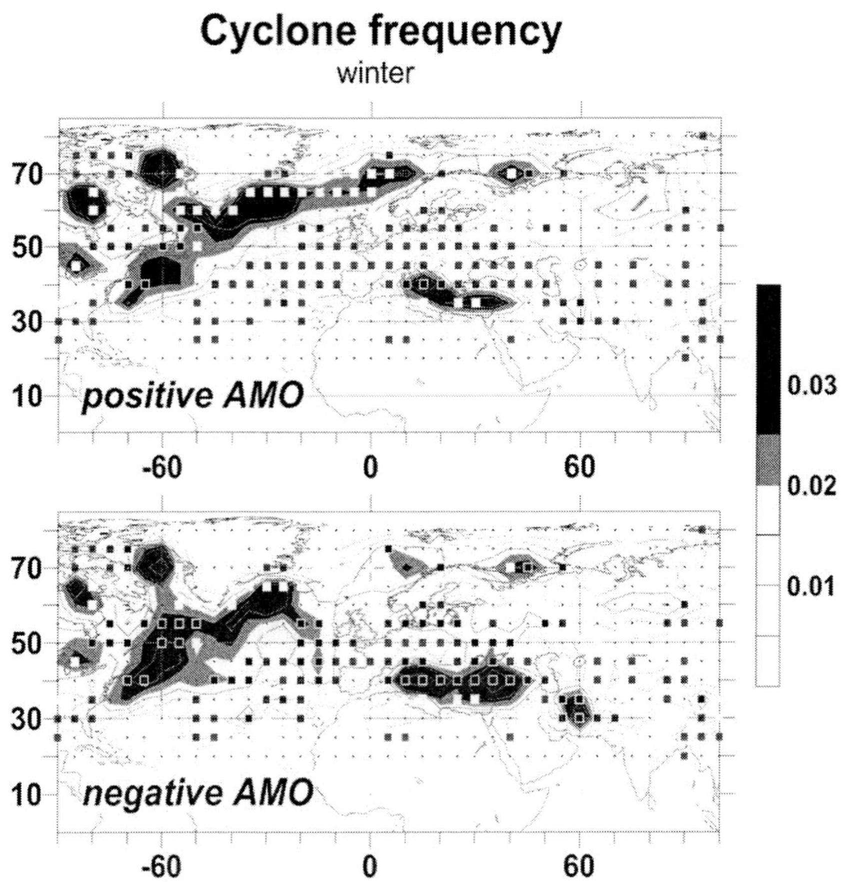

Figure 17. The same as in Figure 7, but for the positive and negative phases of AMO.

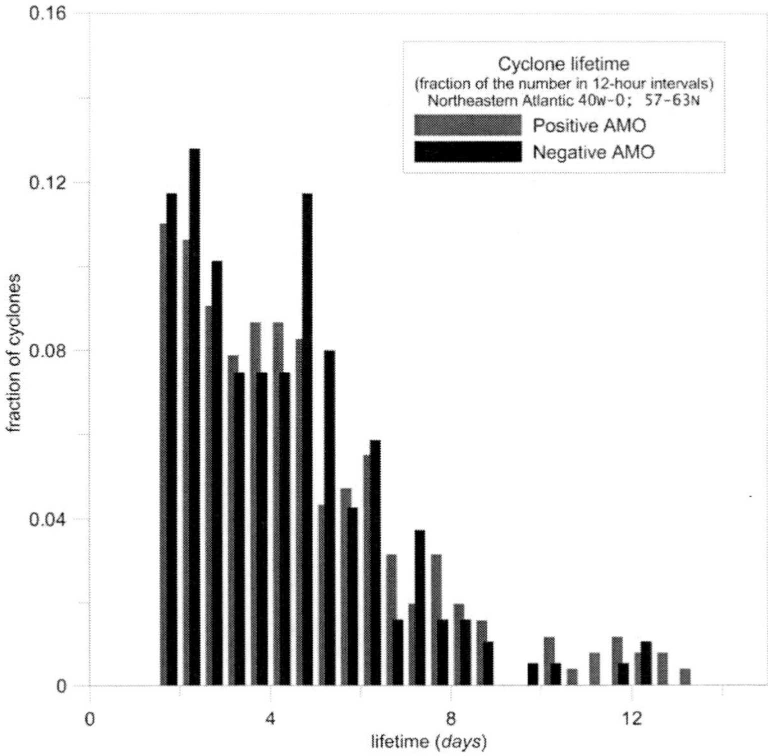

Figure 18. Distributions of the relative lifetime of cyclones with lifetimes of 1.5 days and more in the north-eastern Atlantic region (40W-0; 57-63N) in positive and negative phases of AMO. Relative lifetime is a fraction of all cyclones in the corresponding phase that fall into the half-day category in winter months (DJF) for the negative (1963-1970) and positive (1988-1995) AMO phases. The total number of cyclones in the region for positive and negative AMO phases is 245 and 188, respectively.

The time series of winter cyclone characteristics over the north-eastern part of the Ocean (such as the maxima of the cyclone frequency, cyclogenesis and cyclolysis between Greenland – Iceland, Figure 19) demonstrate pronounced variations on quasi – decadal time scale, especially in the second half of 20th century. These variations are in phase with the NAO index. It seems, this proves the acceptable quality of the data. However, on the multi-decadal timescale the cyclone frequency increasing from the late 1960s (along with the NAO) is not accompanied by rise of the number of cyclogeneses and cyclolyses. Thus, it seems we cannot extract reliably the multi-decadal variations of the cyclone parameters for the first half of the century, since there were two periods of the very fast increasing of number of marine observations after the World Wars, which strongly affects the homogeneity of time series on timescales of 2-3 decades. That is why we consider the time series for the northeast Mediterranean region, where the quite dense routine observational network occurs for more than a century. The series also demonstrate pronounced decadal variability (out-of-phase with NAO). In case of the Mediterranean region, we can draw more reliable conclusions concerning variations on timescales of several decades in the first half of the 20th century. On this timescale we find a decrease in the cyclonic activity in the region since the late 1960, and its increase in the first half of the century, both in agreement with the low-frequency NAO mode (Figure 20).

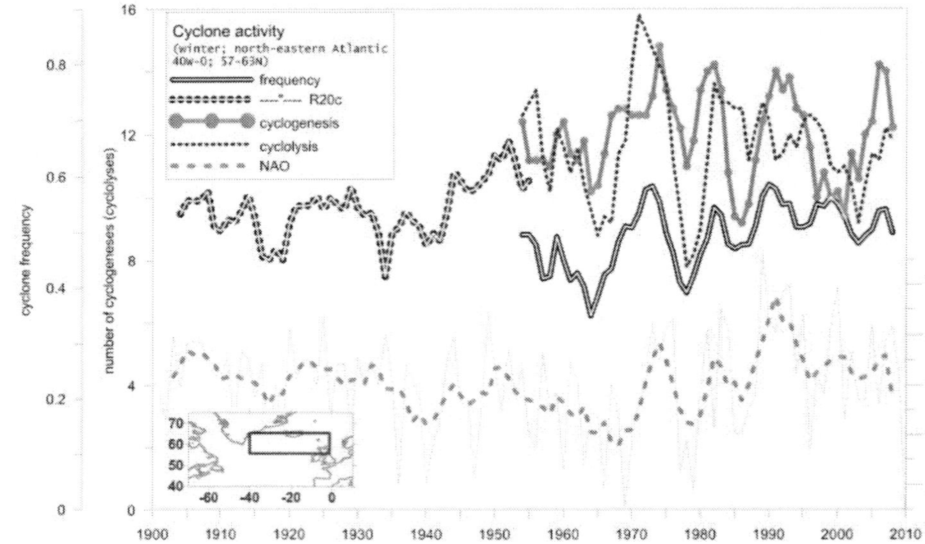

Figure 19. Time series of winter (DJF) characteristics of the cyclonic activity over the "key" region in the north-eastern part of the North Atlantic (40°W-0; 57-63°N): relative frequency and seasonal number of cyclogeneses and cyclolyses. 5-year running means for the entire analyzed region are shown. Data for the second part of 20th century derived from NCEP Reanalysis 1, and for the first half from the R20C (Version 2). Also shown the time series of the seasonal PC-based NAO index (seasonal values and 5-year running means).

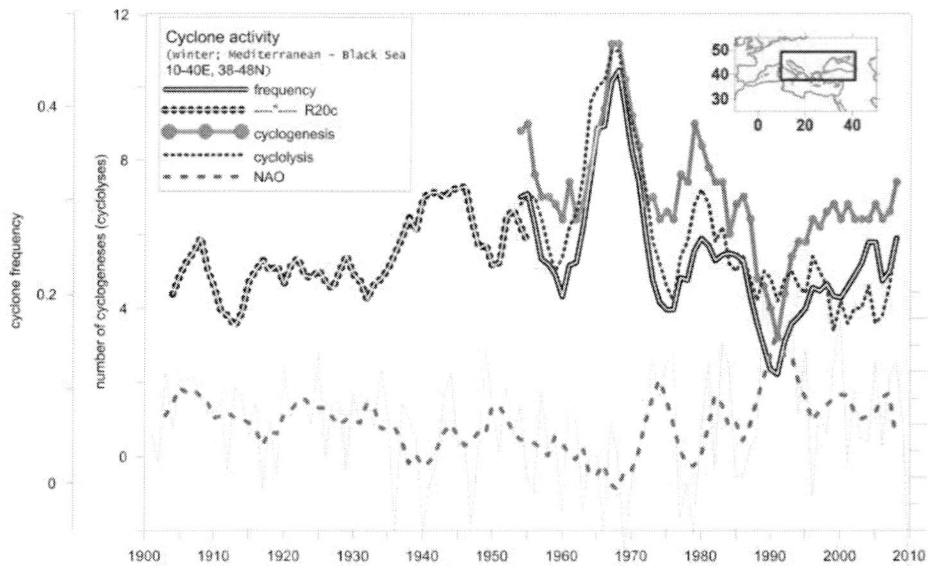

Figure 20. The same as in Figure 19, but for the northeast Mediterranean – Black Sea region (10-40°E; 38-48°N).

We cannot extract the AMO itself from the instrumental observations at the significant level since time series of observations are not sufficiently long. However, practically all reconstructed time series of the NAO index since Medieval period and reconstructed long-term temperature time series reveal the significant spectral peak in the band of periods ~50-70 yrs (see, paragraph ***Atlantic Multi-decadal Oscillation***). Thus, we can conclude that AMO is a significant signal in the coupled system and it manifests itself in the synoptic eddy field over the Northern Hemisphere.

4.2.5 Manifestation of Climate Modes in the Cyclone Activity over Key Regions

To obtain more definite quantitative measure of signal strength, we chose "key" regions where climate signals are supposed to be most pronounced, namely three regions in the North Atlantic, Mediterranean Sea, and three North Pacific regions. For each of them we calculated the mean cyclone frequency and intensity for extreme phases of NAO, NP, ENSO and AMO, and standard deviations for the entire reference period (1958-2010). The annual cycle was subtracted from the monthly regional values prior to the calculation. For the evaluation of signal strength we use a difference of means in the opposite phases. These differences, along with the corresponding standard deviations, are tabulated in Table 4. The signal significance was estimated using standard t-test.

The most intense regional manifestations in the North Atlantic and North Pacific cyclone characteristics are associated with the NAO and NP (PDO) signals, respectively. The winter cyclone frequency increase in the positive NAO phase in the Icelandic Low vicinity by about 0.4 and intensity by 50 *gpm* (in comparison with the negative NAO phase). In the mid-latitudes and subtropics of the North Atlantic, they decrease by up to ~0.3 and 40 *gpm*, respectively. These figures are close to typical figures of seasonal contrasts of the regional cyclonic activity. At the same time, the essential decreasing of cyclonic activity in the positive NAO phase is observed in the Mediterranean region. The AMO signal is also significant, but not so strong (as the NAO one) in the North Atlantic. However, in the Mediterranean region this signal is very strong, especially in cyclone frequency. This is confirmed by early results of the authors ([81, 97], see also, paragraphs ***NAO*** and ***AMO***).

Positive NP phase is accompanied by a dramatic increase of cyclone frequency over the Aleutian Low (up to ~0.5). However, the associated cyclone intensification is not significant here. The North Pacific manifestations of ENSO are significant over Okhotsk Sea and Gulf of Alaska, while over the Aleutian Low they are weak. Remind that we used the combined catalogue of El Niño events of all types, which reduces the significance of the results (see also paragraphs ***North Atlantic Oscillation*** and ***Atlantic Multi-decadal Oscillation***).

Finally, it is worth to discuss briefly the cross-correlation between the analyzed signals. It is well-known that the zero-lag correlation between the NAO and ENSO indices is very weak (e.g., [47]). Lead-lag correlations of NAO/ENSO indices in the range from 1 to 12 months are also mostly insignificant. The same is true for the NAO and NP indices. It does not mean that interrelation between these signals is absent. However, in the simplest case one can consider the NAO-ENSO/NAO-NP pairs as the independent signals. This is supported by data of Table 4. Hence, the differences of regional cyclone patterns produced by Pacific and Atlantic modes and revealed relative importance of these modes are quite reliable. At the same time, the low-frequency NAO mode vs. AMO and NP vs. ENSO indices cannot be considered as statistically independent pairs (in spite of low zero-lag correlations between AMO–NAO indices and ENSO–NP indices). Moreover, the low-frequency NAO mode is one of the characteristics of AMO. As a result, the AMO- and NAO-related regional patterns of cyclone characteristics are similar (see, paragraphs ***NAO*** and ***AMO***). Table 4 confirms that. So, additional analysis needs to separate the impacts of two Atlantic modes on regional cyclonic activity. The same conclusion is true for the two Pacific modes.

Table 4. Differences of winter regional cyclonic activity in opposite phases of the climate modes in 1958-2010. Corresponding standard deviations are presented in parentheses (in italics). *F* is frequency (fraction of observations with cyclone centers within a region), and *I* is intensity (gpm). Values significant at 5%, 1% and 0.1% level are marked by one, two and three asterisks, respectively

	NEN Atl 0-40°W 57-73°N	SWN Atl 63-33°W 40-52°N	SEN Atl 32-10°W 35-55°N	Med Sea 10-40°E 38-48°N	Okh Sea 138-156°E 44-64°N	Aleut Isl 170°E-150°W 42-57°N	Alaska Gulf 125-140°W 50-60°N
NAO							
F	0.377***(0.137)	-0.201***(0.110)	-0.283***(0.088)	-0.171***(0.139)	0.029 (0.108)	-0.049 (0.210)	0.004 (0.061)
I	50.8***(28.5)	-33. 7***(30.3)	-40.3**(48.6)	-15.3***(14.5)	18.9 (38.9)	3.5 (25.4)	13.4 (44.0)
NP							
F	-0.114*(0.177)	0.069*(0.104)	0.022 (0.119)	0.071 (0.132)	-0.193***(0.094)	0.479***(0.130)	-0.099***(0.068)
I	-18.2 (31.7)	-3.6 (30.4)	0.3 (41.4)	4.9 (15.7)	-39. 1***(29.3)	1.4 (30.2)	-20.4 (51.4)
ENSO							
F	-0.049 (0.168)	0.038 (0.123)	0.064 (0.127)	0.061 (0.145)	-0.070*(0.105)	0.011 (0.208)	-0.044*(0.069)
I	8.0 (26.5)	17.1 (33.1)	-3.0 (54.2)	6.5 (18.7)	-11.8 (39.0)	6.4 (26.8)	-16.7 (42.1)
AMO							
F	0.165**(0.160)	-0.069*(0.104)	-0.137**(0.149)	-0.258***(0.139)	-0.006 (0.095)	0.093 (0.203)	-0.025 (0.083)
I	25.3* (34.4)	-17.4*(28.9)	-8.0 (49.4)	-7.4 (18.0)	-3.9 (41.0)	9.7 (24.8)	-14.9 (55.3)

Conclusion

Presented results clearly demonstrate that the principal large-scale modes of the ocean-atmosphere variability, namely NAO and ENSO (at the interannual to sub-decadal scale) and PDO and AMO (at the interdecadal to multi-decadal scale) have the significant signatures in the statistical properties of mid / high -latitude and subtropical cyclones.

The most reliable results on impact of the large-scale modes on the synoptic eddy field in the Northern Hemisphere can be formulated as follows:

- the positive phase of the North Atlantic Oscillation and Atlantic Multi-decadal Oscillation is accompanied by increased/decreased frequency of cyclones in the subarctic/mid-latitude-subtropical Atlantic region, intensification and enlargement of the North Atlantic stormtrack. The speed of movement of cyclones tends to be higher in the positive phase of NAO and AMO. There is strong impact of these two modes on the properties of the Mediterranean cyclones which follow the tendencies of mid-latitude--subtropical cyclones. Both cyclogenesis and cyclolysis in the weak NAO phase are much more widespread over extratropics in the Atlantic – European sector. In the strong NAO phase, they are concentrated within the narrower band in the northern part of the region and in the western Mediterranean – the Near East;
- there are numerous manifestations of the Atlantic ocean-atmosphere signals in the cyclone properties outside the Atlantic – European region, such as changes of the cyclone frequency over the North America, Siberia, Far East and China region, with intensification and enlargement of the cyclones during weak phases of the Atlantic signals in the first three regions and during strong phases in the fourth one;
- El Niño and La Nina events are accompanied by intermittent zonally-oriented bands of increased and decreased cyclone frequency over the Northern Hemisphere. Some Atlantic features of cyclonic activity induced by El Niño-events are reminiscent of the NAO-associated pattern. Last result is mostly due to the more numerous spring El Niño events (or typical El Niño events of Rusmusson and Carpenter). However, its significance decreases as a result of mixture of El Niño events of different types. The comprehensive analysis of regional manifestations of the El Niño events of different types is restricted by insufficient data;
- positive phase of PDO is accompanied by the enhanced frequency of cyclone occurrences, cyclogeneses and cyclolyses, in the Aleutian Low region and by the more intense Pacific stormtrack. Around this region the manifestation of PDO is opposite;
- differences of winter cyclonic activity between positive and negative phases of the Atlantic and Pacific modes in key regions (such as Icelandic and Aleutian Low and Mediterranean region) are the same order as seasonal contrasts of the cyclones' properties here.

Thus, we can conclude that climate variations, associated with interannual to multi-decadal SST anomalies do cause the atmospheric response in the cyclone statistics, resulting in the certain teleconnection patterns. These patterns determine the anomalies of the general atmospheric circulation. Hence the variation of cyclonic activity is one of the crucial elements of feedback in the climatic system which dictates the principal properties of low-frequency variability of GAC.

ACKNOWLEDGMENTS

We thank two anonymous reviewers for valuable comments and suggestions.

REFERENCES

[1] E. N. Lorenz. *The Nature and Theory of the General Circulation of the Atmosphere,* World Meteorological Organization, 218, (1967).
[2] A.S Monin, V.M. Kamenkovich, V.G. Kort, *Variability of the World Ocean,* Leningrad, (1974).
[3] K. Hasselmann, *Tellus,* 28, 473 (1976).
[4] A. B. Polonsky, S.A. Baev, A. Diakite, *Dynamics of Atmosph. and Oceans*, 16, 225 (1992).
[5] M. Yu. Bardin, and A. B. Polonsky, *Izv. Atmoshp. Ocean Physics*, 41, 147 (2005).
[6] J. Bjerknes,*A Coll. Papers*, Moscow, 257 (1969).
[7] A. Polonsky, E. Voskresenskaya, and V. Belokopytov, In book: *Sensitivity to Change: Black Sea, Baltic Sea, North Sea*, Kluwer Academic Publisher, 14 (1997).
[8] A. Czaja, C. Frankignoul, *Geophys. Res.Lett.,* 26, 2969 (1999).
[9] Z. Liu, M. Alexander, *Rev. Geophys.,* 45, 2005, doi:10.1029/2005RG000172 (2007).
[10] M. Latif, T. Barnett, *Science,* 266, 634 (1994).
[11] T.N. Palmer, A. Alessandri, U. Andersen, et.al., *Bull. Amer. Meteorol. Soc.*, 85, 853 (2004).
[12] A. G. Barnston, R. E. Livezey, *Mon. Wea. Rev.,* 115, 1083 (1987).
[13] D. Enfield, A.M. Mestas-Nunez, *J. Climate,* 12, 2719 (1999).
[14] A. Kaplan, M.A. Cane, Y. Kushnir, et al., *J. Geophys. Res.*, 103, 18567 (1998).
[15] E. S. Nesterov, *Russian Meteorology and Hydrology*, 8, 74 (1998).
[16] N.E. Graham, and H.F. Diaz, *Bull. Amer. Meteor. Soc.,* 82, 1869 (2001).
[17] J. Marshall, Y. Kushnir, D. Battisti, et al., *Int. J. Climatology*, 21, 1863 (2001).
[18] A. B. Polonsky, *Physical Oceanography,* 11, 32 (2001).
[19] A. B. Polonsky, D. V. Basharin, E. N. Voskresenskaya, et al., *Physical Oceanography,*14, 38 (2004a).
[20] A. B. Polonsky, D. V. Basharin, E. N. Voskresenskaya, et al., *Pacific Oceanography.,* 2, 52 (2004b).
[21] J.W. Hurrell, and H. van Loon, *Climate Change,* 36, 301 (1997).
[22] J. C. Rogers, *J. Climate,* 10, 1635 (1997).
[23] M. Yu. Bardin, *Russian Meteorology and Hydrology*, 1, 42 (2000).
[24] M. Yu. Bardin, *Russian Meteorology and Hydrology*, 8, 5 (2002).
[25] G.T. Walker, and E.W. Bliss, *Meteorology*, R. Meteorol. Soc., 4 (36), 53 (1932).

[26] J.W. Hurrell,*Science,* 269, 676 (1995).
[27] G. Dzhiganshin, and A. Polonsky, *Izv. Atm. Ocean Physics,* 39, 557 (2003).
[28] G. Eshel, *Mon. Wea. Rev.,* 131, 1018 (2003).
[29] M. Honda, and H. Nakamura, *J. Climate*, 14, 4512 (2001).
[30] M. Honda, H. Nakamura, J. Ukita, et al., *J. Climate,* 14, 1029 (2001).
[31] J.M. Wallance, and D.S. Gutzler, *Mon. Wea. Rev*., 109, 784 (1981).
[32] N.-C. Lau, *J. Atmos. Science,* 45, 2718 (1988).
[33] M. C. Serreze, F. Carse, R. G. Barry, et al., *J. Climate,* 10, 453 (1997).
[34] J. B. Elsner, B. Kocher, *Geophys. Res. Lett.,* 27, 129 (2000).
[35] S.G. Philander, *El Nińo, La Nińa, and the Southern Oscillation,* Academic Press, Inc. (1990).
[36] W. S. Kessler, *Geophys. .Res. Lett.*, 29, 2125, doi:10.1029/2002GL015924 (2002).
[37] Climate glossary –*"Southern Oscilliation Index (SOI)".* Bureau of Meteorology (Australia).(2002). http://www.bom.gov.ua/climate/glossary/soi.shtml.
[38] Ch. Ropelewsky, and D. Halpern, *Mon. Wea. Rev*., 115, 1540 (1987).
[39] D.B. Enfield, D.A. Meyer, *J. Geophys. Res*., 102, 929 (1997).
[40] C. Wang, In book: *Advances in Global Change Research,* Kluwer Academic Publishers, *Dordrecht*, 21 (2004).
[41] G. Hernandez, *The economic implications of an El Nińo,* Washington, March 18 (2002).
[42] B.C. Weare, A.R. Navato, R.E. Newell, *J. Phys. Oceanogr*., 6, 671 (1976).
[43] J. Hsiung, R.E Newell, *J. Phys. Oceanogr*., 13, 1957 (1983).
[44] J. Lanzante, *J. Climate,* 9, 2568 (1996).
[45] R. Saravanan, and P. Chang, *J. Climate.,* 13, 2195 (2000).
[46] K. Fraederich, and. P. Muller, *Int. J. Climatology,* 12, 12 (1992).
[47] E. Voskresenskaya, and A. Polonsky, *Physical Oceanography*, 2, 23 (1992).
[48] A. Polonsky, *Australian Journal of Marine and Freshwater Research*, 45, 700 (1994).
[49] A. Mariotti, N. Zeng, and K-M. Lau, *Geophys. Res. Lett.*, 29, doi:10.1029/2001GL014248 (2002).
[50] R.P. Kane, *Monthly Weather Rev.,* 125, 1792(1997).
[51] K.E. Trenberth, and D.P. Stepaniak, *J. Climate,* 14, 1697 (2001).
[52] J. Xu, and J.C.L. Chan, *J. Climate*, 14, 418 (2001).
[53] T. Horii, K. Hanawa, *Geophys. Res. Let.*,31, 1634 (2004).
[54] N.K Larkin, D.E. Harrison, *Geophys. Res. Lett.*, 32, L13705/doi:10.1029/2005GL022738 (2005).
[55] H. Ashok, and T. Yamagata, *Nature*, 461, 481 doi:10.1038/461481a (2009).
[56] J.S. Kug, F.F. Jin, S.I. An, *J. Climate,* 22, 1499 doi:10.1175/2008JCLI2624.1 (2009).
[57] H.Y. Kao, J.Y. Yu, *J. Climate,* 22, 615 doi:10.1175/ 2008JCLI2309.1 (2009).
[58] E. Voskresenskaya, and N. Mikhailova, *Proceedings of National Academy of sciences of Ukraine*, 4, 54 (2010).
[59] E.M. Rasmusson, and T.H. Carpenter, *Mon. Wea. Rev*., 110, 354 (1982).
[60] K. Ashok, S.K. Behera, S.A. Rao, et al., *J. Geophys. Res.,* 112, C11007, doi:10.1029/2006JC003798 (2007).
[61] D. Gu, and S.G.H. Philander, *Science*, 275, 805 (1997).
[62] D. Neelin, D. Battisti, A.C. Hirst, *J. Geophys. Res*., 103, 14261 (1998).
[63] S.R. Hare, and R. C. Francis, *Can. Spec. Publ. Fish. Aquat. Sci*., 121, 357 (1995).
[64] N.J. Mantua, S.R. Hare, Y. Zhang, et al., *Bull. Amer. Meteor. Soc.,* 78, 1069 (1997).

[65] A. Gershunov, and T.P. Barnett, *Bull. Amer. Meteor. Soc.*, 79, 2715 (1999).
[66] R. D. Garreaud, and D. S. Battisti, *J. Climate*, 12, 2113 (1999).
[67] S. Minobe, *Geophys. Res. Lett.*, 24, 683 (1997).
[68] K.E. Trenberth, and J.W. Hurrell, *Climate Dynamics,* 9, 303 (1994).
[69] J. C. Rogers, *Int. J. Climatology,* 1, 39 (1981).
[70] S.-W. Yeh, Y.-J. Kang, Y. Noh, et al., *J. Climate*, 24, 1170 (2011).
[71] M.Newman, G.P.Compo, M. A.Alexander, *J.Climate,* 16, 3853 (2003).
[72] M. Yu. Bardin, E.N. Voskresenskaya, V.N. Maslova, *Physical Oceanography*, 18, 38 (2008).
[73] Y. Kushnir, *J.Clim.*, 7, 141 (1994).
[74] M. E. Schlesinger, and N. Ramankutty, *Nature,*367,6465 (1994).
[75] T.L. Delworth, and R.J. Greatbatch, *J. Climate,* 13, 1489 (2000).
[76] T.S. Delworth, S. Manabe, R.J. Stouffer, *J. Climate,* 6, 1993 (1996).
[77] A.B. Polonsky, and E. N. Voskresenskaya, *Russian Meteorology and Hydrology*, 7, 89 (1996).
[78] J. Knight, R. Allan, C. Folland, et al., *Proceedings of the Workshop on Decadal Climate Variability*, 12 (2005).
[79] J. Luterbacher, D. Gyalistras, C. Schmitz, et al., *Geophys. Res. Lett.,* 26, 2745 (1999).
[80] C. Eden, T. Jung,*J. Climate*, 14, 676 (2001).
[81] A. B. Polonsky, *Physical Oceanography,* 18, 42 (2008).
[82] F.M. Selten, R.G. Haarsma, J.D. Opsteegh,*J. Climate,* 12, 1956 (1999).
[83] K.M. Zishka, and P.J. Smith, *Mon. Wea. Rev.*, 108, 387 (1980).
[84] L.M. Whittaker, and L.H. Horn, *Int. J. Climatology*, 4, 297 (1984).
[85] M. R. Sinclair,*Mon. Wea. Rev.,* 122, 2239 (1994).
[86] E. Kalnay, et al., *Bull. Amer. Meteor. Soc.,* 77, 437 (1996).
[87] R. Kistler, et al., *Bull. Amer. Meteor. Soc.*, 82, 247 (2001).
[88] G.P. Compo, J.S. Whitaker, and P.D. Sardeshmukh, *Bull. Amer. Met. Soc.*, 87, 175 (2006).
[89] K. Wolter, and M.S. Timlin, *Proceedings of the 17th Climate Diagnostics Workshop*, 52 (1993).
[90] A. Polonsky, *WMO TD-N 1062, JCOMM technical report,* 10, (2003).
[91] S. Worley, et al., *Int. J. Climatol.*, 25, 823 (2005).
[92] R. Otnes, and L. Enochson, *Applied Time Series Analysis, 1: Basic Techniques*, Wiley, New York (1978).
[93] M. Yu. Bardin, *Russian Meteorology and Hydrology*, .11, 24 (1995).
[94] J. R. Key, A.C.K. Chan, *Geophys. Res. Lett.*, 26, 2053 (1999).
[95] T. M. Smith, and R. W. Reynolds, *J. Climate*, 11, 3320 (1998).
[96] A.B. Polonsky, and E.P. Semiletova, *Physical Oceanography*, 12. 28 (2002).
[97] A. Polonsky, M. Bardin, E. Voskresenskaya, *Physical Oceanography,* 17, 47 (2007).

In: Cyclones: Formation, Triggers and Control
Editors: K. Oouchi and H. Fudeyasu

ISBN: 978-1-61942-976-5
©2012 Nova Science Publishers, Inc.

Chapter 9

TROPICAL CYCLONE SIMULATION IN A HIGH-RESOLUTION ATMOSPHERE-OCEAN COUPLED GENERAL CIRCULATION MODEL

Wataru Sasaki[1], Jing-Jia Luo[1,2] and Sebastien Masson[3]
[1]Japan Agency for Marine-Earth Science and Technology, Yokohama, Japan
[2]Centre for Australian Weather and Climate Research, Melbourne, Australia
[3]Laboratoire d'Océanographie Expérimentation et Approches Numériques, Paris, France

ABSTRACT

A global high-resolution coupled general circulation model (CGCM) consisting of a T319 atmosphere general circulation model and an eddy-permitted ocean general circulation model is examined in terms of the reproducibility of the northern hemisphere tropical cyclone (TC) activity as well as the large-scale environmental conditions. The CGCM successfully simulates the realistic TC structure, TC-induced ocean response, and TC genesis frequency. The global TC genesis frequency simulated by the high-resolution CGCM is much closer to the observed, compared that simulated by the medium-resolution (T106) CGCM. In addition, the high-resolution CGCM partially reproduces the bimodal seasonal cycle of the North Indian Ocean cyclogenesis, while the medium-resolution CGCM fails to simulate it. The high-resolution CGCM also reasonably reproduces the environmental conditions favorable for the TC genesis: warm sea surface temperature, low-level cyclonic circulation, weak vertical wind shear, and high relative humidity in the mid-troposphere. The eastward extension of monsoon-trough is well simulated by the high-resolution CGCM as observed, compared to the medium-resolution CGCM. There are, however, still some discrepancies between the modeled and observed TC activity. We discuss about the following two discrepancies from the view point of the simulated large-scale environmental conditions: the high-resolution CGCM fails to reproduce the bimodal seasonal cycle of the Arabian cyclogenesis during the pre-monsoon period, and the western North Pacific TC genesis locations are confined in the southwestern part of the western North Pacific. It is found that less Arabian cyclogenesis during the pre-monsoon period is due to the weak low-level cyclonic circulation in the Arabian Sea during this period, although the weak vertical wind shear is well simulated as observed. For the western North Pacific, less TC genesis in the southeastern part of the western North Pacific is found to be due to the failure to simulate the eastward extension of the monsoon-trough up to the international dateline. Compared to a medium-resolution CGCM, one of the advantages of the high-resolution CGCM is the reproduction of the intense TC. Surface wind speed exceeding 20~40 ms^{-1} is successfully simulated by the high-resolution CGCM, while the TC wind speed simulated by the medium-resolution CGCM is less than 20~30 ms^{-1}. The frequency distribution of TC surface wind speed simulated by the high-resolution CGCM is closer to the observed compared to the medium-resolution CGCM.

1. INTRODUCTION

Tropical cyclone (TC) is one of the most devastating phenomena in the earth's weather system. The primary driver of the variation of TC activity is large-scale climatic events such as El Nino-Southern Oscillation (ENSO) as well as synoptic events such as Madden-Julian Oscillation [1], and monsoon trough. The modulation of the TC activity and the large-scale climatic events has been extensively studied. Relationship between the western North Pacific TC activity and ENSO was studied by Chan [2] and many scientists. Kim et al. [3, 4] investigated the modulation of the Pacific and Atlantic TC activity by the central Pacific warming, in contrast to the east Pacific warming during the conventional El Nino. The North Atlantic TC activity in relation to the Atlantic Meridional Mode [5] and the North Atlantic Oscillation [6] is also studied. Singh [7] reported the relationship between the North Indian Ocean TC activity and Indian Ocean Dipole mode.

High-resolution atmospheric general circulation models (AGCMs) are necessary for realistic TC simulation and TC forecast. Oouchi et al. [8] discussed possible changes in TC activity under global warming conditions by using a 20-km mesh AGCM. Bengtsson et al. [9] also examined future TC activity in a warmer climate condition by using a high resolution AGCM with T319 (~40km) horizontal resolution.

These high-resolution AGCMs indeed have the capability to realistically simulate TCs. However, the stand-alone AGCMs have limitation due to the lack of the coupled process of the atmosphere-ocean interaction. Importance of atmosphere-ocean coupling for TC simulation was investigated based on regional atmosphere-ocean coupled models. Bender and Ginis [10] pointed out that an uncoupled atmospheric model incorrectly predicts too strong TCs, while a coupled model predicts a cooling of sea surface temperature (SST) associated with the vertical mixing and upwelling which in turn lead to reasonable estimation of the central pressure of TC. Thus the atmosphere-ocean coupling process is indispensable for the realistic simulation of TC intensity.

Atmosphere-ocean coupled model is necessary to reproduce the correct air-sea coupling process in real climate. Coupled general circulation models (CGCMs) have been used to study TC activity on a broad range of time-scales [11-13], although the adopted horizontal resolutions are still too coarse to realistic TC simulation. Vitart and Stockdale [11] reported the occurrence of tropical storms simulated by the ECMWF seasonal forecasting system consisting of an AGCM with T63 horizontal resolution and 31 vertical levels and an ocean general circulation model (OGCM) with $2^{\circ} \times 0.5^{\circ}$ horizontal resolution. Matsuura et al. [12] showed the inter-decadal variability of the western North Pacific TC activity simulated by a medium-resolution CGCM consisting of a T106 AGCM with 19 vertical levels and an

OGCM with 1.125°×0.5625° horizontal resolution and 37 vertical levels. Iizuka and Matsuura [13] investigated the response of the western North Pacific TC activity to ENSO simulated by a relatively high-resolution CGCM which has an T213 AGCM with 21 vertical levels and an OGCM with 0.5625°×0.5625° horizontal resolution and 37 vertical levels.

The horizontal resolution of the oceanic component of these CGCMs is coarse to resolve ocean fronts and eddies. Previous study showed that the ocean eddy affects the development of TC intensity [14]. Thus a high-resolution CGCM is expected to realistically simulate the interaction between TCs and ocean fronts and eddies.

In this chapter, a high-resolution CGCM consisting of a T319 (~40km) AGCM and an eddy-permitted OGCM is examined in terms of the reproducibility of the TC activity, particularly in northern hemisphere, as well as the large-scale environmental conditions. As we will show, the high-resolution CGCM realistically reproduces TC structure, TC-induced oceanic response, TC activity, and the large-scale environmental conditions. For comparison, we also examine the performance of a medium-resolution (T106) CGCM in simulating the TC activity.

The rest of this chapter is structured as follows: the CGCMs are introduced in section 2 and we describe the TC detection and tracking method, Sections 3 to 7 presents the TC activity and the large-scale circulation simulated by the high- and medium-resolution CGCMs, and Section 8 provides a summary and discussion.

2. Model Description and TC Detection

2.1. The Cgcms and Observed Data

The high-resolution CGCM introduced here is the version 2 of the Scale Interaction Tropical Experiment-Frontier (SINTEX-F2) CGCM, which has been developed from the SINTEX-F1 CGCM [15]. The atmospheric component of the SINTEX-F2 CGCM is ECHAM5 [16] which has T319 horizontal resolution with 31 vertical levels. The model uses a cumulus parameterization scheme proposed by Tiedtke [17]. A 5-layer land surface thermo-dynamical model is implemented in ECHAM5 [18]. The oceanic component is based on the Nucleus for European Modelling of the Ocean (NEMO) modeling system [19] which includes an ocean model of the Océan Parallélisé (OPA9) and a sea-ice model of the Louvain-la-Neuve Sea Ice Model (LIM2; [20]). The OPA9 and LIM2 employ the ORCA-R025 grid which uses a tri-polar grid with 0.25° zonal resolution and 0.25°cos(latitude) meridional resolution with 46 vertical levels. The atmospheric and oceanic components are coupled with data exchanged every 2 hours, including SST, sea-ice fraction and thickness, water, surface heat, and momentum fluxes, by means of the Ocean Atmosphere Sea Ice Soil (OASIS3) coupler [21] without applying any flux correction. Therefore the feedback of strong surface currents driven by the TC winds on the wind stress forcing has been taken into account. Following Luo et al. [15], the momentum flux of ocean surface current is directly passed to the atmosphere. The model is initialized by setting the temperature and salinity in climatological January equal to the Levitus [22], and a state of the rest without motion. Due to the limited computational resources, the SINTEX-F2 CGCM is integrated for 20 years and the outputs from the model year 6 to year 20 are used for the analysis. We note that the 5-yr model spin-up is still too short to reach the equilibrium states. In fact, the modeled tropical SST has a positive trend which is greater than that observed during 1980-2008 (not shown). Considering that TC activity is influenced by the SST, further model spin-up is necessary. Therefore our results described in this chapter should be viewed as the preliminary results.

We also use the outputs of the SINTEX-F1 CGCM with a medium resolution. The atmospheric component of the SINTEX-F1 is ECHAM4 [23] which has T106 (~100km) horizontal resolution with 19 vertical levels. The oceanic component is the Océan Parallélisé (OPA8.2; [24]). The OPA8.2 employs the ORCA-R2 grid which uses a tri-polar grid with 2° zonal resolution, and a meridional resolution varying from 0.5° at the equator to 2°cos(latitude) pole-ward of 20° with 31 vertical levels. The atmospheric and oceanic components are coupled with data exchanged every 2 hours by means of the Ocean Atmosphere Sea Ice Soil (OASIS2.4) coupler [25] without applying any flux correction. The initialization method is the same as that used for the SINTEX-F2. The SINTEX-F1 CGCM is integrated for 180 years, and the last 15 years are used for the analysis. Readers are referred to [15] for detailed model descriptions.

We employ observed TC data obtained from the National Hurricane Center and the U.S. Joint Typhoon Warming Center. The TC climatology is calculated from the data during 1980-2008 during which observed data are reliable thanks to an introduction of satellite altimeter measurements. To validate the modeled environmental conditions, we use the 1980-2008 mean of the National Center for Environmental Prediction/National Center for Atmospheric Research (NCEP/NCAR) Reanalysis [26]. We also use HadISST [27] which contains observed SST and global sea ice coverage during 1980-2008.

2.2. Method of TC detection

Many different TC detection and tracking schemes have been proposed [28]. Since TCs are characterized by cyclonic circulation associated with strong low-level winds and the warm-core in the mid-troposphere, the features are reflected in most schemes to detect model TC. However, different detection criteria for the threshold in low-level wind speed, low-level vorticity, and warm-core temperature anomaly are employed in individual schemes respectively. In addition, the lifetime of the TC is taken into consideration.

We employ a TC detection and tracking scheme proposed by Iizuka and Matsuura [13]. The scheme employs the surface wind speed threshold (17 ms^{-1}) and surface vorticity threshold (1.5×10^{-4} s^{-1}). The lifetime is required to be longer than 2 days. To exclude the mid-latitude cyclones, the genesis locations are limited within the 30°S-30°N. In addition to their detection scheme, we impose a limitation on the warm-core temperature anomaly. The sum of temperature anomaly at the 700-, 500-and 300-hPa levels is greater than 1.0 K for the high-resolution CGCM, while it is greater than 0 K for the medium-resolution CGCM. The surface wind speed threshold for the high-resolution CGCM is 17 ms^{-1}, while that is changed to 12 ms^{-1} for the medium-resolution CGCM. Walsh et al. [28] proposed a resolution-dependent TC detection and tracking scheme, and recommended to apply the higher surface wind speed threshold for higher-resolution atmospheric models. From their criteria, the surface wind speed threshold for T319 (~40-km) and T106 (~100-km) AGCM is ~17 ms^{-1} and ~15 ms^{-1}, respectively. Our criteria for the high-resolution CGCM satisfy their criteria, while our criteria for the medium-resolution CGCM are not stringent as the recommended threshold. As described in Walsh et al. [28], using a lower threshold leads to an artificial increase in the TC genesis frequency in a simulation. Our low surface wind speed threshold (12 ms^{-1}) may artificially increase the TC genesis frequency in the medium-resolution CGCM. Nevertheless, as we will see, the TC genesis frequency in the medium-resolution CGCM is less than those of observation and high-resolution CGCM simulation. Although we have tried some different combination of the threshold values for the relative vorticity and surface wind speed, the relative relationship of the TC genesis frequency between the SINTEX-F1 and SINTEX-F2 is not unduly changed.

3. Tropical Cyclones Simulated by the High-Resolution CGCM

We briefly describe, by taking as some examples, a TC structure and response of ocean to TCs simulated by the high-resolution CGCM (SINTEX-F2).

In the CGCM, a TC-like vortex occurred at 10.7°N, 142.9°E on 07OCT0018. The TC moved northwestward until 12OCT0018, then moved westward and made landfall on Vietnam (the TC track is shown as red line in Figure 1). The westward propagation of the TC is one of the typical TC tracks during boreal autumn in the western North Pacific. The TC reached its peak strength of the central pressure at 956 hPa with the maximum surface wind speed of 39.8 ms^{-1}. Figure 1 shows the horizontal distribution of sea level pressure, precipitation, and surface wind field of the simulated TC at its mature stage. The decrease in sea level pressure associated with the strong low-level cyclonic circulation is well simulated. The heavy rainfall exceeding 20 mm/hr associated with the TC is simulated. Figures 2a and 2b show the vertical section of zonal and meridional winds of the TC, respectively. The cyclonic structure is maintained up to the 200-hPa level. The weak wind speed at the center of the TC is also simulated. The warm-core exceeding 5°C associated with the strong ascent flow (Figure 2c) is simulated. Although the horizontal resolution of the T319 AGCM is still coarse to resolve the finer structure such as eye wall of a TC, overall characteristics of a modeled TC are in reasonable agreement with observations [29].

The high-resolution CGCM also simulates the realistic ocean response to a TC passage. The black line in Figure 3a shows a track of a simulated TC which occurred at 7.7°N, 137.2°E on 29NOV0019. The TC reached its peak strength of the central pressure at 971.9 hPa, and the surface wind speed was enhanced to 35.6 ms^{-1} at its mature stage. During the passage of the TC, the cyclonically rotating winds (not shown) generated the surface currents which flow nearly the same direction as the surface winds (Figure 3a). Not only the surface currents beneath the TC, but also the strong surface currents behind the TC after the TC passage are simulated (Figures 3a-e). In this case, SST is not drastically cooled as observed, although observations indicate that the SST cooling induced by TCs ranges from 1°C to 6°C [30]. Deep thermocline/mixed layer in the tropical western Pacific may prevent the cooling of SST. Instead, a significant vertical motion in the subsurface induced by the TC is seen in the model. Figures 3e-h show the simulated vertical current velocity during 02DEC0019-05DEC0019. The vertical motion induced by the TC consists of two components: the upwelling associated with the Ekman pumping forced by the surface winds and near-inertial oscillations characterized by a pattern of cells of upwelling and downwelling behind the TC. Figure 3h clearly shows the TC-induced upwelling beneath the TC and the alternate upwelling and downwelling behind the TC. The simulated ocean currents induced by the TC are consistent with the observed and previous modeled results [31, 32].

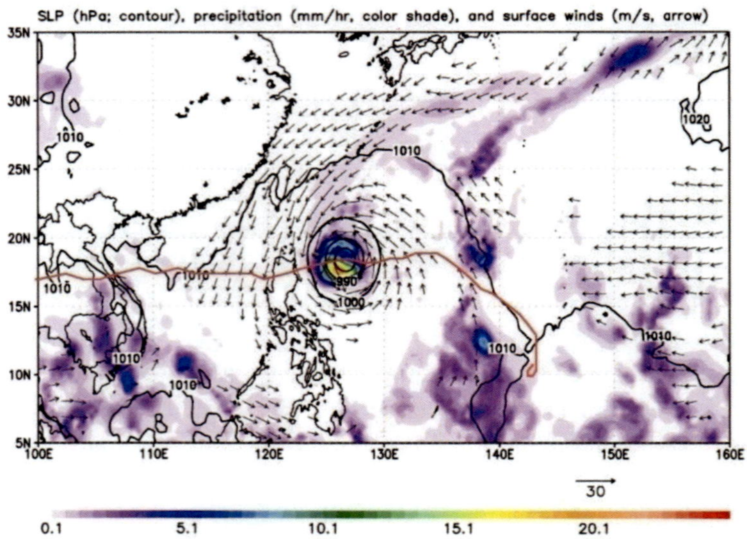

Figure 1. Sea level pressure (contour; hPa), precipitation (color shade; mm hr^{-1}), surface wind vectors (arrow; ms^{-1}) of a TC-like vortex simulated by the SINTEX-F2. The contour interval is 10-hPa. The surface wind vectors are shown for the magnitude exceeding 8 ms^{-1}. Red line denotes the track of the TC.

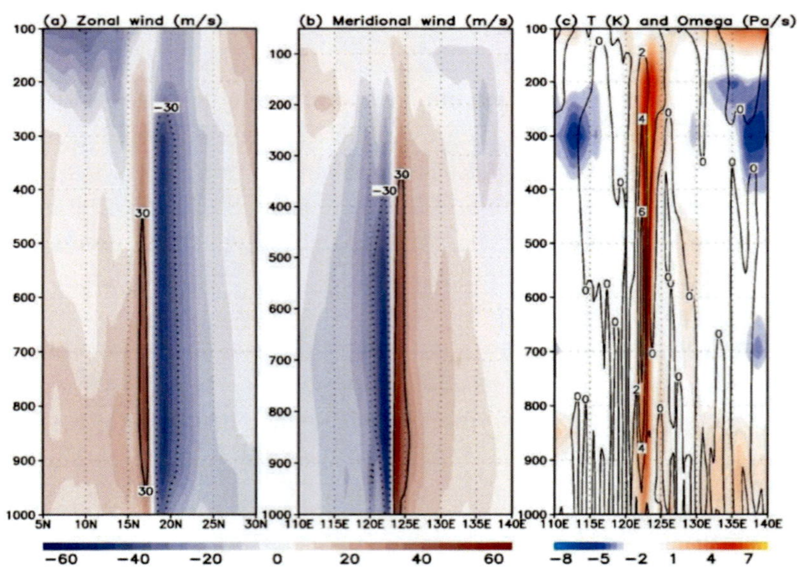

Figure 2. Vertical section of (a) zonal wind (ms^{-1}), (b) meridional wind (ms^{-1}), (c) temperature anomaly (color; K) and vertical wind speed (contour, Pa s^{-1}) associated with a TC in Figure 1. The contour in (a) and (b) shows the wind speed exceeding 30ms^{-1}.

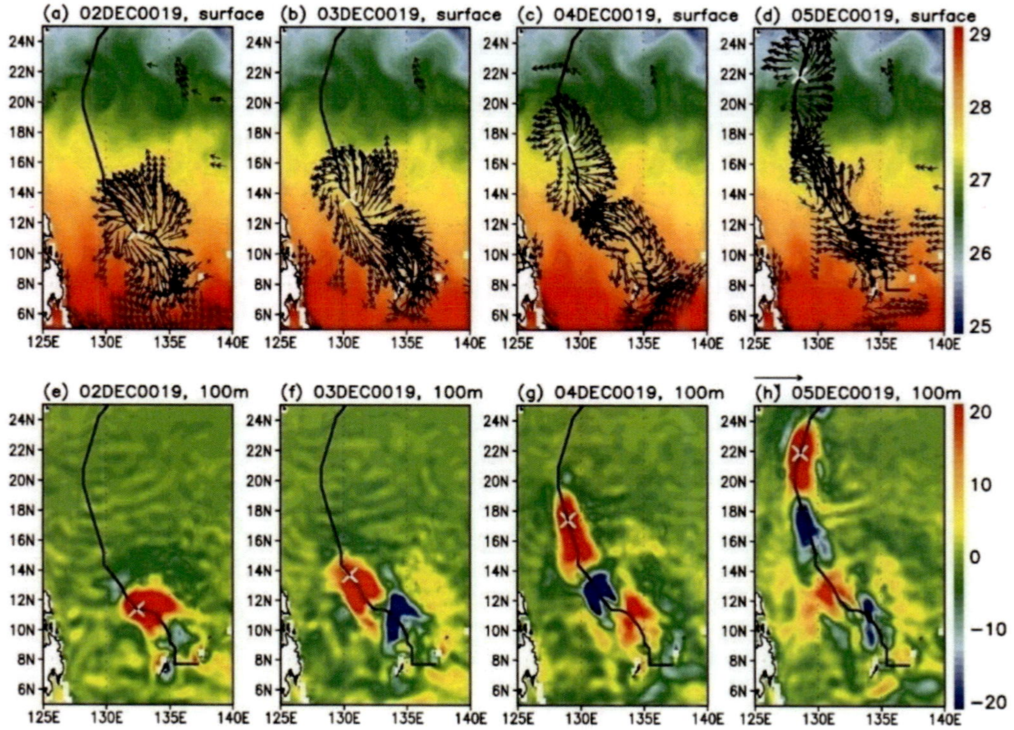

Figure 3. (a-d) SST (color shade; °C) and surface currents (arrow; ms^{-1}) associated with a TC passage during 02DEC0019-05DEC0019 simulated by the SINTEX-F2. The surface current velocity less than 0.2 ms^{-1} is omitted. Bold line denotes the TC track. Cross mark shows the position of the TC. (e-h) As in (a-d), but for the vertical current velocity (color shade; m day^{-1}) at the 100-m depth.

Figure 4 shows changes in SST associated with the passage of a simulated TC in the eastern North Pacific during 10SEP0015-16SEP0015 in the high-resolution CGCM. The SST is reduced by the TC passage particularly on the right hand side of the moving direction (Figures 4a-d). Different from the less SST cooling during the TC passage from 02DEC0019 to 05DEC0019 in the western North Pacific (Figure 3), SST cooling associated with the TC passage during 10SEP0015-16SEP0015 is remarkable. The strong SST cooling could be caused by both the TC passage near the SST front and a shallow thermocline in the eastern North Pacific. On the other hand, Wu et al. [14] showed that the ocean eddy affects the development of TC intensity. These are good examples that a high-resolution CGCM has a capability to simulate the interaction between TCs and ocean fronts and eddies.

Thus, the high-resolution CGCM has a capability to simulate realistically the atmosphere-ocean phenomenon associated with TCs.

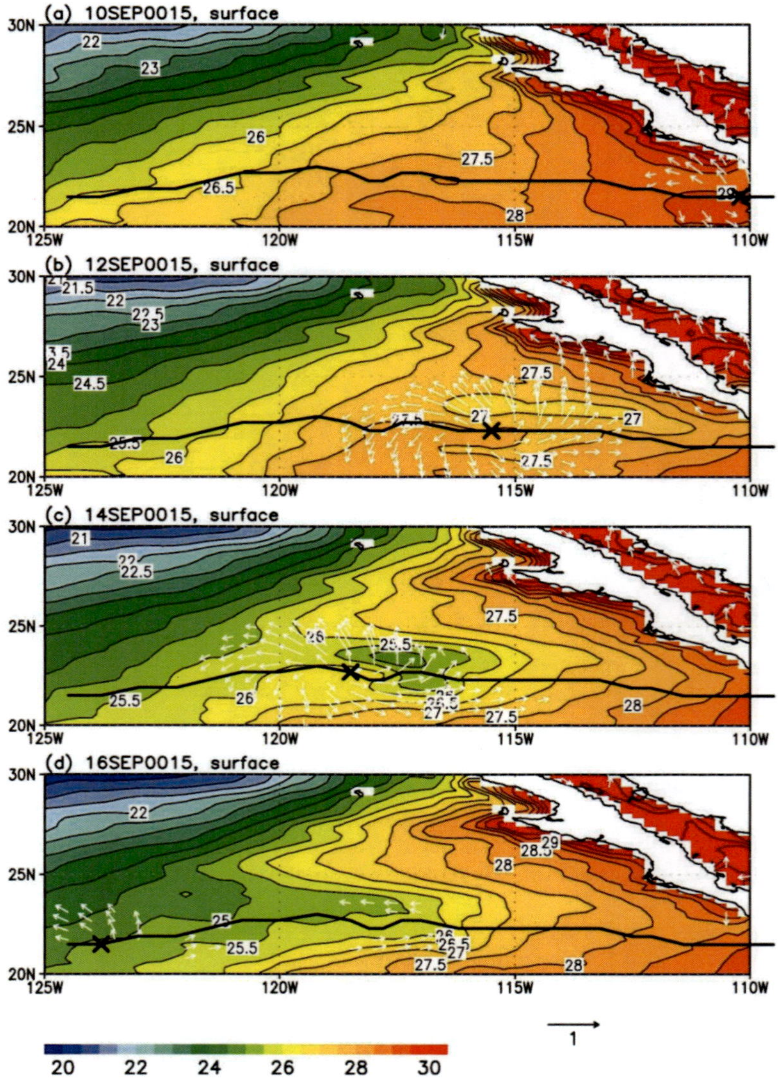

Figure 4. SST (color shade; °C) and surface currents (arrow; ms^{-1}) associated with a TC passage during 10SEP0015-16SEP0015 simulated by the SINTEX-F2. The surface current velocity less than 0.2 ms^{-1} is omitted. Bold line denotes the TC track. Cross mark shows the position of the TC.

4. Environmental Fields Simulated by the High-Resolution CGCM

Gray [33, 34] showed that warm SST, high vorticity in the low-troposphere, weak vertical wind shear, and high humidity in the mid-troposphere are favorable large-scale conditions for TC genesis. In this section, we evaluate the performance of the high-resolution CGCM in terms of capturing these large-scale environmental conditions.

Figure 5a shows SST averaged for the TC peak season in the northern hemisphere (June-November) simulated by the SINTEX-F2 CGCM. The horizontal distribution of the SST is in general agreement with that observed (Figure 5c). The SST warmer than 26°C, which is a favorable condition for TC genesis, in the tropical oceans is well simulated as observed (Figure 5a). Meanwhile, the Pacific cold tongue SST in the SINTEX-F2 is approximately 1°C warmer than the observed (Figure 5a). The warm SST bias may be due to the weak equatorial upwelling associated with the weak easterlies in the model when compared to those observed (Figures 6a and 9.6c). For the tropical Atlantic, the modeled SST to the west of African continent is warmer than that observed (Figure 5a). The warm SST bias in the eastern equatorial Atlantic is one of the common problems for most of current CGCMs [35] to be solved. The medium-resolution CGCM (SINTEX-F1) also generally simulates the SST as observed, although there is a warm SST bias in the North Indian Ocean (Figure 5b).

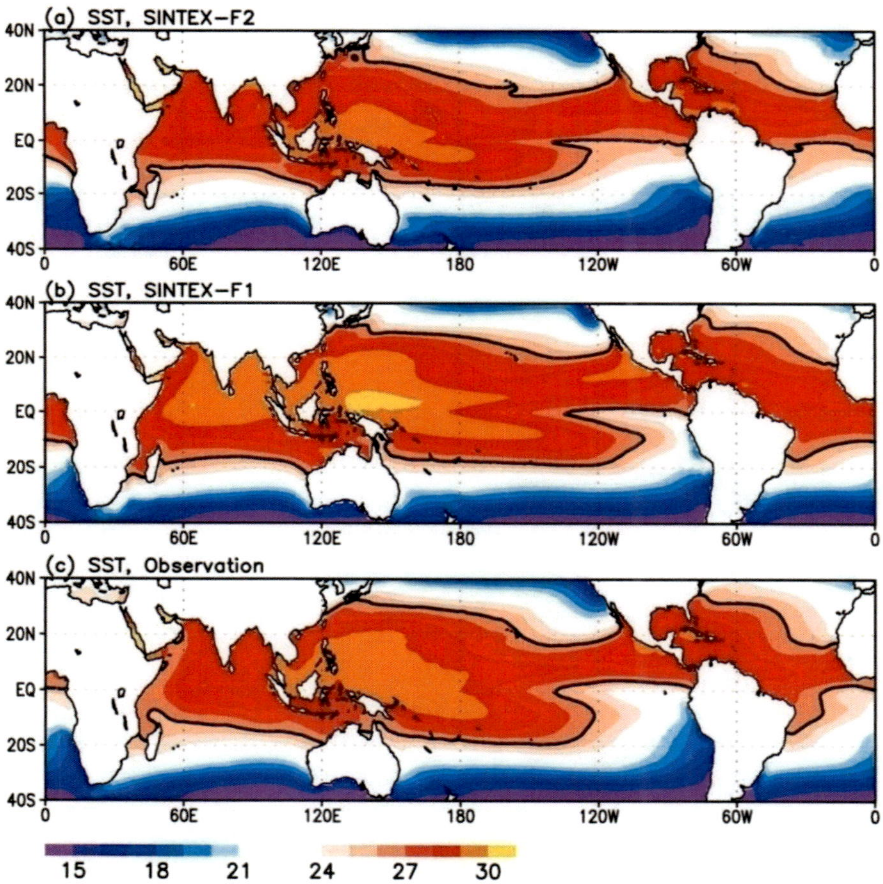

Figure 5. Climatological Jun-November mean of SST (color shade; °C) from the (a) SINTEX-F2, (b) SINTEX-F1, and (c) HadISST. Bold line denotes the 26°C isotherm.

The modeled wind fields at the 850-hPa level also accord with those observed (Figure 6). In particular, the eastward extension of monsoon westerly to the east of the Philippines in the SINTEX-F2 is well simulated as observed (Figures 6a and 9.6c). We however note that the modeled easterly extends slightly westward compared to the observed. On the other hand, the eastward extension of the monsoon westerly in the SINTEX-F1 is weaker than that observed (Figures 6b and 9.6c). The eastward extension of the monsoon-trough is one of the key factors for the western North Pacific cyclogenesis [36]. We further examine the seasonal cycle of the western North Pacific TC genesis frequency and the large-scale circulation simulated by the CGCMs in Section 6.

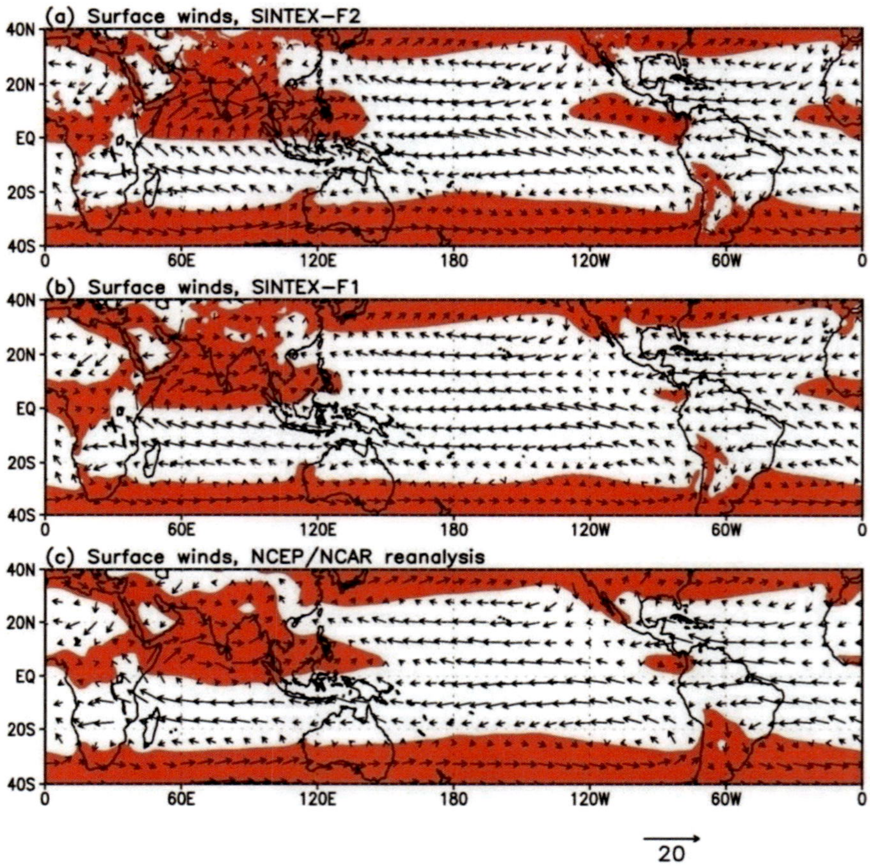

Figure 6. Climatological June-November mean of wind vectors (ms^{-1}) at the 850-hPa level from the (a) SINTEX-F2, (b) SINTEX-F1, and (c) NCEP/NCAR reanalysis. Red denotes the positive zonal surface wind speed.

Figure 7 shows a comparison of the relative vorticity at the 850-hPa level simulated by the CGCMs and that observed. The modeled strong cyclonic circulation in the Arabian Sea, Bay of Bengal, western North Pacific, eastern North Pacific, and tropical North Atlantic is in good agreement with that observed. However, the low-level vorticity in the eastern North Pacific simulated by the SINTEX-F2 extends further northward compared to that observed (Figures 7a and 9.7c).

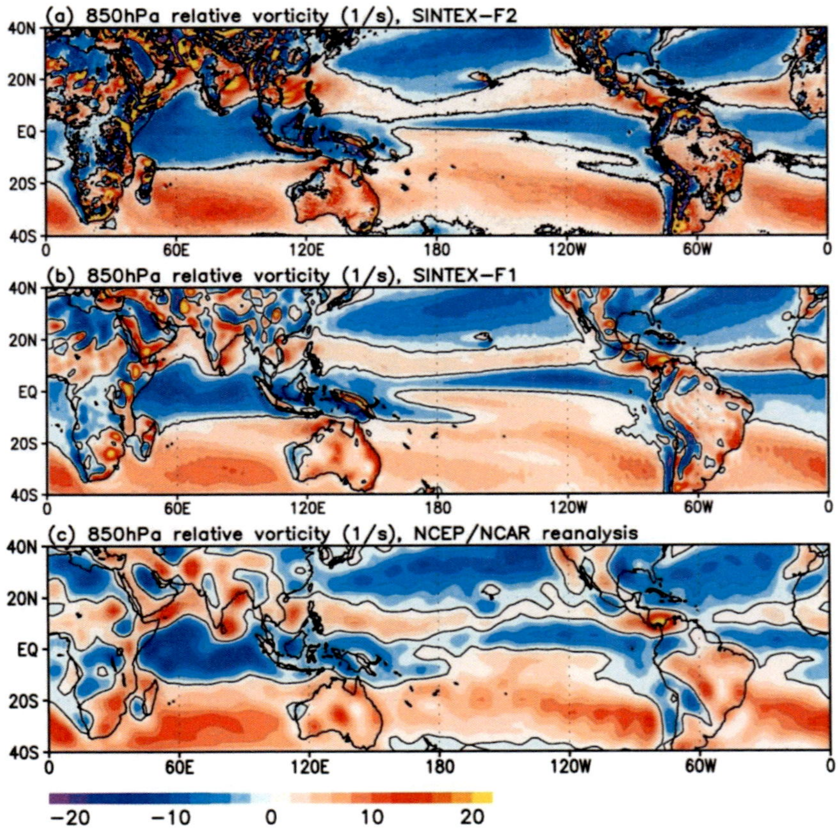

Figure 7. Climatological June-November mean of the relative vorticity at the 850-hPa level (color shade; $\times 10^{-6}$ s^{-1}) from the (a) SINTEX-F2, (b) SINTEX-F1, and (c) NCEP/NCAR reanalysis. The contour denotes the zero line.

The modeled and observed relative humidity in the lower troposphere has a good agreement in the spatial pattern: relatively high humidity in the region of tropical convection and relatively low humidity in the mid-latitudes (Figure 8). However, there is a discrepancy between the modeled and observed relative humidity in the region of tropical convection. The relative humidity at the 700-hPa level in the region of tropical convection simulated by the CGCMs is greater than that of the NCEP/NCAR reanalysis (Figure 8). Trenberth and Guillemot [37] pointed out that the NCEP/NCAR reanalysis has a substantial dry bias in the region of tropical convection associated with the weak Hadley circulation. Annamalai et al. [38] suggested the tendency for the Arakawa-Schubert convection scheme to dry the tropical atmosphere. The difference in the convection scheme between the NCEP/NCAR reanalysis AGCM (Arakawa-Schubert convection scheme) and the CGCMs (Tiedtke convection scheme [17]) may lead the difference in the magnitude of the humidity in the region of tropical convection. The relative humidity simulated by the SINTEX-F2 is much greater than that of the SINTEX-F1 (Figures 8a and 8b). This is consistent with too strong tropical convective precipitation in the SINTEX-F2.

In: Cyclones: Formation, Triggers and Control
Editors: K. Oouchi and H. Fudeyasu

ISBN: 978-1-61942-976-5
©2012 Nova Science Publishers, Inc.

Chapter 10

Multi-Scale Dynamics of Tropical Cyclone Formations in an Equilibrium Simulation Using a Global Cloud-System Resolving Model

Wataru Yanase[1], Masaki Satoh[1,2], Shin-ichi Iga[3], Johnny C. L. Chan[4], Hironori Fudeyasu[5], Yuqing Wang[6] and Kazuyoshi Oouchi[2]

[1]The University of Tokyo, Kashiwa, Japan
[2]Japan Agency for Marine-Earth Science and Technology, Yokohama, Japan
[3]RIKEN, Kobe, Japan
[4]City University of Hong Kong, China
[5]Yokohama National University, Yokohama, Japan
[6]University of Hawaii at Manoa, Honolulu, Hawaii, US

Abstract

An equilibrium July simulation was attempted to examine how well a global cloud-system resolving model (GCRM) simulates the multi-scale dynamics of tropical cyclone (TC) formations. Results for the eastern North Pacific were analyzed to elucidate the capability of the model in capturing the multi-scale dynamics of TC formation in the basin where both the TCs and the large-scale environmental fields were simulated the best. The lifetimes and intensities of the simulated TCs were comparable to the best-track data, which is attributed to the model resolution marginally high enough to represent essential dynamics within a TC inner-core, such as a precipitation-free eye and eye-wall updrafts. Consistent with previous observational studies, some TCs were transformed from synoptic-scale tropical easterly waves that propagated from the North Atlantic. The model also showed meso-scale vortex dynamics related to TC genesis, such as ITCZ-breakdown, vortex merger, and Rossby-wave energy dispersion. Thus, the results of the 14-km simulation encourage us to use GCRMs for understanding the multi-scale dynamics of vortices responsible for TC formations, and provide a useful benchmark for the longer-time or higher-resolution GCRM simulations in the future.

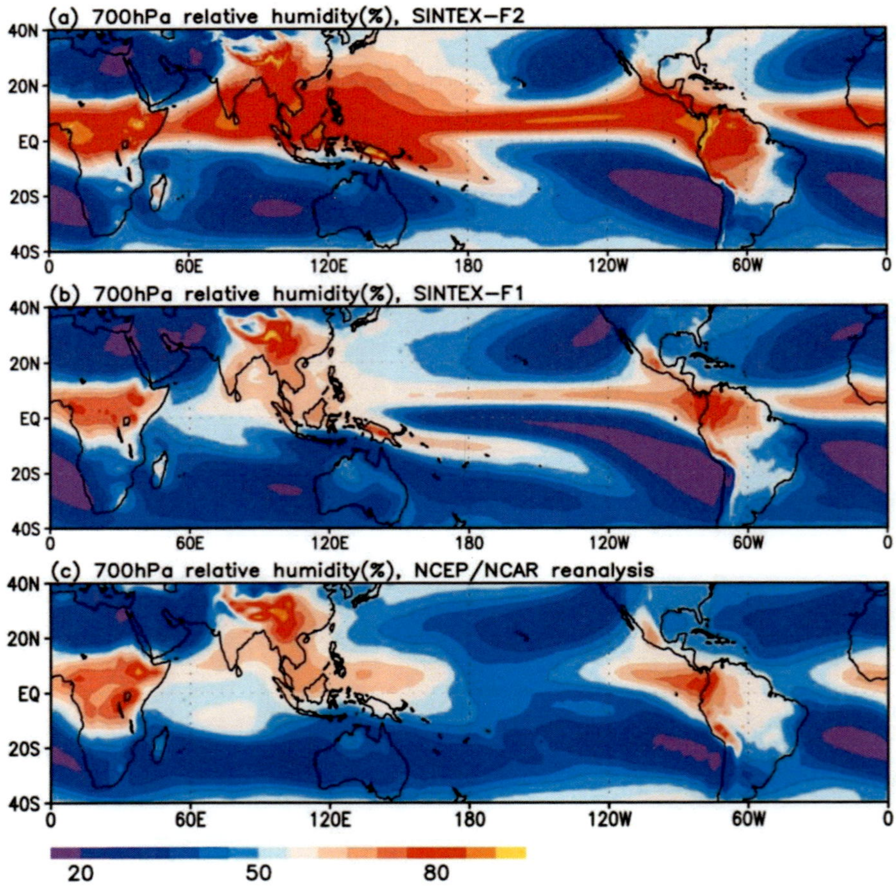

Figure 8. Climatological June-November mean of the relative humidity (%) at the 700-hPa level from the (a) SINTEX-F2, (b) SINTEX-F1, and (c) NCEP/NCAR reanalysis.

Figure 9 shows the climatological June-November mean of the magnitude of the vertical wind shear from the CGCMs and the NCEP/NCAR reanalysis. The simulated vertical wind shear, calculated as the difference of the winds between the 850- and 200-hPa levels, accords with that observed. The vertical wind shear in the eastern North Pacific simulated by the SINTEX-F2 is slightly stronger than that of the SINTEX-F1 (Figures 9a and 9b). This indicates that the eastern North Pacific cyclogenesis in the SINTEX-F2 is less than that of the SINTEX-F1. However, as we will see, the eastern North Pacific cyclogenesis in the SINTEX-F2 is greater than that of the SINTEX-F1 (Table 1). The higher low-level vorticity and mid-level humidity in the SINTEX-F2 relative to the SINTEX-F1 (Figures 7 and 8) may enhance the eastern North Pacific cyclogenesis in the SINTEX-F2.

The vertical wind shear in the North Indian Ocean simulated by the SINTEX-F1 is weaker than that of the SINTEX-F2, indicating that the North Indian TC genesis in the SINTEX-F1 may be greater than that of the SINTEX-F2. We however note that the North Indian cyclogenesis in the SINTEX-F1 is less than that of the SINTEX-F2 (Table 1). The less North Indian cyclogenesis in the SINTEX-F1 may be due to the lower relative vorticity and relative humidity when compared to the SINTEX-F2 (Figures 7 and 8).

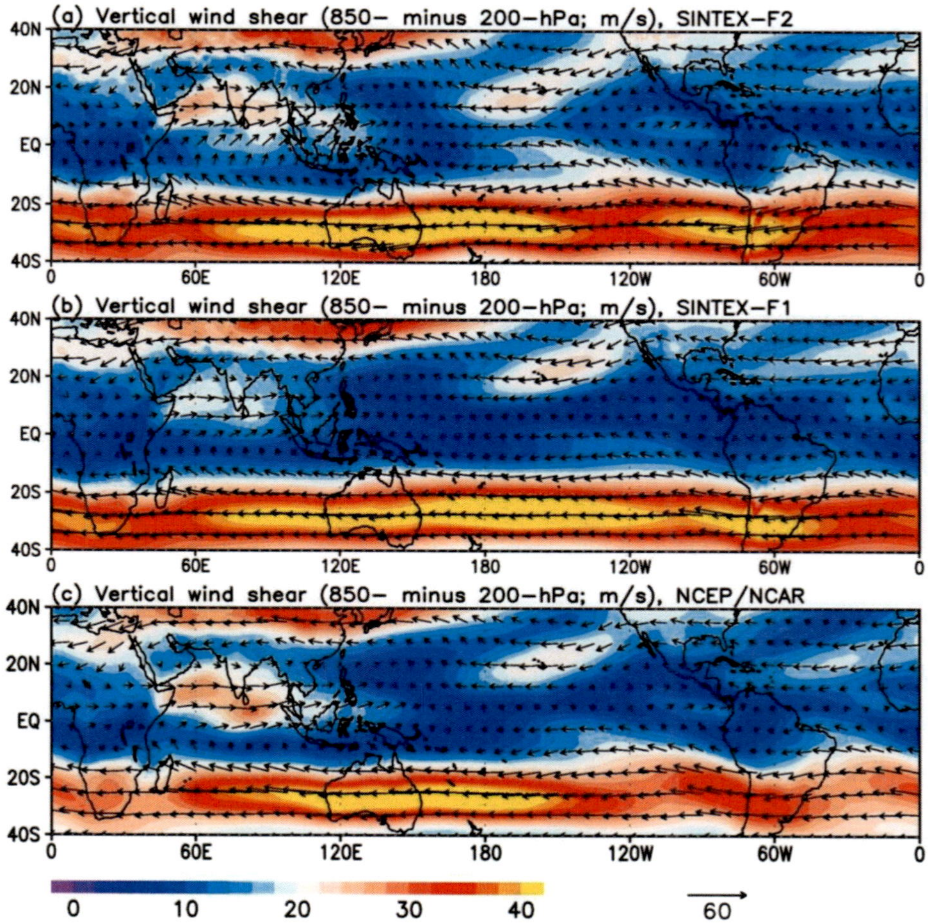

Figure 9. Climatological June-November mean of the magnitude of the vertical wind shear (color shade; ms^{-1}) and vertical wind shear vector (arrow) from the (a) SINTEX-F2, (b) SINTEX-F1, and (c) NCEP/NCAR reanalysis. The vertical wind shear is calculated as the difference in wind speed between the 850- and 200-hPa levels.

Table 1. The annual mean of observed and simulated TC genesis frequency

	Global	Western North Pacific	Eastern North Pacific	North Indian Ocean	North Atlantic
SINTEX-F2	78.0	21.5	15.5	3.7	8.3
SINTEX-F1	46.6	10.3	5.2	2.4	3.5
Observation	81.4	27.6	15.6	4.6	9.6

5. Climatological TC Genesis Frequency

In this section we examine the reproducibility of the climatological TC genesis frequency simulated by the SINTEX-F2 and SINTEX-F1.

It has been well known that there are four major TC genesis basins in the northern hemisphere: the western North Pacific, eastern North Pacific, North Indian Ocean, and North Atlantic. Table 1 shows the annual mean of the modeled and observed TC genesis frequency over the whole globe as well as the four TC basins. The annual mean of the global TC genesis frequency simulated by the SINTEX-F2 (78) is comparable to that observed (81.4). The regional TC frequency is also generally simulated by the model, although the model slightly underestimated the TC frequency in the western North Pacific and North Indian Ocean (Table 1). In contrast, the SINTEX-F1 considerably underestimates the TC genesis frequency for all TC basins.

Figure 10 shows a comparison of the horizontal distribution of the climatological annual mean of observed and modeled TC genesis frequency. The horizontal distribution of the cyclogenesis simulated by both of the CGCMs is in general agreement with that observed. In particular, the SINTEX-F2 successfully simulates the frequent TC genesis in the western North Pacific and eastern North Pacific as observed (Figures 10a and 10b). The SINTEX-F2, however, underestimates the South Indian Ocean cyclogenesis, while the SINTEX-F1 simulates well. The unsuccessful reproduction of South Indian cyclogenesis in the SINTEX-F2 may result from the less westward extension of the monsoon westerly in the South Indian Ocean during the austral summer (not shown). The weak westerly may suppress the horizontal shear which is favorable condition for TC genesis.

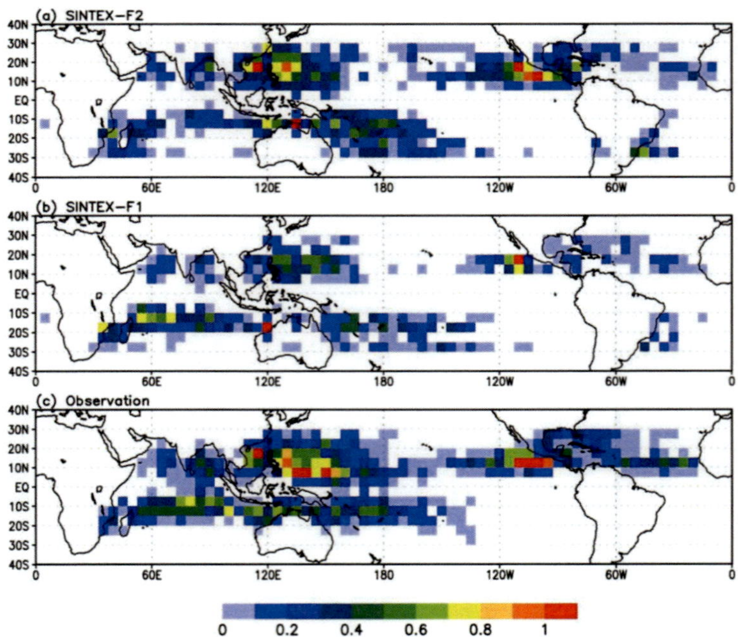

Figure 10. Horizontal distribution of the annual mean of TC genesis frequency from (a) SINTEX-F2, (b) SINTEX-F1, and (c) observation.

Although the SINTEX-F2 well simulates the annual mean of the North Atlantic TC genesis frequency as observed (Table 1), there is a difference in the horizontal distribution of the cyclogenesis between the model and observation. The North Atlantic cyclogenesis is observed primary between 10°N and 20°N (Figure 10c), termed the main development region (MDR; Goldenberg and Shapiro [39]) where African waves propagate westward from Africa across the tropical North Atlantic and Caribbean Sea. The MDR has favorable conditions for the TC genesis: positive low-level relative vorticity (Figure 7c), and weak vertical wind shear (Figure 9c). The SINTEX-F1 and SINTEX-F2 well simulate the high relative vorticity and weak vertical wind shear in the MDR (Figures 7 and 9). However, the modeled TC genesis frequency in the MDR is less than the observed (Figure 9.10). The underestimates of the North Atlantic cyclogenesis frequency is also simulated by the previous medium-resolution CGCM (Fig 17 in [40]). Our results indicate that just increasing the spatial resolution of the CGCM is not enough to improve the location of the North Atlantic cyclogenesis.

Both of CGCMs simulate the TC genesis in the South Atlantic (Figures 10a and 10b), although the cyclogenesis has not been observed in that basin (Figure 10c). The reason for the South Atlantic cyclogenesis in the CGCMs is unclear. The observed South Atlantic SST exceeds 26°C and the vertical wind shear in the basin is relatively weak during the austral summer, while the relative vorticity in the low-troposphere is positive (not shown). This suggests that the anti-cyclonic circulation suppresses the South Atlantic cyclogenesis. The model well simulates not only the warm SST and weak vertical wind shear but also the anti-cyclonic circulation in the South Atlantic. Despite the successful simulation of the South Atlantic large-scale conditions, the model simulates the TC genesis in the basin. The unrealistic South Atlantic cyclogenesis is also simulated by previous medium-resolution CGCMs (Figure 17 in [40]). Further investigation is required to understand this common bias of current models.

Figure 11 shows the seasonal cycle of the TC genesis frequency in the four TC basins in the northern hemisphere. The western North Pacific has a cyclone season during July-October (black line in Figure 11a). The increase in TC genesis during this period is generally reproduced by the two CGCMs (colored lines in Figure 11a). However, there are some discrepancies in the western North Pacific cyclogenesis between the CGCMs and the observations. The SINTEX-F2 overestimates the TC genesis frequency during November-December compared to the observations. In addition, the amplitude of the modeled seasonal cycle of the cyclogenesis is weaker than that of observed. In particular, both of CGCMs underestimate the TC frequency during the peak season (Figure 11a).

The eastern North Pacific experiences cyclones frequently during boreal summer (black line Figure 11b). The seasonal cycle of the eastern North Pacific cyclogenesis is also well simulated by the two CGCMs, although the amplitude of the modeled seasonal cycle, particularly for the SINTEX-F1, is weaker than the observed.

As shown in Figures 11a and 11b, the seasonal cycle of the TC genesis frequency in the tropical Pacific simulated by the SINTEX-F2 is closer to the observed, compared to that simulated by the SINTEX-F1. In particular, the underestimates of the TC genesis frequency in the SINTEX-F1 are much improved by the SINTX-F2.

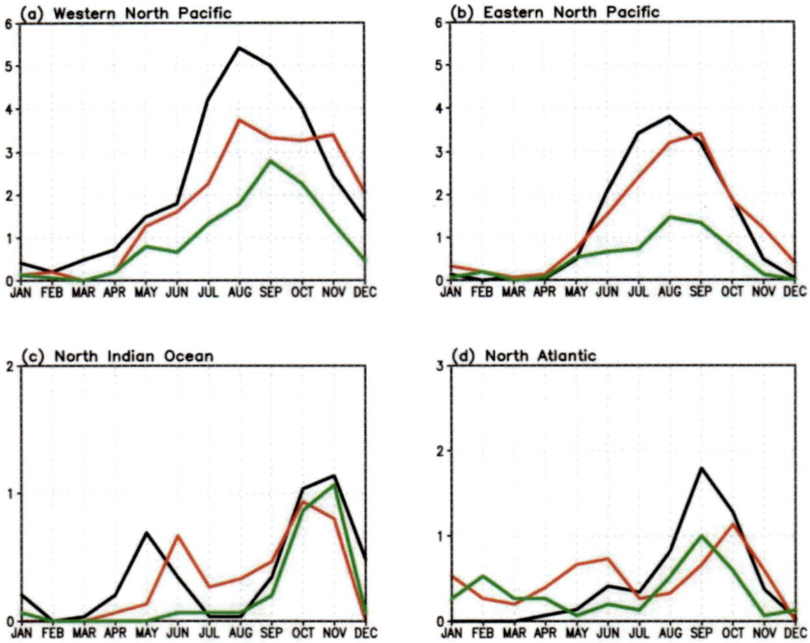

Figure 11. Climatological seasonal cycle of TC genesis frequency over the (a) western North Pacific, (b) eastern North Pacific, (c) North Indian Ocean, and (d) North Atlantic. Black, red, and green denotes the observation, SINTEX-F2, and SINTEX-F1, respectively.

The North Indian Ocean has two TC genesis peaks during the pre- and post-monsoon periods (black line in Figure 11c). The North Indian cyclogenesis reduces during the monsoon period due to the strong vertical wind shear of the monsoonal circulation [41]. Instead, weaker than TC systems known as monsoon depressions are formed during the monsoon period. The bimodal seasonal cycle of the North Indian cyclogenesis is reproduced by the SINTEX-F2 to some extent: the modeled TC frequency in June is well simulated while that in May is not (Figure 11c). Meanwhile, the SINTEX-F1 fails to simulate the North Indian cyclogenesis during the pre-monsoon period (Figure 11c). In section 6, we further discuss the possible reasons for the North Indian cyclogenesis during the pre-monsoon period from a view point of the large-scale environmental conditions.

Seasonal cycle of the North Atlantic cyclogenesis is qualitatively well simulated by the SINTEX-F1, although the modeled amplitude is weaker than that of observed (Figure 11d). In contrast, the SINTEX-F2 tends to produce the two cyclone seasons in May and September-October (red line in Figure 11d).

6. Seasonal Cycle of Environmental Conditions

As we have presented in the Section 5, the high-resolution CGCM reproduces the overall feature of the TC climatology in terms of the genesis frequency, horizontal distribution, and seasonal cycle, although there are some discrepancies between the modeled and observed TC climatology. In this section we further validate the seasonal cycle of TC genesis frequency from a viewpoint of the large-scale environmental conditions with a particular focus on the North Indian Ocean and the western North Pacific.

Figure 12c shows the climatological seasonal cycle of the observed SST averaged over the latitudinal band 5°N-25°N, together with the observed climatological TC genesis frequency summed up over the same latitudinal band. As Gray [33] revealed, most of observed TCs are generated over the region where the SST exceeds 26°C. Both of CGCMs successfully simulates TC genesis over the region where the SST exceeding 26°C (Figures 12a and 12b).

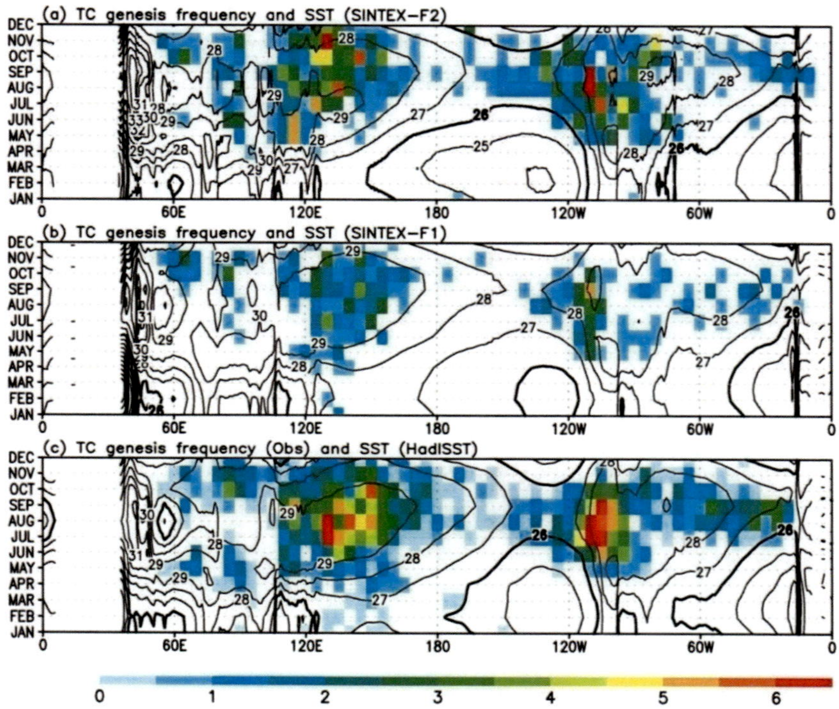

Figure 12. Seasonal cycle of the climatological TC genesis (×10; color) summed up over the latitudinal band 5°N-25°N, and SST (contour; °C) from the (a) HadISST, (b) SINTEX-F2, and (c) SINTEX-F1. TC genesis frequency for (a) is calculated from the observed TC data of the NHC and JTWC. Bold line denotes the 26°C isotherm.

A significant difference between modeled and observed seasonal cycle of the North Indian Ocean TC genesis frequency is the less modeled cyclogenesis during the pre-monsoon period (Figure 11c). Furthermore, the SINTEX-F1 simulates no TC genesis in the Arabian Sea and less TC genesis in the Bay of Bengal during the pre-monsoon period compared to that observed (Figure 12b). In contrast, the SINTEX-F2 simulates the TC genesis in the Arabian Sea and Bay of Bengal during the pre-monsoon period (Figure 12a). We however note that the modeled Arabian cyclogenesis is much less than that observed, while the TC genesis frequency in the Bay of Bengal is greater than that observed (Figures 12a and 12c). These results suggest that the bimodal seasonal cycle of the North Indian Ocean cyclogenesis in the SINTEX-F2 (Figure 11c) results from the unrealistic increase in the TC genesis in the Bay of Bengal during the pre-monsoon period.

The Arabian cyclogenesis in the SINTEX-F2 appears to lag behind by one month compared to that observed (Figures 12a and 12c). Evan and Camargo [41] showed that an early monsoon onset favors Arabian storms in May, while late monsoon onset is typically associated with June storms. This suggests that the less early monsoon onset in the model may suppress the modeled Arabian TC genesis in May.

Gray [42] identified the bimodal seasonal cycle of the cyclogenesis frequency in the Arabian Sea, and attributed the depression of the cyclogenesis in July and August to strong vertical wind shear. The contour in Figure 13c shows the seasonal cycle of the observed vertical wind shear averaged over the latitudinal band 5°N-25°N. The vertical wind shear over the Arabian Sea and Bay of Bengal is relatively weak during the pre- and post-monsoon periods (less than 15 ms^{-1}). On the other hand, the vertical wind shear is strong during the monsoon period (Figure 13c). The observed data clearly show that the bimodal seasonal cycle of the Arabian cyclogenesis during the pre- and post-monsoon periods coincides with the periods of the weak vertical wind shear over the Arabian Sea (Figure 13c). The SINTEX-F2 and SINTEX-F1 successfully reproduce the bimodal seasonal cycle of the vertical wind shear in the Arabian Sea, while the modeled Arabian cyclogenesis during the pre-monsoon period is less than that observed (Figure 13). The SINTEX-F2 simulates a few Arabian TCs (Figure 13a), while no Arabian cyclogenesis during the pre-monsoon period is simulated in the SINTEX-F1 (Figure 13b). These results suggest that the vertical wind shear does not explain the less Arabian cyclogenesis during the pre-monsoon period in the CGCMs.

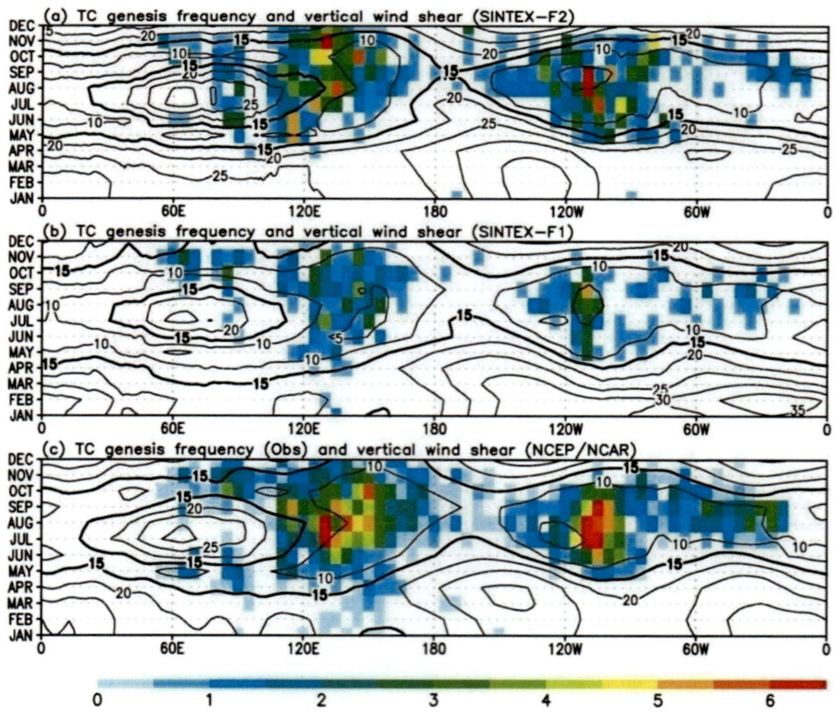

Figure 13. As in Figure 12, but for the vertical wind shear (contour; ms^{-1}) calculated as the difference in the wind speed between the 850- and 200-hPa levels averaged over the latitudinal band 5°N-25°N from the (a) NCEP/NCAR reanalysis, (b) SINTEX-F2 and (c) SINTEX-F1. Bold line denotes the 10 ms^{-1}.

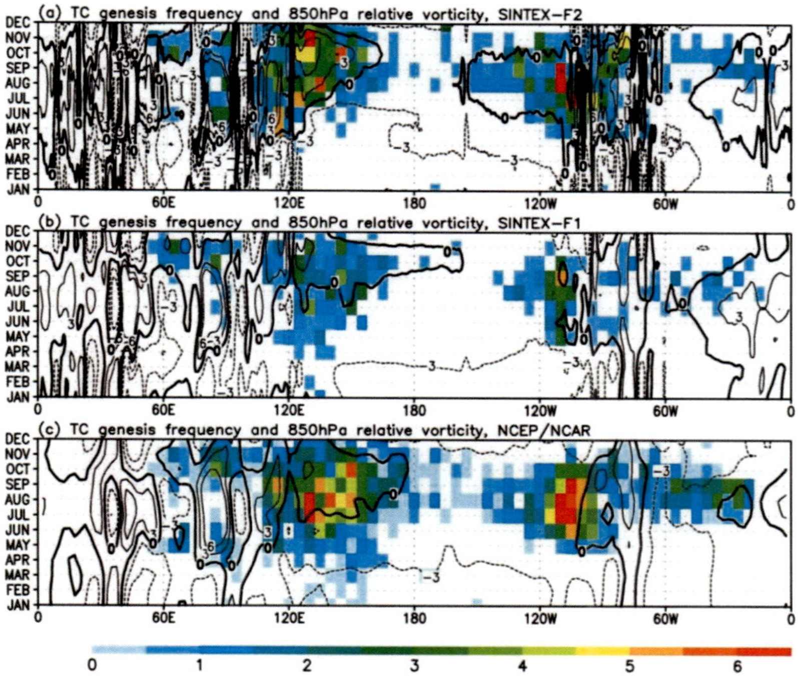

Figure 14. as in Figure 12, but for the relative vorticity at the 850-hPa level ($\times 10^{-6}$ s^{-1}; contour) from the (a) NCEP/NCAR reanalysis, (b) SINTEX-F2 and (c) SINTEX-F1. Bold denotes the zero line.

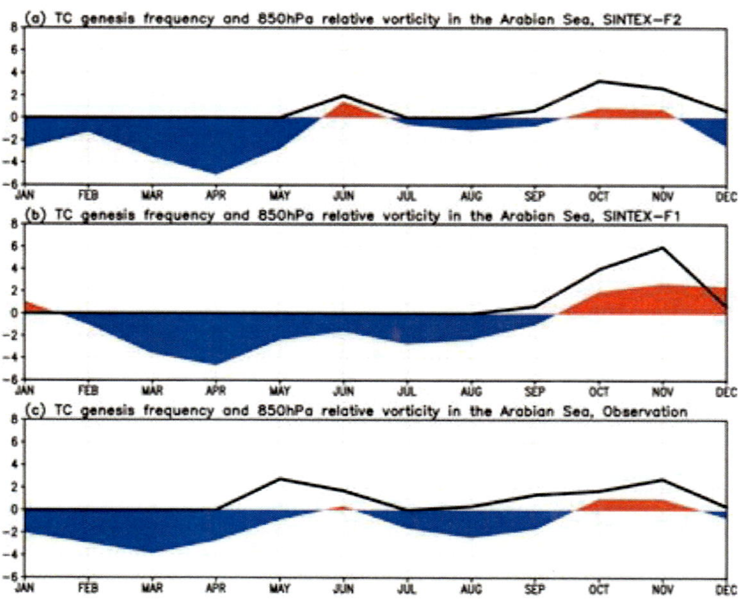

Figure 15. Climatological annual cycle of TC genesis frequency ($\times 10$; black line) and the relative vorticity ($\times 10^{6}$ s^{-1}; color) at the 850-hPa level over the Arabian Sea from the (a)SINTEX-F2, (b) SINTEX-F1, and (c)observations. Red and blue denote the positive and negative relative vorticity, respectively.

The low-level cyclonic circulation is also related to the Arabian cyclogenesis activity. Color shade in Figure 14a shows the climatological seasonal cycle of the observed relative vorticity at the 850-hPa level averaged over the latitudinal band 5°-25°N. In the Arabian Sea, there is a good agreement with the increase in observed TC genesis frequency and the enhanced low-level cyclonic circulation (Figure 14c). In contrast, the positive vorticity during the pre-monsoon period in the SINTEX-F2 is ambiguous, while during the post-monsoon period it is significant (Figure 14a). Figure 15 shows the seasonal cycle of the modeled and observed relative vorticity at the 850-hPa level in the Arabian Sea. The relative vorticity from the observation and SINTEX-F2 has the bimodal peaks in the pre- and post-monsoon periods (Figures 15a and 15c). The less Arabian TC genesis during the pre-monsoon period in the SINTEX-F2 may result from the weak positive low-level vorticity in the basin. In case of SINTEX-F1, there is no significant positive vorticity during the pre-monsoon period (Figure 15b). No Arabian cyclogenesis during the pre-monsoon period in the SINTEX-F1 agrees with the low-level anti-cyclonic circulation in the Arabian Sea during this period.

For the western North Pacific, although the high-resolution CGCM well simulates the climatology of TC genesis, the modeled TC genesis frequency is underestimated (Table 1). In addition, if looking carefully, it turns out that modeled TC genesis locations are too much confined to east of the Philippines compared to the observations (Figure 12). Observed data shows that the western North Pacific cyclogenesis locations extend eastward up to the international dateline during the peak season (Figure 12c). During the same period, the monsoon-trough characterized by the weak vertical wind shear and low-level cyclonic circulation also expands eastward to the dateline during the peak season (Figures 13c and 14c). The eastward extension of the weak vertical wind shear is generally simulated by both of the CGCMs. However, the SINTEX-F1 simulates the unrealistic eastward extension of the positive low-level vorticity zone during the boreal autumn (Figure 14b). The unrealistic eastward extension of the positive vorticity zone during boreal autumn simulated by the SINTEX-F1 is improved by the SINTEX-F2. The SINTEX-F2 successfully simulates the eastward extensions of the weak vertical wind shear and low-level cyclonic circulation (Figures 13a and 14a). However, the SINTEX-F2 fails to simulate the eastward extension of the monsoon-trough up to the international dateline, compared to that of the observed. In particular, observed positive vorticity extends to the large portion of the western North Pacific during July-October, while the modeled positive vorticity zone does not extend as far east as the observed (Figures 14a and 14c). The modeled weak vertical wind shear also does not extend as far east as the observed during the TC peak season compared to that of the observed (Figures 13a and 13c). Thus, the less eastward extensions of the low-level cyclonic circulation and the weak vertical wind shear in the SINTEX-F2 lead to the less TC genesis west of the dateline. The less modeled TC genesis in the southeastern part of the western North Pacific may lead to the less western North Pacific TC genesis frequency (Figure 10).

Figure 14 shows that the observed low-level cyclonic circulation in the western North Pacific extends eastward rapidly in July, while the rapid eastward extension of the cyclonic circulation is not well simulated by the both of CGCMs. The rapid eastward extension of the low-level cyclonic circulation may be related to a rapid increase in the western North Pacific cyclogenesis in July (black line in Figure 11a).

7. TC INTENSITY

Bengtsson et al. [9] showed that higher horizontal resolution AGCM simulates more frequent occurrence of intense TCs. In this section we compare the frequency distribution of the simulated TC surface wind speed between the SINTEX-F2 and SINTEX-F1.

Figure 16 shows a comparison of the modeled and observed frequency distribution of TC surface wind speed in the four TC basins. Over all the TC basins, the TC wind speed simulated by the SINTEX-F1 is less than 20~30 ms^{-1}, while that simulated by the SINTEX-F2 exceeds 30~40 ms^{-1}. The frequency of high wind speed (> 20 ms^{-1}) simulated by the SINTEX-F2 is much greater than that simulated by the SINTEX-F1, while the frequency of the low wind speed (< 15 ms^{-1}) of the SINTEX-F2 is less than that of the SINTEX-F1. As a result, the frequency distribution of TC surface wind speed simulated by the high-resolution CGCM is closer to that of observation compared to the medium-resolution CGCM.

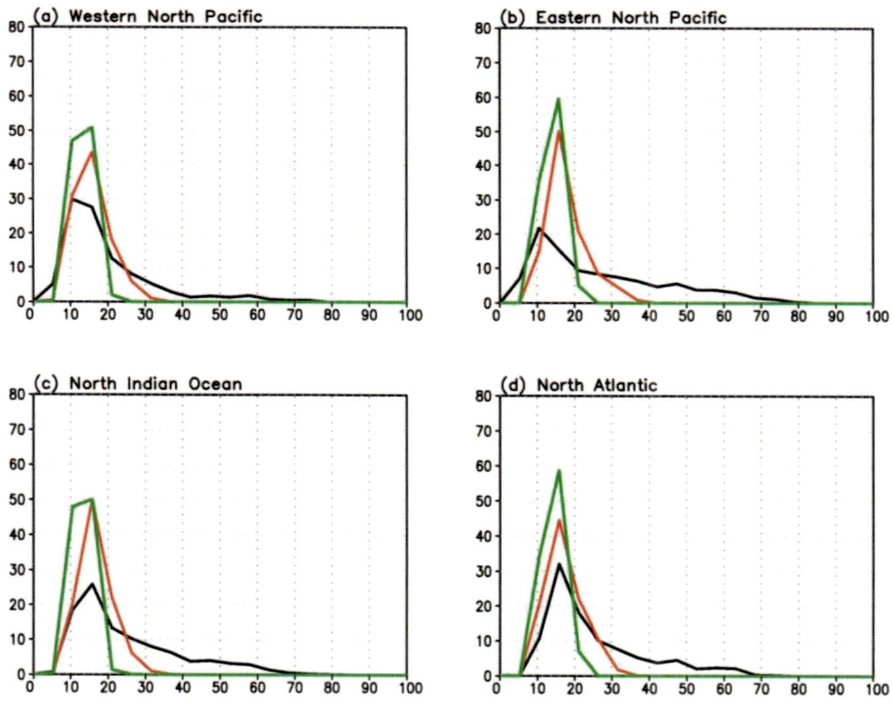

Red and green denote the SINTEX-F2 and SINTEX-F1, respectively. Black shows the observation.

Figure 16. Normalized frequency distribution of modeled and observed surface wind speed of TCs. The unit for the vertical and horizontal axis is % and ms^{-1}, respectively.

However, the modeled TC intensity even with the high resolution (about 40 km atmosphere) is far from the success. The modeled TC wind speed is less than ~40 ms^{-1}, while the observed one exceeds 60 ms^{-1} (Figure 16). In particular, the difference in the modeled and observed wind speed is large at the mature stage of the TCs. In fact, there is a large difference in the surface wind speed at the latitudinal band of 10°N-30°N where TCs develop at its mature stage (not shown). There may be several reasons for the weak TCs in the model compared to those observed. The observed surface wind speed is the maximum surface wind speed for 10 minutes at 6-hour interval, while the modeled surface wind speed is the 6-hour average of the surface wind speed. This suggests that modeled surface wind speed may be reasonably underestimated due to the time-average compared to the observed maximum values. Besides, the model analysis period (15-yr) may be too short to exactly compare with the 29-yr observed TC data. However, the most important factor for the underestimated TC intensity could be due to the model resolution. The horizontal resolution of the SINTEX-F2 (T319) is still too coarse to resolve the inner structure of TCs; the spatial average of surface wind at model grid box may also reduce the simulated TC wind, compared to point-measurement of the observed.

SUMMARY AND DISCUSSION

We have examined the global high-resolution CGCM consisting of the T319 AGCM and eddy-permitted OGCM in terms of the climatological TC activity as well as the large-scale environmental conditions favorable for the TC genesis. The overall feature of the climatological TC genesis frequency as well as the environmental conditions is reasonably reproduced by the CGCM. The global TC genesis frequency simulated by the high-resolution CGCM is much closer to the observed, compared that simulated by the medium-resolution (T106) CGCM. In addition, the high-resolution CGCM partially reproduces the bimodal seasonal cycle of the North Indian Ocean cyclogenesis, while the medium-resolution CGCM fails to simulate it. There are, however, still some discrepancies between the modeled and observed TC activity. For instance, although the high-resolution CGCM partially reproduced the bimodal seasonal cycle of the North Indian Ocean cyclogenesis, the model failed to reproduce the Arabian cyclogenesis during the pre-monsoon period. The western North Pacific TC genesis locations are found to be confined in the southwestern sub-domain of the western North Pacific. We discussed the cause of the discrepancy from the view point of the environmental conditions. It has been found that the low-level cyclonic circulation in the Arabian Sea during the pre-monsoon period is relatively weak compared to that observed, although the vertical wind shear is well simulated as observed. For the western North Pacific, we have found that the modeled monsoon-trough associated with the low-level cyclonic circulation and weak vertical wind shear does not extend eastward far enough compared to that observed. These results suggest that the better reproduction of the large-scale conditions is indispensable for the successful simulation of TC activity. The TC activity can also be simulated reasonably well by the medium-resolution model. Compared to the coarse-resolution model, one of the advantages of the high-resolution CGCM is the reproduction of the intense TC. Surface wind speed exceeding 30~40 ms^{-1} is successfully simulated by the high-resolution CGCM, while the TC wind speed simulated by the medium-resolution CGCM is less than 20~30 ms^{-1}.

We again note that a 5-yr model spin-up for the SINTEX-F2 is insufficient to reach the equilibrium states. Further integration should be needed for the complete simulation and analysis.

Further realistic simulations of TC intensity should be necessary for both the better reproduction of the mean-state of climate and societal applications. Fedorov et al. [43]

showed that enhanced vertical mixing associated with enhanced TC activity leads to mix the subsurface ocean waters in the extra-tropical Pacific and thereby increase the temperature of subsurface waters that have been subducted off the coast of California and Chile. This eventually leads to a warming of the Pacific cold tongue SST which in turn enhances TC activity in the region. This ocean-TC feedback hypothesis indicates that the realistic simulation of TC intensity may change the climate conditions as well as TC activity in CGCMs.

Bengtsson et al. [9] showed more frequent intense TC would occur under the global warming condition. Correctly reproducing the TC strength with higher-resolution CGCM would be beneficial to not only TC simulation but also the societal application. Previous studies identified that the U.S. societal damage by a hurricane is proportional to the 3rd-9th power of wind speed at landfall [44]. Although there is uncertainty about how changes in wind speed are related to corresponding changes in the societal damage, the societal damage is no doubt sensitive to the TC intensity. The correct estimation of the TC intensity will be required to make a quantitative assessment of societal damage from TCs, particularly for the assessments of relatively long-term (seasonal-decadal-century timescale) climate impact on society.

REFERENCES

[1] R. A. Madden, P. R. Julian, *Mon. Wea. Rev.*, 122, 814 (1994).
[2] J. C. L. Chan, *Mon. Wea. Rev.*, 113, 599 (1985).
[3] H.-M. Kim, P. J. Webster and J. A. Curry, *Science*, 325, 77 (2009).
[4] H.-M. Kim, P. J. Webster, and J. A. Curry, *J. Climate*, 24, 1839 (2011).
[5] J. P. Kossin and D. J. Vimont, *Bull. Amer. Meteor. Soc.*, 88, 1767 (2007).
[6] J. W. Hurrell, *Science*, 269, 676 (1995).
[7] O. P. Singh, *Current Science*, 94, 29 (2008).
[8] K. Oouchi, J. Yoshimura, H. Yoshimura, R. Mizuta, S. Kusunoki, and A. Noda, *J. Meteorol. Soc. Japan*, 84, 259 (2006).
[9] L. Bengtsson, K. I. Hodges, M. Esch, N. Keenlyside, L. Kornblueh, J.-J. Luo, and T. Yamagata, *Tellus*, 59A, 539 (2007).
[10] M. A. Bender and I. Ginis, *Mon. Wea. Rev.*, 128, 917 (2000).
[11] F. Vitart, and T. N. Stockdale, *Mon. Wea. Rev.*, 129, 2521 (2001).
[12] T. Matsuura, M. Yumoto, and S. Iizuka, *Clim. Dym.*, 21, 105 (2003).
[13] S. Iizuka and T. Matsuura, *Clim. Dyn.*, 30, 815 (2008).
[14] C.-C. Wu, C.-Y. Lee, and I.-I. Lin, *J. Atmos. Sci.*, 64, 3562 (2007).
[15] J.-J. Luo, S. Masson, E. Roeckner, G. Madec, and T. Yamagata, *J. Climate*, 18, 2344 (2005).
[16] E. Roeckner, and Coauthors, "The atmospheric general circulation model ECHAM5, Part I: Model description", Max-Planck-Institut für Meteorologie, Rep. 349, 127 pp (2003).
[17] M. Tiedtke, *Mon. Wea. Rev.,* 117, 1779 (1989).
[18] J. -P. Schulz, L. Dumenil, and J. Polcher, *J. Appl. Meteorol.*, 40, 642 (2001).

[19] G. Madec, NEMO ocean engine, Note du Pole de Modelisation, Institut Pierre-Simon Laplace (2006).

[20] T. Fichefet, M. A. Morales Maqueda, *J. Geophys. Res.*, 102, 12609 (1997).

[21] S. Valcke, A. Caubel, R. Vogelsang, D. Declat, OASIS3 ocean atmosphere sea ice soil user's guide, Technical Report TR/CMGC/04/68, CERFACS, Toulouse, France (2004).

[22] S. Levitus, Climatological atlas of the world ocean, Rep. NOAA Prof. Pap. 13 (1982).

[23] E. Roeckner and Coauthors, "The atmospheric general circulation model ECHAM-4: Model description and simulation of present-day climate", Max-Plank-Institut fur Meteorologie Rep. 218, 90pp (1996).

[24] G. Madec, P. Delecluse, and C. Levy, "OPA 8.1 ocean general circulation model reference manual, LODYC/IPSL Tech. Rep. Note 11, 91pp (1998).

[25] S. Valcke, L. Terray, and A. Piacentini, "The OASIS coupler user guide version 2.4", CERFACS Tech. Rep. TR/CGMC/00-10, 85pp (2000).

[26] E. Kalnay and Coauthors, *Bull. Amer. Meteor. Soc.*, 77, 437 (1996).

[27] N. A. Rayner and Coauthors, *J. Geophys. Res.*, 108, D14, 4407 (2003).

[28] K. J. E. Walsh, M. Fiorino, C. W. Landsea, K. L. Mcinnes, *J. Climate*, 20, 2307 (2007).

[29] W. M. Frank, *Mon. Wea. Rev.*, 105, 1119 (1977).

[30] J. F. Price, *J. Phys. Oceanogr.*, 11, 153 (1981).

[31] M. A. Bender, I. Ginis, and Y. Kurihara, *J. Geophys. Res.*, 98, 23245 (1993).

[32] T. Dickey and Coauthors, *Mon. Wea. Rev.*, 126, 1195 (1998).

[33] W. M. Gray, "Tropical cyclone genesis", Department of Atmospheric Science paper no. 234, Colorado State University, Fort Collins, CO, 121pp (1975).

[34] W. M. Gray, *J. Meteorol. Soc. Japan,* 55, 465 (1977).

[35] J.-L. Lin, *J. Climate*, 20, 4497 (2007).

[36] P. A. Harr, and R. L. Elsberry, *Mon. Wea. Rev.*, 123, 1225 (1995).

[37] K. E. Trenberth, and C. J. Guillemot, *Clim. Dyn.*, 14, 213 (1998).

[38] H. Annamalai, J. M. Slingo, K. R. Sperber, and K. Hodges, *Mon. Wea. Rev.*, 127, 1157 (1999).

[39] S. B. Goldenberg and L. J. Shapiro, *J. Climate.*, 9, 1169 (1996).

[40] S. Yokoi, Y. Takayabu, and J. C. L. Chan, *Clim. Dyn.*, 33, 665 (2009).

[41] A. T. Evan, and S. J. Camargo, *J. Climate*, 24, 140 (2011).

[42] W. M. Gray, *Mon. Wea. Rev.*, 135, 3990 (2007).

[43] A. V. Fedorov, C. M. Brierlrey, and K. Emanuel, *Nature*, 463, 1066 (2009).

[44] R. A. Pielke Jr., *Phil. Trans. R. Soc. A*, 365, 2717 (2007).

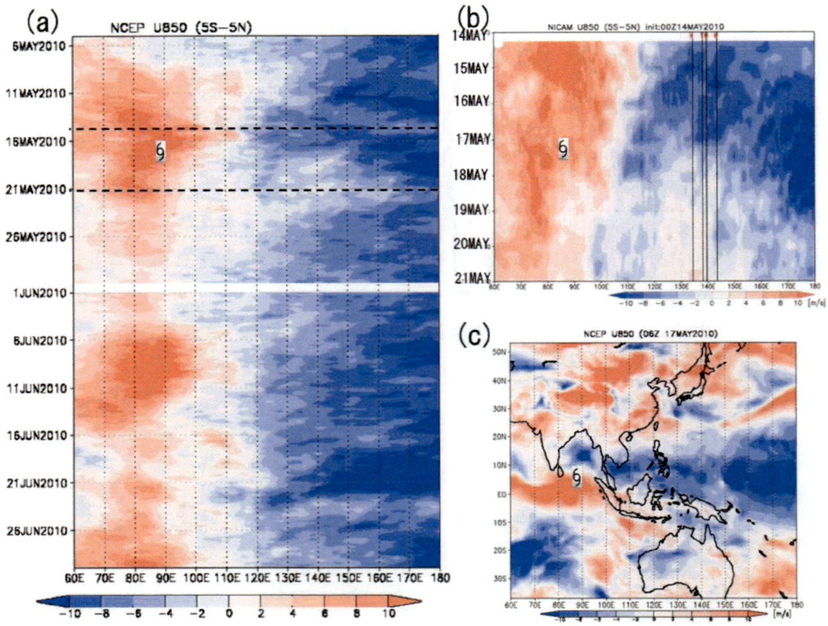

Figure 6. Hovmoller diagram of zonal velocity at 850 hPa averaged over 5^0S - 5^0N for (a) observation from NCEP-NCAR Reanalysis from May 6th to June 30th, (b) NICAM forecast from May 14th to 21st; (c) Horizontal distribution of zonal velocity [ms^{-1}] at 06Z, May 17th, 2010.

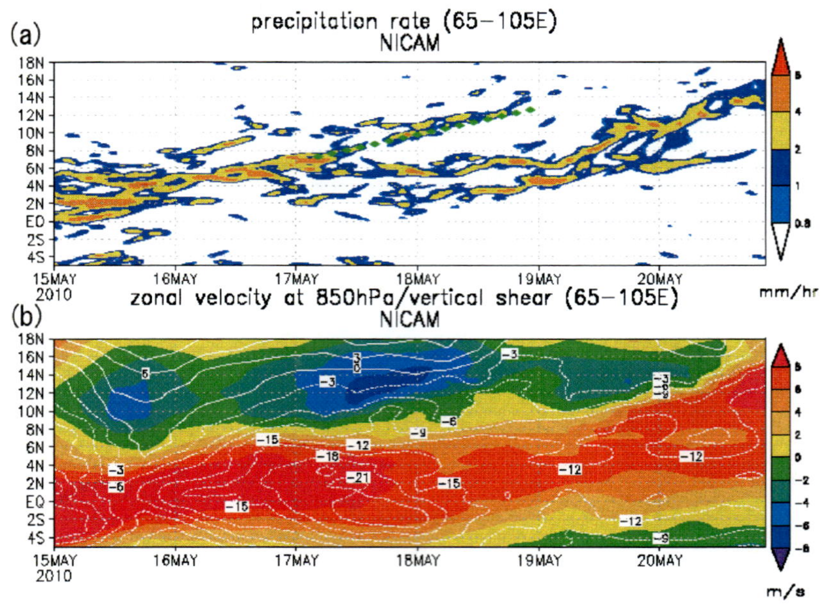

Figure 7. Time-meridional section for zonally averaged (65-105^{0}E) (a) precipitation rate, and (b) zonal velocity at 850hPa (shade) and vertical shear (line) as defined horizontal velocity at 200 hPa minus that at 850 hPa. Green dotted line in (a) denotes signal corresponding to TC LAILA.

1. INTRODUCTION

Recent progress of computing technology has made it possible to afford cloud-system resolving (~10-km-grid) simulations using a global atmospheric model. The global cloud-system resolving model (GCRM) has been shown to be able to simulate TCs which formed within large-scale atmospheric dynamics in the absence of any artificial lateral boundaries [1-4]. In previous numerical studies of actual TC events, observational data were used as initial conditions. With further increases in computing capability, GCRMs can be used to study the climatology and inter-annual variability of TCs and to assess the impact of global warming on TCs. Since the initial information would be lost in such long time simulations, a GCRM has to be able to represent all of the dynamics associated with TC formations by itself. Although the simulation longer than several years is difficult at present even for the current state-of-the-art computer, a month-long equilibrium simulation can give us useful information as a first effort. Although Yamada et al. [5] performed 5-month simulations using a 14-km GCRM to examine the TC distribution under the present climate and global warming, it focused little on the individual dynamics of TC formations.

This chapter presents results from a month-long equilibrium simulation to examine how well a GCRM at about 14-km resolution simulates the multi-scale dynamics of TC formations. It should be emphasized that all of the atmospheric process is simulated in the global equilibrium experiment, which is different from the short-time simulation using observational data for the initial condition, and is also different from the regional simulation using observational data for the boundary condition. The following two questions will be addressed: (1) does the model reproduce the lifecycles of TCs with realistic intensity and structure in the simulated large-scale environment? and if so, (2) what is the multi-scale dynamics associated with TC formation in the simulation?

2. METHODOLOGY

A global non-hydrostatic model, NICAM, employs an icosahedral-based quasi-uniform grid system [6]. The horizontal grid interval is about 14 km. The model has 40 vertical layers with the model top at 38 km in a terrain-following coordinate. The cloud microphysics scheme with simple liquid/ice categories developed by Grabowski [7] was used to represent all moist and convective processes explicitly without the use of any cumulus parameterization in the model. Note that some previous global (or large domain) models of 10~20-km-grid obtained slightly better TC simulations when they did not use any cumulus parameterization [2, 8]. Paulis and Garner [9] showed some encouraging insensitivity of the largest convective cores to resolution up to 16 km in a radiative-convective equilibrium simulation. Therefore, we decided to use no cumulus parameterization in the 14-km-grid GCRM this time, although we do not think it is a perfect approach. It should be also noted that the 14-km-grid NICAM simulated the large-scale organization of the cloud systems of MJO, TCs and meso-scale convective features such as the diurnal cycle of precipitation [10-12].

In the perpetual July experiment, the following boundary conditions were fixed for climatological July mean: sea surface temperature, sea ice, and solar insolation. After a 120-day spin-up period, the simulation reached a quasi-equilibrium state where most of the information of initial conditions seems to be lost. The following 30-day equilibrium simulation was used in this chapter. The detailed experimental design was given as CS4L100 experiment in the report by Iga [13].

In most of the previous studies, the definition of a TC was arbitrarily adjusted such that the total global number of TCs in the model became close to the observed. Walsh et al. suggested that objective criteria of TCs should be used to diagnose the actual model performance [14]. They showed that the horizontal resolution of about 10 km was sufficient to capture the TC intensity even without any extra artificial definition. Therefore, we examined how the non-artificial threshold works for TC detection in our global simulation by employing the operational wind speed threshold (~17.5 ms^{-1} at 10 m above the sea level). To exclude the extratropical cyclones and short-lived disturbances in the analysis, additional thresholds were used to define a TC in this study: a warm core structure (temperature anomaly more than 2 K between 700 hPa and 300 hPa), tropical origins of TC formations (30°S -30°N), and lifetime of more than 2 days. The simulated tracks are compared with the best-track data provided on the web-site of the Unisys Corporation (http://weather.unisys.com/hurricane/). The five July (155-days from 1995 to 1999 which include both El Nino and La Nina years) TCs over the globe was used as July observations for a proxy of climatology.

3. RESULT

The model reproduced TC formations over the eastern North Pacific the best among the tropical oceans. Figure 1 show the TC tracks (black curves) in the 30-day simulation and in the 155-day best-track observation. Since the five-year July TC data were used for the observational tracks to remove the interannual variability, the number of TCs in Figure 1b is expected to be about five times that in Figure 1a if the model reproduces the realistic TC number. In the 30-day simulation, several TCs over the eastern North Pacific formed off the west coast of North America and moved westward, in good agreement with the observations. Figure 1 also shows distributions of the genesis potential index (GPI [15]; color shading) in the 30-day simulation and that from 20-year July reanalysis data of ERA40 [16], respectively. The GPI considers the following four environmental factors responsible for cyclogenesis: the lower-tropospheric absolute vorticity, mid-tropospheric relative humidity, potential intensity, and vertical shear. The pattern of the simulated GPI over the eastern North Pacific is similar to the observation: the large GPI extends westward from North America to the Pacific along ~10°N with decreasing magnitude. There was a bias in the genesis locations over the western North Pacific, and significant underestimation over the North Atlantic (not shown). These characteristics are consistent with the bias in the simulated GPI, which results from the humidity bias related to the monsoon over the western North Pacific and vertical shear bias related to Walker circulation over the North Atlantic. These results indicate that a correct simulation of the large scale fields is necessary to simulate TC formations. Therefore, we present the detailed dynamics of TC formations over the eastern North Pacific in this chapter.

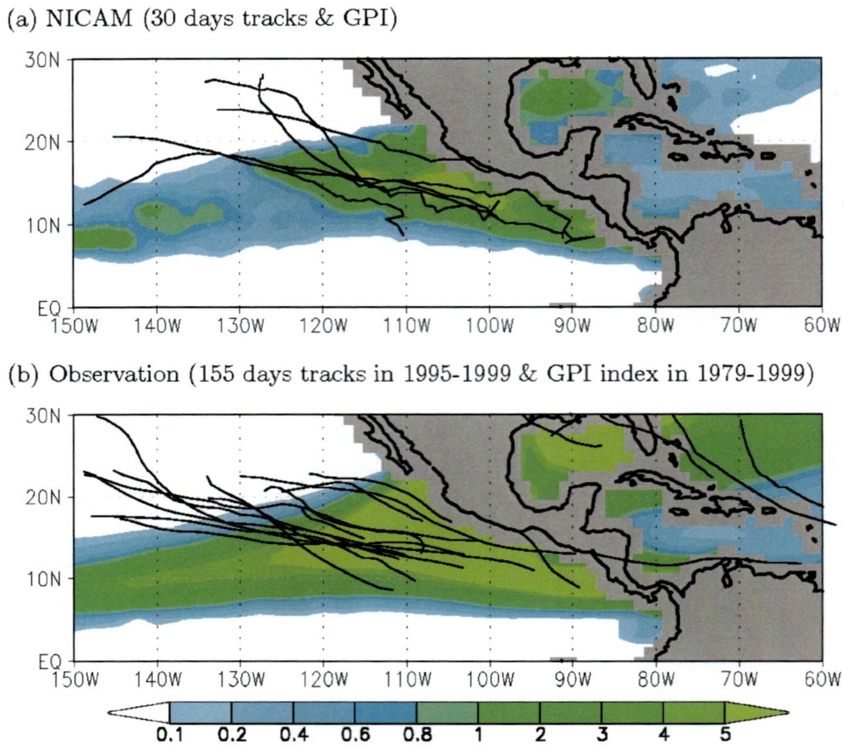

Figure 1. Tracks of TCs (black curves) and GPI (shading) over the eastern North Pacific calculated from the given period-mean fields. (a) NICAM simulation. (b) Observation.

Next, we compare the lifetime and intensity of TCs over the eastern North Pacific in the simulation with those in the five-July best-track for reference (Figure 2). The simulated lifetimes were about one week with a variance of a few days, which is in good agreement with those in the best-track. Note that the average and standard deviation of the life time of the observed tropical storms that lasted more than 2 days are 5.5 and 2.4 days, respectively, for the July cases between 1991 and 2010. The simulated maximum wind speeds of ~50 ms^{-1} at most were also comparable to the best-track although some underestimations are evident in the simulation. Since the simulated maximum wind speeds were much larger than the operational threshold of ~17.5 ms^{-1}, the TC tracking algorithm worked well even without any artificial definitions (Figure 1a). In order to test how well the lower-resolution data can capture the TC intensity, we constructed a 112-km-grid dataset by averaging the 8 × 8 grids of the original 14-km-grid dataset and recalculated the maximum wind speeds (the blue lines in Figure 2). It should be noted that this coarse-grid dataset is different from the result of a "coarse-grid simulation," because the latter may not represent the realistic dynamics of TC development. Now the maximum wind speeds reduced to ~30 ms^{-1} at most, and some TCs only marginally exceeded the wind speed threshold. This demonstrates that the TC intensity in ~10-km grid spacing is remarkably comparable to the observation; the ordinary ~100-km grid spacing is too coarse to resolve the TC intensity, even if the sub-grid-scale dynamics can be parameterized in some way.

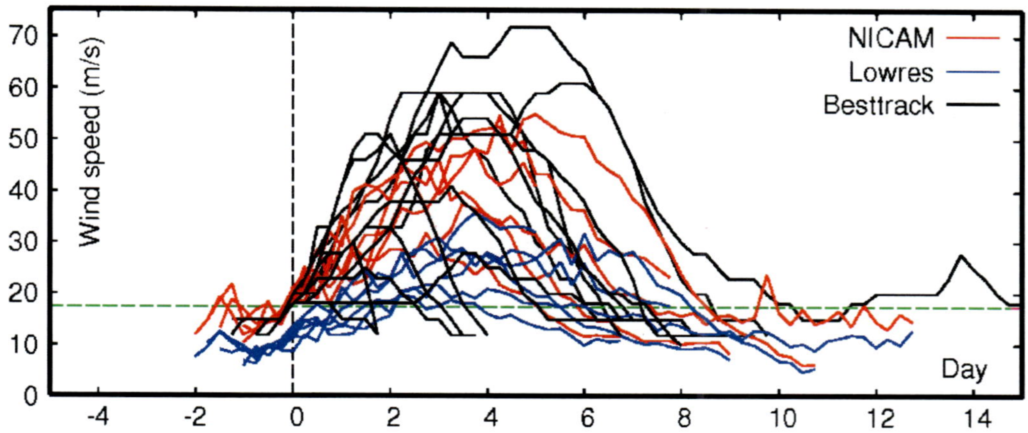

Figure 2. Time series of the cyclone's low-level maximum wind speed over the eastern North Pacific in the NICAM simulation (red lines) and in the best-track (black lines). The cyclone's maximum wind speed of low resolution NICAM data (blue lines) was calculated by averaging the 64 (8 × 8) grids of the original model output. The horizontal axis is the time relative to the cyclogenesis which was defined by the low-level maximum wind speed exceeding 34 kt (~ 17.5 m s^{-1}; a green dashed line).

The acceptable TC intensities simulated with 14-km-grid result from the realistic representation of the inner-core structure of TCs. Figure 3 illustrates the structure of a mature TC simulated over the eastern North Pacific. The outgoing longwave radiation and precipitation (Figure 3a) showed that the model resolved the spiral cloud pattern and a precipitation-free eye at the TC center. The radial-vertical cross-section of the azimuthally-averaged tangential wind field (Figure 3b) shows the radius of maximum wind of about 50 km in the simulated TC. The vertical wind field (Figure 3c) shows strong updrafts in the eyewall near the radius of maximum wind and weak subsidence in the central precipitation-free eye. Therefore, the 14-km-grid model resolved the essential dynamics within a TC inner-core reasonably well, and thus simulated the TC intensity comparable to the observation.

One of the merits to use a GCRM for long simulation is the absence of horizontal boundaries in the computation domain so that the origin of TCs can be retraced without any ambiguity. Figure 4 shows the longitude-time Hovmöller diagram of the anomalous meridional wind at 700 hPa (anomalies from the 30-day mean). It can be seen that some cyclogenesis over the eastern North Pacific were triggered by synoptic-scale tropical easterly waves, which had their origin in North Africa and propagated all the way westward across the North Atlantic and North America and finally arrived at the eastern North Pacific. This relation between tropical easterly waves and cyclogeneses is quite similar to the observational analysis [17]. It should be noted that the model reproduced the observed characteristics of faster wave propagation over the North Atlantic than over the eastern North Pacific [17].

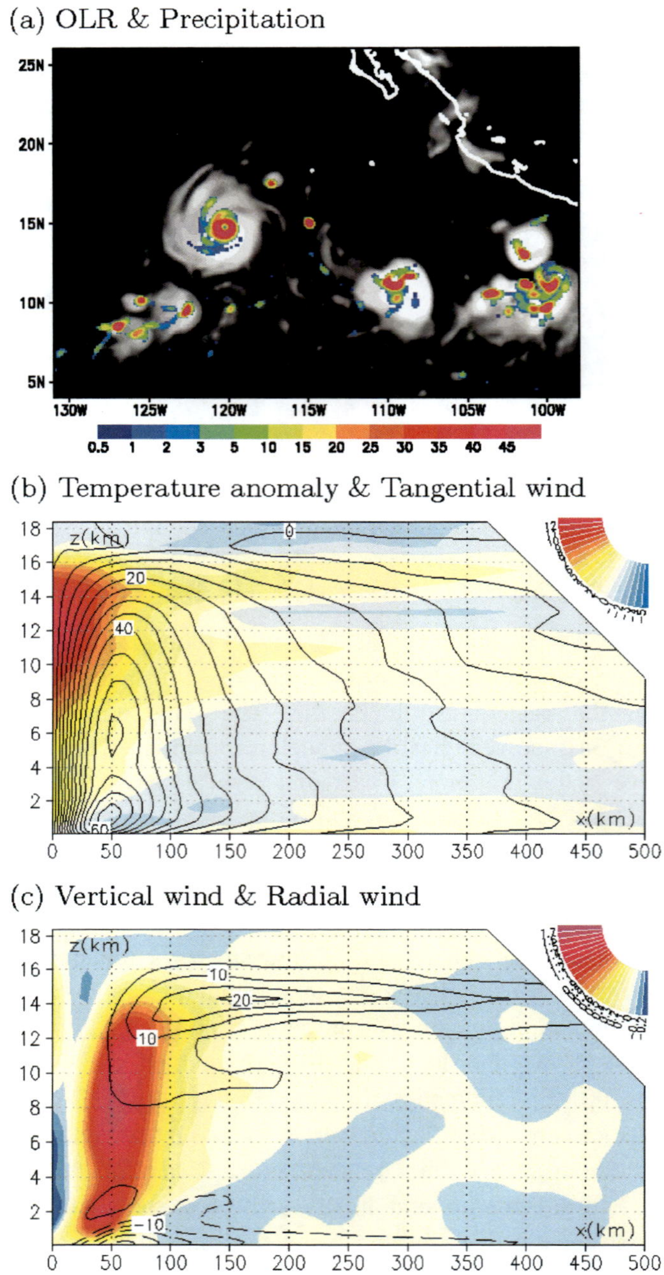

Figure 3. **Structure of a simulated TC over the eastern North Pacific. (a) Outgoing longwave** radiation (gray) and precipitation (color; the unit is mm h^{-1}) at the mature stage (at 1800UTC on the 26th day). (b) Radial-vertical cross-section of the azimuthal mean temperature anomaly (shade; the unit is K), and tangential wind (black contours with contour interval of 5 m s^{-1}). (c) Radial-vertical cross-section of azimuthal mean vertical wind (shade; the unit is m s^{-1}), and radial wind (black contours with contour interval of 5 m s^{-1}; the contour of 0 ms^{-1} is omitted).

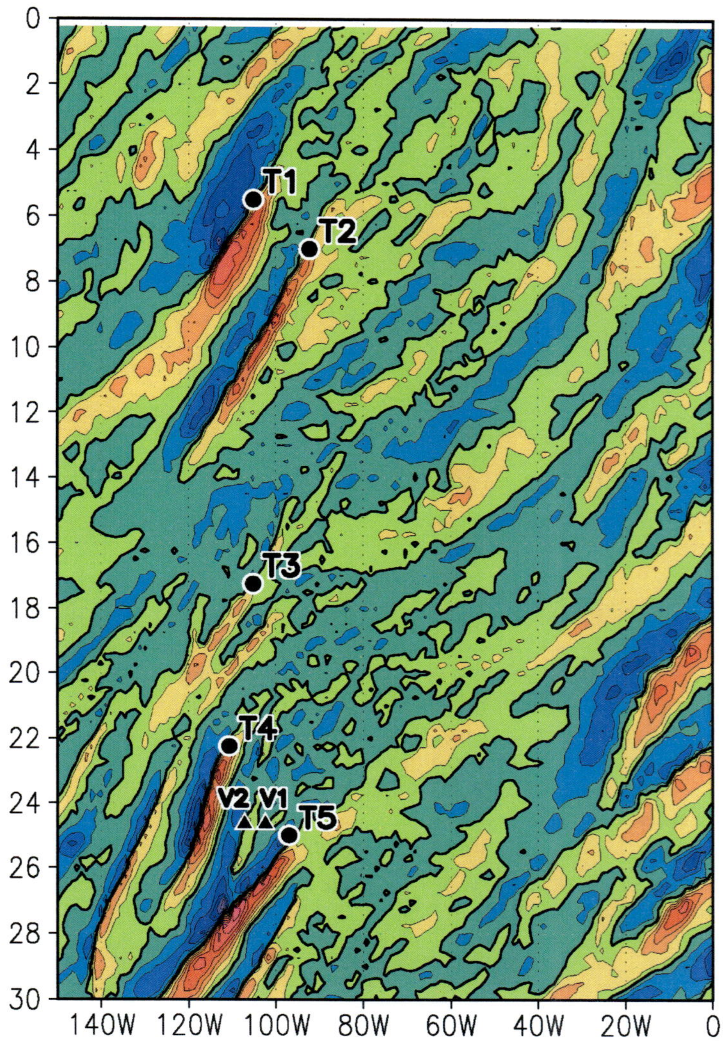

Figure 4. Longitude-time Hovmöller diagram of meridional wind anomaly at 700 hPa (averaged between 10°N and 15°N). The contour intervals are 2 m s^{-1}, and the thick contours indicate 0 m s^{-1}. The circles indicate the genesis location of TCs (T1 -T5), while the triangles indicates the vortices V1 and V2 in Figure 5.

Dunkerton et al. [18] proposed a simple conceptual framework for tropical cyclogenesis by the interaction between a closed circulation associated with easterly waves and deep moist convection. Figure 5 shows the precipitation and streamlines of horizontal flow at 850 hPa relative to the westward-moving disturbance associated with the cyclogenesis of T5 in Figure 4. As Dunkerton et al. showed, tropical cyclogenesis occurred within the closed stream lines accompanied by active convection, which formed near the intersection of the trough axis (black curve) and isopleths of zero relative zonal flow (purple curve). Also note that the center of the closed circulation is located to the south of the negative gradient of absolute vorticity (or negative effective beta), which are similar characteristics to Tropical Storm Fabio (2000) [18]. The zero line of meridional gradient of absolute vorticity implies the necessary condition of barotropic instability [19].

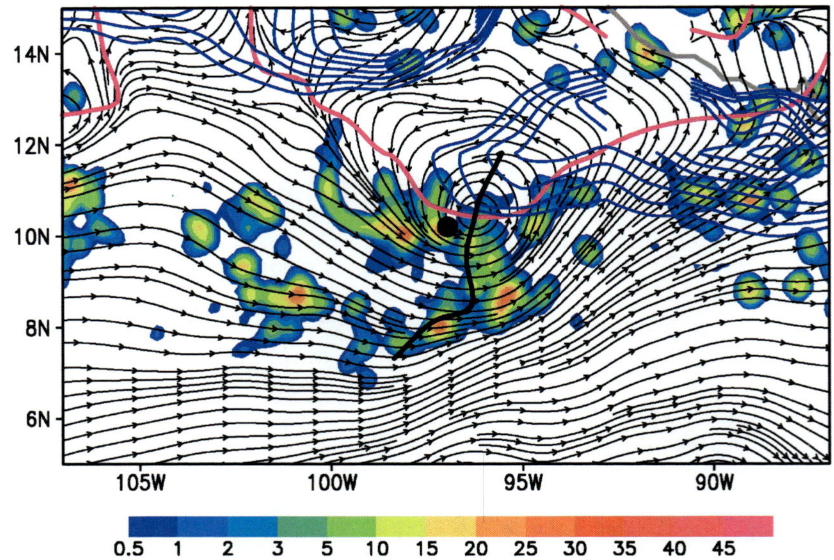

Figure 5. Stream lines of horizontal flow at 850 hPa and 1-day averaged precipitation (color; the unit is mm h^{-1}) at 0000 UTC on 26th day. The horizontal flow is relative to the westward moving frame at the speed of the disturbance (~7 ms^{-1}) which was estimated from the longitude-time Hovmöller diagram. Purple curve indicates isopleth of zero zonal flow relative to the moving disturbance. The black curve shows wave's trough axis for reference. The negative gradient of the absolute vorticity is calculated using the 30-day averaged horizontal flow (blue contours with contour interval is 10^{-11} $m^{-1}s^{-2}$). The black circle indicates the genesis location of TC T5.

Furthermore, the 14-km-grid model resolved some meso-scale dynamics of vortices during the cyclogenesis over the eastern North Pacific. For example, some TCs and meso-scale vortices (e.g. T4, v1 and v2 in Figure 4) form locally near the zone of barotropic instability around 10°~15°N, which is associated with the intertropical convergence zone (ITCZ) [19]. Previous studies show that TCs and meso-scale vortices form through the instability associated with the ITCZ-breakdown over the eastern North Pacific [20-22], some of which interact with each other. Figure 6 shows the horizontal wind field at 850 hPa from 25th to 28th day. TC T5 was originally propagated from the North Atlantic (Figure 4). On 25th day, to the west of TC T5, meso-scale vortices V1 and V2 formed locally as has been mentioned. From 25th to 28th day, TC T5 moved northwestward and merged vortex V1 and then vortex V2. Wang and Magnusdottir [20] found two types of cyclogenesis process associated with ITCZ-breakdown, vortex rollup and westward propagating disturbance; tropical cyclones originated from westward propagating disturbances have a better chance to merge the smaller disturbances and tend to move northwestward (e.g. Hurricane Fausto, 2002), which is consistent with the characteristics of TC T5 in our simulation. Figure 7 shows the zonal-vertical cross section at 10°N on 25th day. The vorticity of TC T5 is upright with the peak in the lower troposphere, which is typical structure of a TC. On the other hand, meso-scale vortex V1 has a vertically-tilted structure with the vorticity maximum in the mid-troposphere, which is accompanied by diabatic warming (cooling) above (below) it (not shown). This characteristic is similar to the meso-scale convective vortex (MCV) in the observational and modeling analyses [21, 22], which merged and developed into Tropical Storm Eugene (2005).

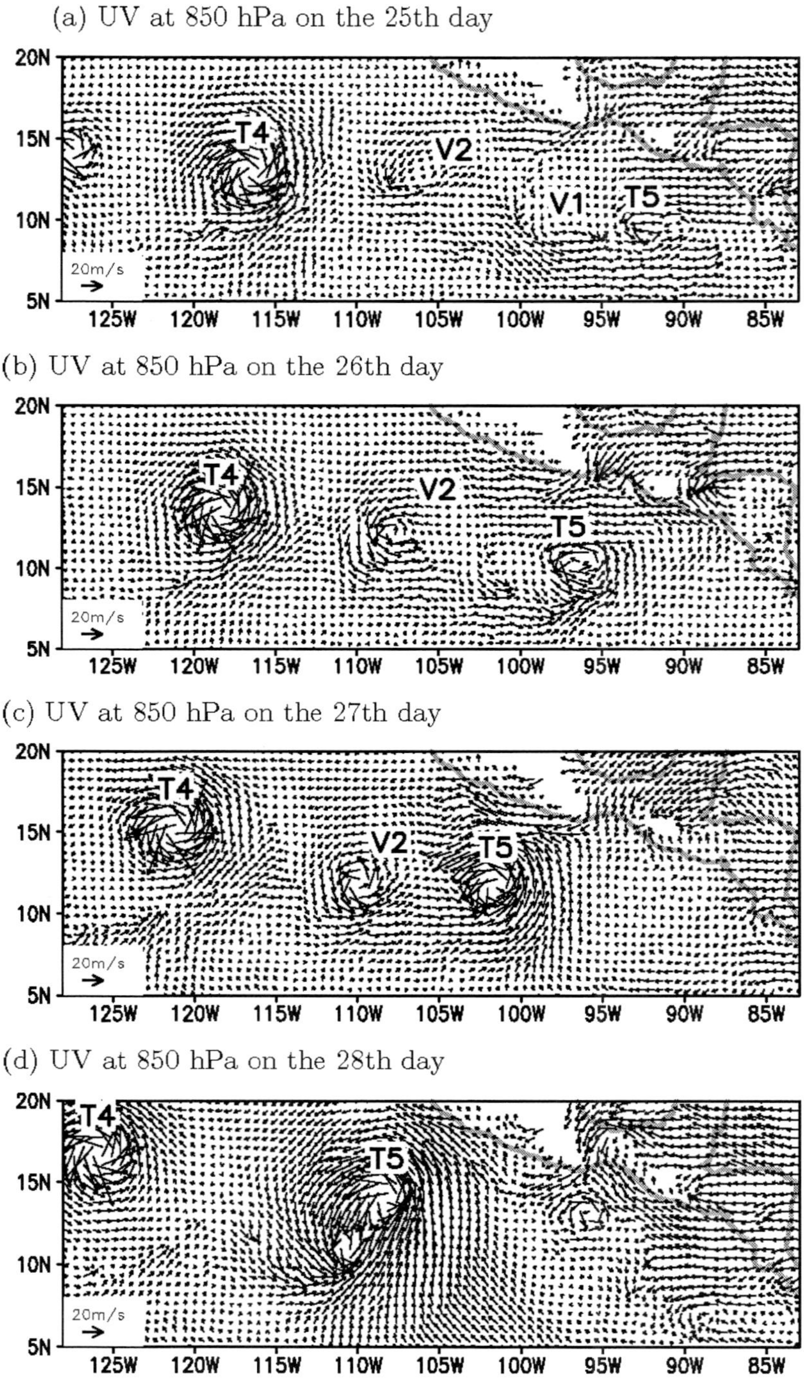

Figure 6. Horizontal wind (anomaly from the 30-day average) at 850 hPa at 0000 UTC on (a) the 25th day, (b) the 26th day, (c) the 27th day, and (d) the 28th day.

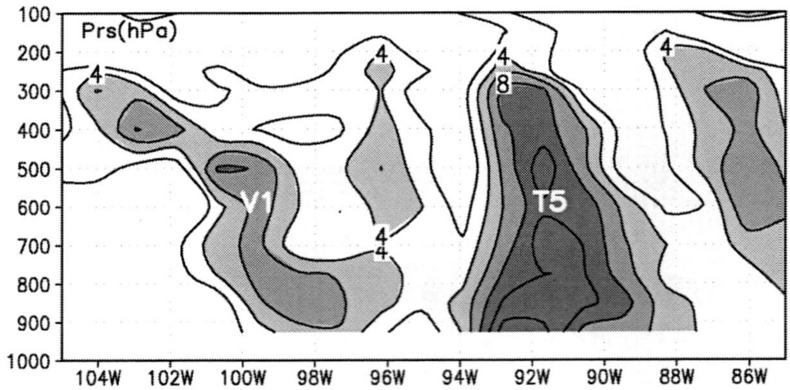

Figure 7. Positive relative vorticity at 0000 UTC on the 25th day on the longitude-height cross section at 10°N. The contour intervals are 2 x 10^{-5} s^{-1}.

Another example of vortex interaction was related to TC T2, which developed just after the previous TC T1 (Figure 4). The atmospheric fields at 850 hPa during the cyclone formations are shown in Figure 8. The horizontal wind at 850 hPa (Figure 8a) shows a wave-train emanated from TC T1 east-southeastward to T2. This is similar to the pattern associated with the Rossby wave energy dispersion [23]. In fact, an E-vector (an energy propagation vector) analysis at 850 hPa illustrates the east-southeast-ward propagation of Rossby wave energy from T1 (Figure 8b). Thus, the GCRM with ~14-km-grid can simulate various meso-scale interactions of vortices during the TC genesis such as ITCZ-breakdown, vortex merger and Rossby wave energy dispersion.

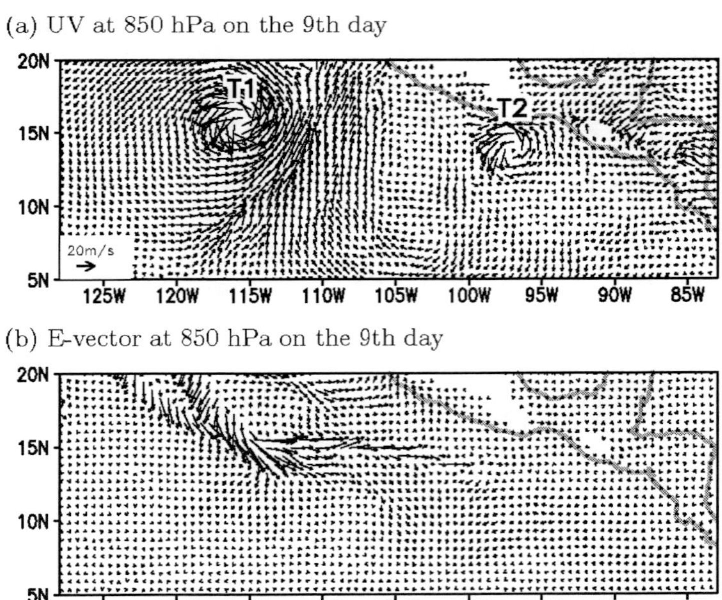

Figure 8. Wind field associated with TCs T1 and T2 at 0000 UTC on the 9th day. (a) Horizontal wind (anomaly from the 30-day average) at 850 hPa. (b) E-vector ([−u'u'+v'v']/2, [−u'v']) at 850 hPa, where [x] indicates the time average of x over a 15-day period centered on the 9th day, and x' indicates the anomaly of x from the time average.

Conclusion

In order to examine how well a GCRM simulates the multi-scale dynamics of TC formations, we performed an equilibrium July experiment using the 14-km-grid NICAM model. In this chapter, we focused on the eastern North Pacific, because the model simulated several TC formations and the environmental field which are similar to the observation.

The model produced the lifetimes and intensities of TCs comparable to the best-track analysis. Since the cyclone intensity reached about 50 m s^{-1} at most, the cyclone tracking algorithm worked sufficiently well even in the absence of any artificial definitions. This is attributed to the GCRM approach using a 14-km grid, which represented the essential dynamics of a TC inner-core including a precipitation-free eye and eye-wall convection. Note that even a 20-km-grid global model required an artificial threshold to capture the simulated tropical cyclones [24]. Thus, a 14-km-grid GCRM could be quite attractive in the sense of the model resolution and computational cost for long-term simulation.

A GCRM represented comprehensively the multi-scale vortices associated with TC genesis over the eastern North Pacific, which has been shown in observational studies [18, 20, 21]. Tropical easterly waves originated over Africa propagated all the way westward across the North Atlantic and entered to the eastern North Pacific to trigger some cyclogenesis. Camargo et al. [25] showed that low resolution models produced few TCs over the eastern North Pacific, which was attributed to the poor representation of easterly waves. In the future work, the GCRM can be useful to investigate what factor determines the success or failure of easterly waves to develop to TCs. Furthermore, it was demonstrated that the model can simulate the dynamics of meso-scale vortex including ITCZ-breakdown, vortex merger, and Rossby wave energy dispersion. The 14-km-grid resolution seems to represent well the meso-scale process, particularly in case organized convective systems are dominantly controlled by the vortex dynamics rather than by detailed cumulus dynamics. The detailed cumulus dynamics in meso-scale process may become more realistic, if an appropriate cumulus parameterization suitable for the cloud-system resolving model is incorporated [26, 27].

Thus, we have demonstrated that the GCRM could be a useful tool to investigate (1) TC intensity and life-cycle, (2) origins of TCs, and (3) meso-scale vortex dynamics associated with TC formation, including ITCZ-breakdown and vortex merger. This chapter should be informative to the future work for higher resolution GCRM and for longer-time simulations focusing on interannual variability of TCs and potential impact of global warming on the activity of TCs.

Acknowledgments

We appreciate the comments by two anonymous reviewers. The simulation discussed in this chapter was performed on the Earth Simulator at the Japan Agency for Marine-Earth Science and Technology. This research was supported by KAKENHI 20740266 and KAKENHI 23740349.

References

[1] B. W. Shen, R. Atlas, J. D. Chern, O. Reale, S. J. Lin, T. Lee, and J. Chang, *Geophys. Res. Lett.*, 33, doi:10.1029/2006GL024594 (2006).

[2] B. W. Shen, R. Atlas, O. Reale, S. J. Lin, J. D. Chern, J. Chang, C. Henze, and J. L. Li, *Geophys. Res. Lett.*, 33, doi:10.1029/2006GL026143 (2006).

[3] H. Miura, M. Satoh, H. Tomita, A. T. Noda, T. Nasuno, and S. Iga, *Geophys. Res. Lett.,* 34, doi:10.1029/2006GL027448 (2007).

[4] H. Fudeyasu, Y. Wang, M. Satoh, T. Nasuno, H. Miura, and W. Yanase, *Geophys. Res. Lett.*, 35, doi:10.1029/2008GL036003 (2008).

[5] Y. Yamada, K. Oouchi, M. Satoh, H. Tomita, and W. Yanase, *Geophys. Res. Lett.*, 37, doi:10.1029/2010GL042518 (2010).

[6] M. Satoh, T. Matsuno, H. Tomita, H. Miura, T. Nasuno, and S. Iga, *J. Comp. Phys.*, 227, 3486,doi:10.1016/j.jcp.2007.02.006 (2008).

[7] W. W. Grabowski, *J. Atmos. Sci.*, 55, 3283 (1998).

[8] T. R. Knutson, J. J. Sirutis, S. T. Garner, I. M. Held and R. E. Tuleya, *Bull. Amer. Meteor. Soc.*, 88, 1549 (2007).

[9] O. Paulis and S. Garner, *J. Atmos. Sci.*, 63, 1910 (2006).

[10] H. Miura, M. Satoh, T. Nasuno, A. T. Noda and K. Oouchi, *Science*, 318, 1763 (2007).

[11] K. Oouchi, A. T. Noda, M. Satoh, B. Wang, S. P. Xie, H. G. Takahashi and T. Yasunari, *Geophys. Res. Lett.*, 36, doi:10.1029/2009GL038271 (2009)

[12] T. Sato, H. Miura, M. Satoh, Y. N. Takayabu and Y. Wang, *J. Clim.*, 22, 4809 (2009)

[13] S. Iga, H. Tomita, Y. Tsushima, and M. Satoh, *Geophys. Res. Lett.*, 34, doi: 10.1029/2007GL031048 (2007).

[14] K. J. E. Walsh, M. Fiorino, C. W. Landsea, and K. L. Mcinnes, *J. Climate*, 20, 2307 (2007).

[15] S. J. Camargo, K. A. Emanuel, and A. H. Sobel, *J. Climate*, 20, 4819 (2007).

[16] S. Uppala, et al., *Quart. J. Roy. Meteor. Soc.*, 131, 2961 (2005).

[17] J. Molinari, D. Vollaro, S. Skubis, and M. Dickinson, *Mon. Wea. Rev.*, 128, 125 (2000).

[18] T. J. Dunkerton, M. T. Montgomery and Z. Wang, *Atmos. Chem. Phys.*, 9, 5587 (2009).

[19] R. N. Ferreira and W. H. Shubert, *J. Atmos. Sci.*, 54, 261 (1997)

[20] C.C. Wang and G. Magnusdottir, *Mon. Wea. Rev.*, 134, 1405 (2006).

[21] C. Q. Kieu and D. L. Zhang, *J. Atmos. Sci.*, 65, 3419 (2008).

[22] C. Q. Kieu and D. L. Zhang, *J. Atmos. Sci.*, 66, 1980 (2009).

[23] G. J. Holland, *Meteor. Atmos. Phys.*, 56, 57 (1995).

[24] K. Oouchi, J. Yoshimura, H. Yoshimura, R. Mizuta, S. Kusunoki and A. Noda, *J. Meteor. Soc. Japan*, 84, 259 (2006).

[25] S. J. Camargo, A. G. Barnston, and S. E. Zebiak, *Tellus*, 57A, 589 (2005).

[26] M. Yamasaki, *J. Meteor. Soc. Japan*, 85B, 145 (2007).

[27] M. Yamasaki. *J. Meteor. Soc. Japan*, 88, 755 (2010).

In: Cyclones: Formation, Triggers and Control
Editors: K. Oouchi and H. Fudeyasu

ISBN: 978-1-61942-976-5
©2012 Nova Science Publishers, Inc.

Chapter 11

A Prototype Quasi Real-Time Intra-Seasonal Forecasting of Tropical Convection over the Warm Pool Region: A New Challenge of Global Cloud-System-Resolving Model for a Field Campaign

Kazuyoshi Oouchi[1*], Hiroshi Taniguchi[2], Tomoe Nasuno[1], Masaki Satoh[1,3], Hirofumi Tomita,[1,4], Yohei Yamada[1], Mikiko Ikeda[1], Ryuichi Shirooka[5], Hiroyuki Yamada[5], and Kunio Yoneyama[5]

[1]Japan Agency for Marine-Earth Science and Technology, Yokohama, Japan
[2]International Pacific Research Center, University of Hawaii at Manoa, Honolulu, Hawaii, US
[3]The University of Tokyo, Kashiwa, Japan
[4]RIKEN, Kobe, Japan
[5]Japan Agency for Marine-Earth Science and Technology, Yokosuka, Japan

Abstract

A new prototype of quasi real-time forecast system for tropical weather events with timescales spanning across diurnal-to-intra-seasonal ranges is developed. Its hallmark is the use of a global cloud-system resolving model (GCRM) that avoids uncertainties arising from cumulus convection schemes inherent to traditional hydrostatic models. This allows forecasting of multi-scale convective disturbances such as tropical cyclones and super cloud clusters that are particularly associated with Madden-Julian Oscillation (MJO), even more straightforwardly, with unprecedented details, and disseminate detailed forecast products to *in-situ* observation networks. The system can operate, for example, in an observation campaign in the tropical ocean, where observation network is

* E-mail address: k-ouchi@jamstec.go.jp

in synchronous with modeling components to investigate the atmospheric and atmosphere-ocean interaction processes controlling tropical disturbances such as MJO.

This chapter overviews the prototype system and reports a first pilot forecast example of tropical cyclone LAILA that happened over the Indian Ocean in the late May 2010. The forecast system with the finest mesh of 14 km over the target region projected the overall track and evolution of LAILA quite decently prior to its landfall with up to about five days lead time. LAILA featured north-westward movement after its genesis in association with a propagation of the predicted MJO. The MJO traversed the Indian Ocean up about 5-7 days before the genesis of LAILA, according to an MJO index analysis from observation. A key ingredient of the encouraging forecast is inferred to be the GCRM's capability of representing a mechanistic link between LAILA and MJO. Additionally, the forecast system captured a northward migration of precipitation embedded in a synoptic-scale low-level westerly intrusion over the Indian Ocean region, which signals the onset of southwesterly monsoon for this particular year.

These findings, albeit from only one specific forecast experiment, lend some reliability to the GCRM-based research forecast of MJO-associated convection events to assist on-site observation planning. Moreover, this suggests a potential pathway of future GCRM-based forecast system development towards extended seamless intra-seasonal forecast of MJO-associated weather events that would threaten the tropical warm-pool and neighborhood regions.

1. Introduction

The aim of this chapter is to introduce a quasi real-time forecasting system of tropical multi-scale convection events that has its unique strength of being driven by a global non-hydrostatic model, along with an overview of a forecast product from its first pilot experiment. *Multi-scale* is a key [1] but loosely used term to broadly characterize convection residing over the warm pool region in the tropical Indian Ocean and western Pacific. These regions indeed experience a wide variety of convective systems such as those accompanied by monsoon depression, equatorial waves [2] and tropical cyclones. Multi-scale convection is, in this sense, a hybrid manifestation of organized convection influenced by these disturbances across various scales over this particular area. The warm pool is also known to be critical for genesis and development of Madden-Julian Oscillation (MJO; [3,4]). As extensively reviewed by [5], MJO is a strong intra-seasonal mode of tropical variability. Its impact on prominent weather phenomena including tropical cyclogenesis is not only significant over the Indian Ocean and western Pacific region [6-8], but stretched eastward out to Atlantic region [9], covering nearly whole of the tropics. Seasonal prediction of tropical cyclones was suggested to be vastly improved by using a methodology based on the phase of MJO [10]. In addition, over some part of the Indo-Asian region, we also experience a quasi-periodic intra-seasonal variation (ISV) of local weather, which sometimes constitutes a sub-component of monsoon intra-seasonal oscillation (MISO; [11]). This is generally accompanied by northward migration of disturbance, and regarded as another leading intra-seasonal mode controlling the weather [12,13] and its predictability [11] in the boreal summer monsoon-prone area. It is therefore rational to direct our attention to prediction and understanding of *multi-scale* convective events over these basins in terms of impacts by MJO/ISV on intra-seasonal time frames, in line with a suggestion from observational studies (*e.g.*, [14-18]).

A major impediment in clarifying these multi-scale convective events has been complex interaction involved across various spatial and temporal scales; observation cannot cover enough area with sufficiently high temporal and spatial resolutions, and the traditional hydrostatic modeling approach also suffered from the similar problems. With the emergence of cloud-(system)-resolving model (CRM), many of these problems were nearly resolved (*e.g.*, [19,20]), and the CRM concept behind these efforts culminated in the development of Multiscale Modeling Framework [21,22] and global CRM (GCRM), with the Nonhydrostatic ICosahedral Atmospheric Model (NICAM; [23,24]) being the first prototype of GCRM. A key new finding from these CRM approaches is that interactions involved in MJO/ISV are controlled by mesoscale convection of O(10km) or its ensemble, a most basic organized form of cumulus convection in tropical organized convective disturbances [25]. Importantly, mesoscale convection plays not passive but active role in organization of MJO-associated convection and larger-scale features as was suggested in observational [1], conceptual [26] and cloud-resolving modeling [20] studies. Some crucial insights into upscale control of mesoscale processes such as mesoscale gravity wave, cold pool, and cumulus momentum transport associated with MJO were provided already before the Earth Simulator's Era (*e.g.*, [26-29]). Yet, its observational verification remains to be seen. These accumulated evidences point to the need for intensive advanced observation and its collaboration with mesoscale-resolving modeling calculations to advance our understanding of MJO and MJO-associated multi-scale convection.

A plan is in progress to meet this issue; an intensive observation project, CINDY2011(Cooperative Indian Ocean experiment on intra-seasonal variability in the Year 2011)/DYNAMO (Dynamics of the MJO) (http://www.jamstec.go.jp/iorgc/cindy/), will start in 2011. It will aim at collecting the dataset of atmosphere and ocean over the Indian Ocean for 4 months to improve our understanding of the genesis of MJO, and related variability and processes in atmosphere and ocean under international collaboration endorsed by WCRP (World Climate Research Program)/CLIVAR (Climate Variability and Predictability Project). Across this oceanic basin, observation and modeling teams with state-of-the-art observational equipments and models will collaborate closely toward the goal. Among them, a JAMSTEC/Mirai observation team and a global cloud-resolving modeling (NICAM) team have started a closer collaboration to prepare for the project. One of the tasks in the NICAM team is to construct and provide quasi real-time prediction of convection events associated with MJO to assist *in-situ* observation. At the same time, we expect, through this collaboration, to promote our modeling and forecast skill over this challenging region [30].

Mirai is a research vessel (R/V) of JAMSTEC that has provided valuable outputs for researches on tropical atmospheric and oceanic phenomena over the world Oceans during various research cruises including the recent MISMO project (Mirai Indian Ocean cruise for the Study of the MJO-convection Onset; MISMO, [31]). Using the research vessel, an attempt was made this summer of 2010 to collaborate the global cloud-resolving model (NICAM) team in JAMSTEC with Mirai group during a research cruise, PALAU (Pacific Area Long-term Atmospheric Observation for Understanding Climate Change) 2010 over the tropical western Pacific region (Main observation sites include 5^0N, 139.5^0E; intensive observation period: May 16th to June 22nd). One of the aims of PALAU2010 campaign was to clarify the mechanisms responsible for northward propagation of the summertime ISV and associated tropical cyclogenesis in and around the observation site. The aim of the NICAM team was to test the newly developed GCRM-based quasi-real-time forecast system, assess its performance and check if it facilitates observation planning of specific atmospheric cloud disturbance on site. NICAM has exhibited its reasonable performance in capturing MJO [32], associated multi-scale convective phenomena over the tropical warm-pool region [33] and monsoon region [30], as well as tropical cyclone development embedded in the MJO-associated synoptic-scale control ([34-37]). Given the amount of scientific scrutiny to which case study of convective event with NICAM has been subjected, it is expected to derive a new insightful synergy through its collaboration with the observation and other modeling groups.

While the PALAU2010 observation was conducted over the western Pacific, a tropical cyclone LAILA attacked India far out over the Mirai observation site, and caused serious damages over some Indian regions in late-May, 2010. It was the first tropical cyclone forecasted after our quasi real-time forecast system started its operation. The main north Indian Ocean tropical season runs from May to November, and TC-LAILA was the first tropical cyclone in the season this year. TC-LAILA developed on May 17th in the Bay of Bengal from persistent convection bands. Strengthening as it tracked northwestward, it became a severe storm on May 19th, made landfall on the eastern coast of Indian subcontinent on May 20th. It caused flooding and damages along its path, and killed at least 23 people after making landfall. Apart from the original objective of observation-modeling collaboration, TC-LAILA is, as it happened, an excellent target for assessing the performance of quasi-real-time CGRM calculation over the Indian Ocean, the venue of the CINDY2011/DYNAMO project. This sets a valuable example of studying a possible link between tropical cyclogenesis and MJO, which is one of our forecast strategies in the project.

The aim of this article is to overview the world's first GCRM-based, quasi-real-time forecast system, and to show some initial output results concerning TC-LAILA, by which we try to clarify what role and strategy the GCRM-based forecast system can take in the tropical observation project, such as CINDY2011/DYNAMO. We can report only one forecast case study here and many more experiments are necessary, but this is a good demonstration of performance of the system, whereby we can start addressing modeling issues that need to be handled for productive collaboration between observation and modeling groups in the project. The article is organized as follows: Section 2 explains the flow of the quasi-real-time forecast; Section 3 highlights examples of tropical cyclone phenomena predicted reasonably in the system during the observation period; Section 4 provides summary and remarks.

2. Overview of Quasi-Real-Time Forecast during JAMSTEC's PALAU2010 Field Campaign

The quasi real-time forecast system is shown in Figure 1(a), along with its output example displayed on a dedicated web server in Figure 1(b). As a test of the system, the system was operated during JAMSTEC's PALAU2010 Field Campaign during May-June, 2010. The operation of the forecast system was collaborated with the *in-situ* observation team during PALAU 2010 project. The PALAU is an observation project which has been conducted in the Republic of Palau since 2000 [38]. The project, which is led by the Research Institute for Global Change/Japan Agency for Marine-Earth Science and Technology (RIGC/JAMSTEC), is designed to explore characteristics of the atmospheric and oceanic variability in the western Pacific, and some events of tropical convection including tropical cyclones.

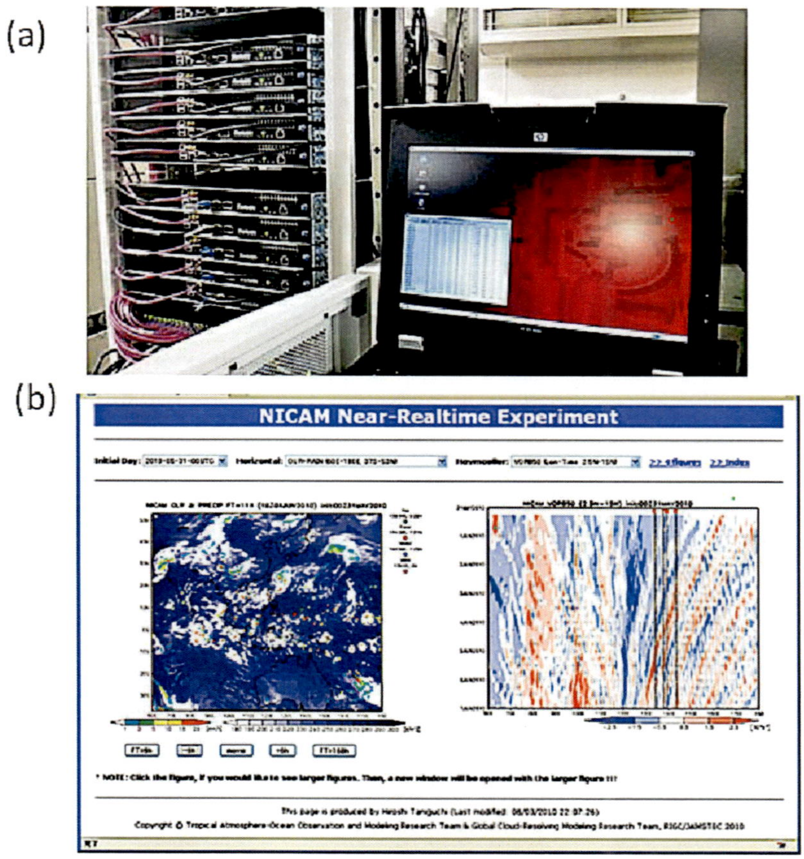

Figure 1. (a) Quasi real-time forecast system (b) An example of forecast results displayed on the dedicated web server.

The quasi real-time forecast for the PALAU 2010 project uses a stretched grid system [39] applied to NICAM [23, 24]. The physical schemes and the manner of constraint of boundary conditions are the same as in [35] except for sophisticated boundary layer (MATSIRO, [40]), new microphysical scheme that predicts six water categories (water vapor, cloud water, rain, cloud ice, snow, and graupel; [41]), and a slab ocean scheme that handles thermal feedback between the atmosphere and the underlying ocean surface layer. The stretched grid covers the finest mesh area at 14-km over the warm water pools of Indian Ocean and western Pacific, and coarser meshes outside stretched up to 56 km. The stretched configuration minimizes computational cost by reducing grid points while retaining the horizontal resolution for the target region of interest as finest as possible. The validity of the use of 14-km mesh in the simulation of an MJO event [42], MJO-associated tropical cyclogenesis over the Indian Ocean [36,37] and western Pacific [35] has been confirmed by other case studies with NICAM.

The initial conditions of atmospheric variables are provided by objective analysis datasets from Final Operational Global Analysis data of the National Centers for Environmental

Prediction (NCEP). The model is run with the realistic boundary condition of topography and the baseline sea surface temperature (SST) toward which predicted SST is nudged at the time scale of 5 days. The baseline SST is the temporally and spatially interpolated OI-SST data provided by National Oceanic and Atmospheric Administration (NOAA).

The flow of the quasi-real-time forecast during PALAU 2010 project is overviewed in Figure 2. Each calculation has been performed for one-week forecast unit initiated with the atmospheric objective analysis dataset for 00UTC of the start date that was derived from the dataset server. Neither specific initialization processing nor nudging of the atmospheric fields was applied throughout the time integration. It took about 1.7 day wall clock time for completing online calculation of the one-week forecast, and some hours of post-processing calculations such as coordinate transformation. Once the gridded output data were created, it can be processed to plot necessary figures of the disturbances, mean field, and disturbance metrics on a variety of demands from the *in-situ* observation sites. Most of these procedures can be automatically processed via cron commands and shell scripts on UNIX platform, depending on specification of the calculation resources. Details of the forecast system are planned to be reported elswhere.

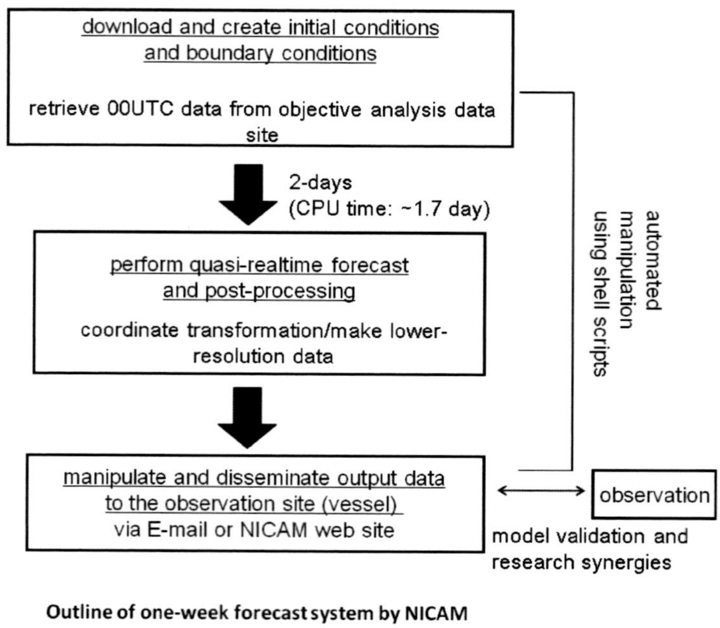

Figure 2. Flowchart of the quasi-real-time forecast.

3. An Example of Forecast - Tropical Cyclone LAILA over the Indian Ocean

Among various outputs for the observation period during PALAU 2010, we highlight Tropical Cyclone (TC) LAILA whose development and movement were captured quite reasonably in the real-time forecast system. TC-LAILA made landfall on the northeastern coast of India late May 20th. 2010. The TC, being developed over the Indian Ocean, was by no means a target of our concern at the time when we started to run our new-built real-time forecast system with the initial condition of May 14th; neither did we imagine it could develop into such a devastating TC bringing heavy rain and gale force winds on the coastal areas of eastern India 6 days later (on May 20th). Yet, the GCRM-based system forecasted its overall behaviors as it turned out, and it is worth highlighting this specific case.

Figure 3 shows an infrared image of TC-LAILA a couple of hours before making landfall on the eastern coast of the Indian subcontinent on about 00UTC, May 20th. We can see the circle-shaped cloud region of TC-LAILA measures about 5 degree in diameter, and not very large but it caused substantial damages on some regions until it decays about a few days after its landfall. Its track is plotted in Figure 4 for simulation (blue) and best track from observation with IBTrACS (red). The dots on each track line represent positions of TC at 6-hour interval. The simulation forecast starts on 00UTC May 14th, and the model-simulated tropical cyclone attains TC intensity on 06Z May 17th. The movement of LAILA in northwestward direction and its landfall on the eastern coast of India looks typical of those for summer storms in the Indian Ocean [43]. Comparison between both tracks indicates a quite good consistency in its northwestward movement and path from about 06UTC, May 18th to 00UTC, May 20th. Both simulated and observed paths coincide with high sea surface temperature waters (about and more than 304 K) during this period.

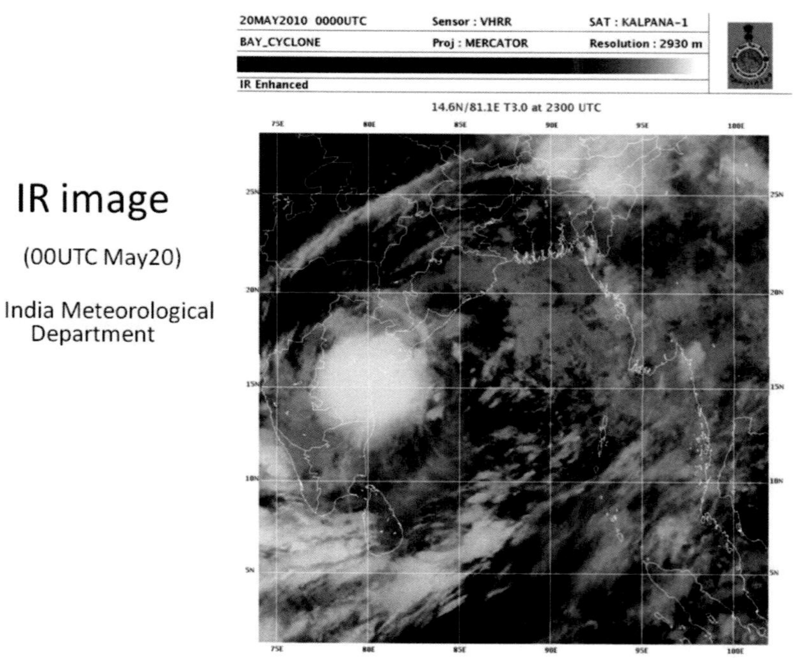

Source: Indian Meteorology Department.

Figure 3. Infrared temperature imagery of tropical cyclone LAILA on May 20th 00UTC.

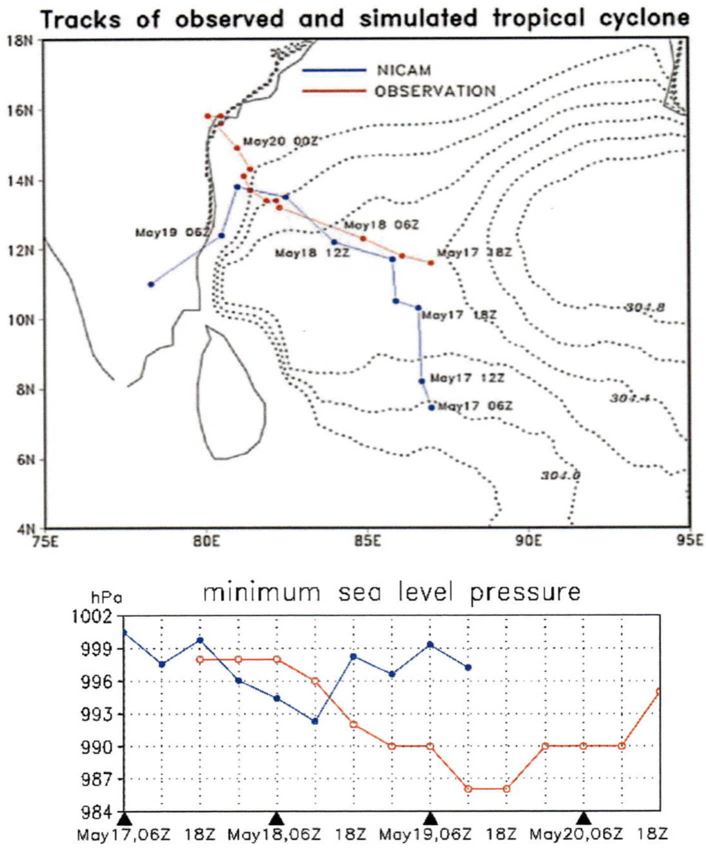

Figure 4. (upper) Tropical cyclone tracks for forecast and observation (IBTrACS:International Best Track Archive for Climate Stewardship). Dotted contours plot sea surface temperature averaged over May 17th to 20th. (lower) Time series of the minimum sea-level pressure of LAILA for IBTrACS and NICAM.

While promising in its simulated movement, we cannot trace the model-simulated TC thoroughly back to its genesis stage and argue its mechanism as the time integration started May 14th. We can infer, however, that propagation of MJO plays some role in its genesis. Figure 5 displays phase diagram of real-time multivariate MJO index from Apr 8th to May 17th created from observation dataset by Dr. Wheeler [44]. The diagram demonstrates that clear eastward propagation of MJO exists across the Indian Ocean and the maritime continent region during early-to-mid May. The signal of eastward propagation of MJO can also be seen Figure 6 (a), which plots Hovmoller diagram of observed zonal velocity at 850 hPa over the equatorial region (5^0S - 5^0N). A region of westerly propagates eastward from about May 11th to May 16th when it reaches the maritime continent (around 110^0E). The period between the broken lines correspond to the target period of the real-time forecast which is displayed in Figure 6 (b). We can see that eastward progression of moderate-to-strong westerly (exceeding 6ms-^{1}) region propagates over 70^0-100^0E toward May 16th and from May 17th to May 18th. These are likely to be associated with model-simulated propagation of MJO - basically in consistent with the observed counterpart. As demonstrated in [34,35], these low-level westerly accompanied behind the MJO is a key precursor to MJO-associated tropical cyclogenesis in NICAM. Under such condition, the tropical North Indian Ocean is subject to cyclonic shear composed of equatorial westerly and off-equatorial easterly, which is evident in the snapshot in Fig 11.6 (c).

To identify how TC-LAILA developed in association with the enhanced equatorial westerly, time-meridional plots of precipitation and zonal velocity are shown in Figure 7. Precipitation is concentrated near the equator around May 15th (one day after the start of the time integration), and moves northward with time. On May 17th, the northward moving precipitation is accompanied with a cyclonic vortex (not shown) that attains the tropical cyclone intensity around May 17th (See. Figure 3). The series of evolution of these atmospheric fields reminds us the typical scenario of MJO-associated tropical cyclogenesis in NICAM [35], and evidences in this case also suggest that it is very likely that TC-LAILA developed in association with the MJO.

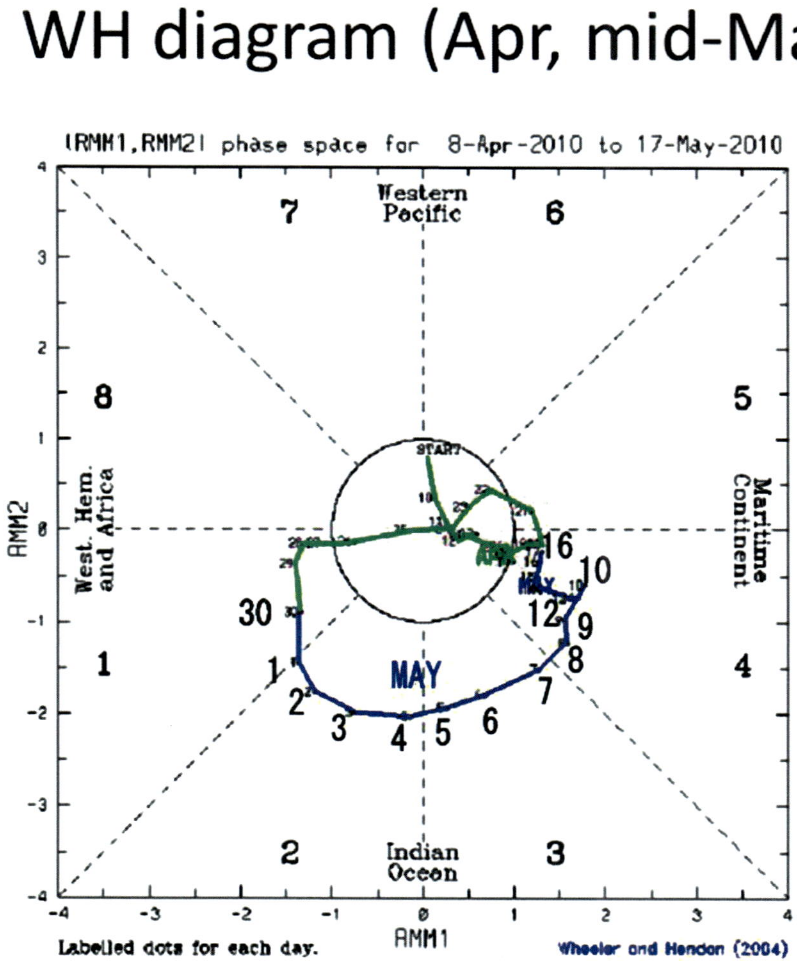

Figure 5. Phase space points each day from Apr 8th to May 17th for real-time multivariate MJO index ([44]). Eight defined regions of the phase space are labeled that are considered to signify MJO activity.

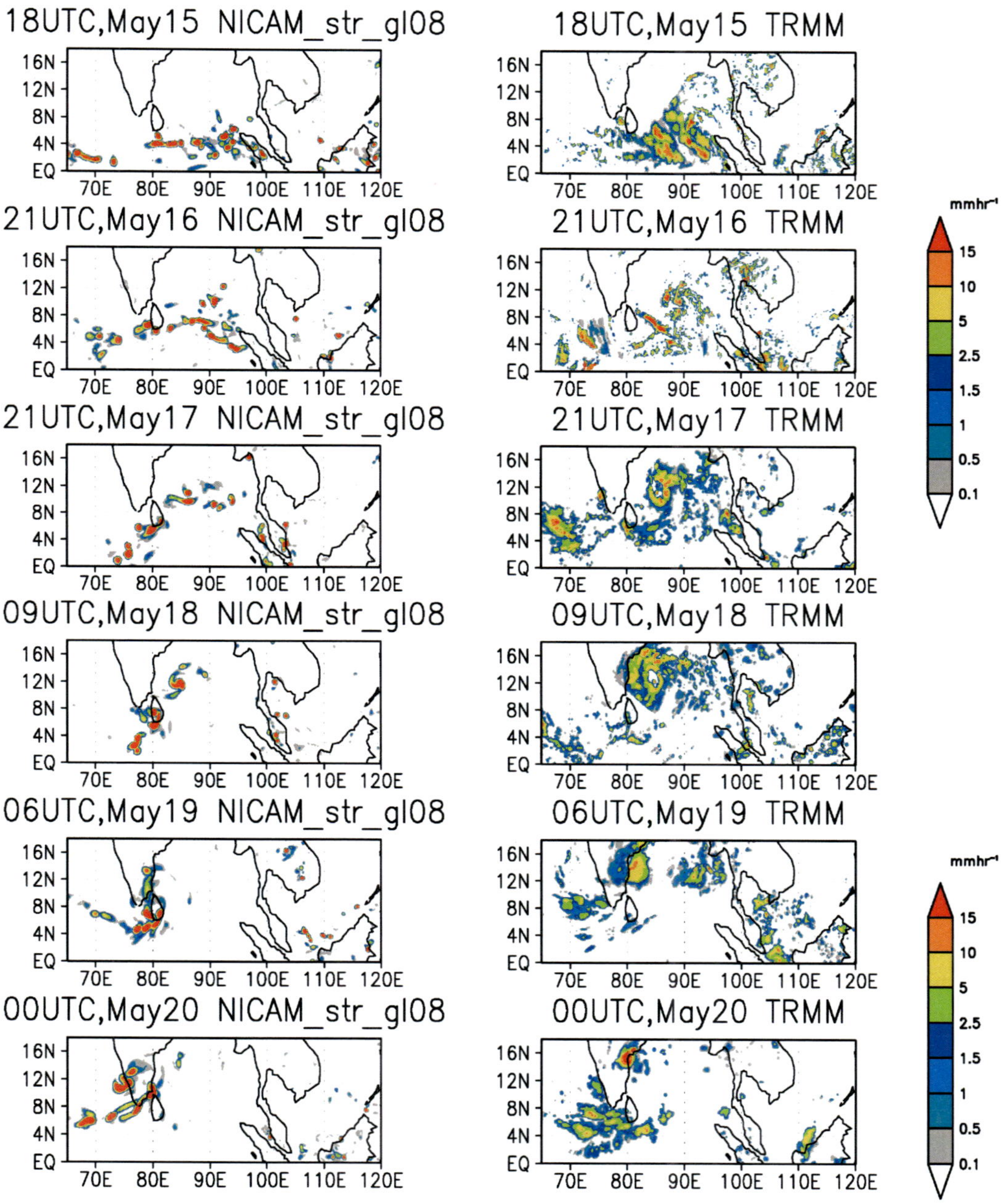

Figure 8. Time evolution of precipitation in and around Tropical Cyclone LAILA for (left) forecast and (right) observation from TRMM 3B42: 18UTC May 15th; 21UTC May 16th, 21UTC May 17th, 09UTC May 18th, 06UTC May 19th, 00UTC May 20th.

Another notable and interesting feature in Figure 7 is that equatorial (0-6^{0}N)/off-equatorial (north of about 6^{0}N) region exhibits a regime shift from rainy/drier to drier/rainy situations toward the end of May. Actually, it was reported that, in association with the formation of TC-LAILA over Bay of Bengal, southwest Indian monsoon set in over Andaman Sea around 17th May, 3 days before its normal date (India Meteorological Department, 2010). The onset of southwest monsoon is characterized by appearance of monsoon low-level flow over Andaman Sea, and change from drier to rainy spell as simulated here. This can be an interesting coincidence - propagation of MJO, tropical cyclogenesis and southwest monsoon break – and it is encouraging that GCRM forecasted a partial trend of this suite of atmospheric events reasonably.

Time evolution of TC-LAILA is shown in Figure 8 (left panels), which illustrates snapshots of precipitation selected at various stages of evolution (18UTC, May 15th - 00UTC, May 20th). Precipitation from TRMM observation at the corresponding time is also displayed (Figure 8; right panels). There exist clusters of precipitation region near the equator on 18UTC, May 15th both for the forecast and the observation. They are associated with the low-level westerly and precipitation areas as seen in Figure11.7. The precipitation area gradually becomes organized until 21UTC, May 17th, and takes on spiral form at 09UTC, May 18th. The northwest-to-southeast aligned concentric precipitation regions for 21UTC, May 16th and 21UTC, May 17th are basically consistent between forecast and observation. The location of the forecasted cyclone is generally more southward after 06UTC, May 19th, and so is its landfall location. Overall, it is fair to say that forecast system predicted pattern of evolution and main path of TC-LAILA pretty consistently with observation.

SUMMARY AND ISSUES

This article overviews a prototype quasi real-time forecast system of tropical convective events that is built on the global cloud resolving model, NICAM [23,24]. Our immediate aim will be to provide forecast information in the CINDY2011/DYNAMO project in timely and collaborative framework among the international observation and modeling networks under the endorsement of WCRP/CLIVAR. The system, including a suite of options of physical schemes, will continue to be improved and tuned to provide reliable forecast information in the project. Main target of the project includes MJO genesis, development, and related atmosphere and ocean processes centered over the Indian Ocean. Our overriding goal is to make full use of the GCRM-based forecast system with observation team of JAMSTEC in particular, and achieve utmost synergies from the combination of the state-of-the-art observation and cloud-system modeling. This kind of collaboration has been long awaited in the community, and it is expected to spark the ideas, new understanding, and improvement of predictive skills on MJO and associated disturbances.

As an initial pilot study, the GCRM-based forecast system was operated in parallel with the PALAU2010 project that includes the JAMSTEC observation research cruise with R/V Mirai over the Northwestern Pacific during a May-June period in 2010. We obtained an encouraging forecast result: the system forecasted the movement almost before its landfall and the overall evolution of TC-LAILA over the Bay of Bengal reasonably - a few to 5 days in advance. The tropical cyclone was forecasted after a likely passage of MJO, and northward migration of precipitation and intrusion of background low-level westerly accompanied with MJO were captured quite reasonably, which set a good example of forecast demonstration for MJO-associated convective events for the upcoming CINDY2011/DYNAMO project. It is unfortunate that only limited output variables were obtained from this experiment and in-depth investigation into TC-LAILA was impossible. Yet, the result is likely to reinforce the recognition from some hindcast experiments with NICAM that GCRM is capable of capturing MJO [32], MJO-associated tropical cyclogenesis and its northward movement quite reasonably across the warm ocean basins, including North Indian Ocean [36,37], South Indian Ocean [34] and Northwestern Pacific [35]. Although more samples need to be accumulated to assess the robustness of the forecast system quantitatively, the success of the forecast of TC-LAILA points to a key strategy of GCRM-based forecast - to capture development of precursor atmospheric convection events by focusing on control from intra-seasonal disturbances - which can be shared among the participating observation network in the CINDY2011/DYNAMO campaign. More systematic forecasting of tropical cyclognesis will be achievable when we perform detailed analysis on the mechanism of tropical cyclogenesis and its linkage with MJO as demonstrated by recent studies [45-47]

It was demonstrated, besides the forecast information on the case of TC-LAILA over the Indian Ocean, that the forecast system provided useful information that assisted the *in-situ* observation over the western North Pacific during the PALAU2010 project. Collaborated research output from both observation and modeling efforts will be reported elsewhere. The only regret in the project was that the western Pacific over which PALAU2010 focused was quiescent during the period when the forecast system was in operation, and neither tropical cyclones nor MJO-associated strong convective events were observed, at odds with our expectation. Nonetheless, the GCRM output will serve as a valuable dataset for comparison and some analysis of intra-seasonal tropical phenomena with the collected observation dataset, and in turn, the observation dataset is expected to help validate GCRM simulation output. Collaboration from both sides will be useful for pursuing specific roles and issues to be identified, and progress can be made on both fronts.

We are elaborating the prediction system so that it can be operated for flexible purposes for the *in-situ* observation of CINDY2011/DYNAMO and future observation projects. Examination into output interval and spatial coverage of finer-mesh forecast will be crucial, in addition to planning of inter-comparison among observation and other models. The horizontal mesh of 14-km used in the current version to cover the warm-pool region can be, arguably, too coarse to express interaction between mesoscale convection processes and larger-scale disturbances. As specification of computing resources is a primary limiting factor that determines experimental design, earlier brainstorming on target resolution will be crucial. The hope is that forecasts will be improved all the more as model resolutions begin to be higher, and resolve processes on ever finer spatial scales. An implementation of effective data assimilation routine into the forecast process will also be required to provide more skillful forecast information in future.

Improvement of physical schemes in the model should be also critical for providing more reliable forecast outputs. Ongoing efforts include incorporation of ocean-atmosphere interaction process and possible monitoring of the surface budgets, which can be significant for improving prediction of intra-seasonal variability and mean state in the Indian region, given the suggestion and hypothesis from previous studies [48,30,49]. We have now a solid observational evidence that relevant atmospheric response of MJO is closely linked with mixed-layer characteristics of the ocean [50], which therefore strengthens the recognition. Tuning of cloud microphysical process that represents details of stratiform precipitation characteristics is also significant. It was revealed, from a study on mesoscale convective system over the western Pacific detected with R/V Mirai, that the both the stratiform cloud dynamics and environmental moisture contrast caused by larger-scale disturbance are important to maintain evolution of the convective system [51]. Thus, further coordination is necessary towards observation campaign to foster a consensus of collaborative plans and ideas about target phenomena of interest to be shared among the participating communities. Moreover, it is exciting to see how this new big collaboration would bring about new findings, deeper understanding of tropical weather, and greater interactions among previously separate research communities.

ACKNOWLEDGMENTS

We appreciate the permissions of Dr. Rastogi of India Meteorological Department and Dr. Matt Wheeler of the Centre for Australian Weather and Climate Research at the Bureau of Meteorology to use Figure11.3 and 11.5, respectively. The source of Figure 5 is contained in the following URL created and maintained by Dr. Matt Wheeler: http://cawcr.gov.au /bmrc/clfor/cfstaff/matw/maproom/RMM/index.htm. Some resources for research are provided under the framework of the Innovative Program of Climate Change Projection for the 21st century (KAKUSHIN) project funded by Ministry of Education, Culture, Sports, Science and Technology, Japan, and by the Core Research for Evolutional Science and Technology (CREST) of the Japan Science and Technology Agency. We appreciate Profs. Taro Matsuno and Fujio Kimura of Japan Agency for Marine-Earth Science and Technology for supports and discussion that motivated model development and construction of the quasi real-time forecast system. We thank past and present NICAM team members for their efforts on model developments.The authors thank anonymous reviewers for their careful review of the manuscript.

REFERENCES

[1] C.-H. Sui and K.-M. Lau, *Mon Wea. Rev.* 120(3), 407 (1992).
[2] T. Matsuno, *J. Meteorol. Soc.Japan*, 44, 25 (1966).
[3] R. A. Madden and P. R. Julian, *J. Atmos. Sci.*, 28, 702 (1971).
[4] R. A. Madden and P. R. Julian, *J. Atmos. Sci.*, 29, 1109 (1972).
[5] C. Zhang, *Rev Geophys.* 43: RG2003 doi:10.1029/2004RG000158 (2005).
[6] T. Nakazawa, *J. Meteor. Soc. Japan,* 66, 823 (1988)
[7] B. Liebmann, H. H. Hendon and J.D. Glick, *J. Meteorol. Soc. Japan*, 72, 401 (1994).
[8] S.J. Camargo, *Bulletin of American Meteorological Society*, 89, S74 (2008).
[9] P.J. Klotzbach, *J. Climate*, 23, 282 (2010).
[10] F. Vitart, *Geophys. Res. Lett.*, 36, L15802, doi:10.1029/2009GL039089. (2009).
[11] X. Fu, B. Wang, Q. Bao, P. Liu and J.-Y. Lee, *Geophys. Res. Lett.* 36, L08801 doi:10.1029/2009GL037166 (2009).

[12] B. Wang and H. Rui, *Meteor. Atmos. Phys.*, 44(1-4), 43 (1990).
[13] K.Kikuchi and B. Wang, *J. Meteor. Soc. Japan*, 88, 475 (2010).
[14] T. Yasunari, *J. Meteor. Soc. Japan*, 57, 227 (1979).
[15] B. Wang, P. Webster, K. Kikuchi, T. Yasunari and Y. Qi, *Clim. Dyn.*, 27, 661 (2006).
[16] X. Jiang, X., T. Li and B. Wang, *J. Climate*, 17, 1022 (2004).
[17] C. Zhou and T. Li, *J. Climate*, 23, 5738 (2010).
[18] P.C.,Hsu and T. Li, *J. Climate*, 24, 942 (2011).
[19] W.W.Grabowski and M. W. Moncrieff, *Quart. J. Roy. Met. Soc.*,127, 445 (2001).
[20] K. Oouchi and M. Yamasaki, *J.Meteor. Soc. Japan*, 79, 201 (2001).
[21] W.W. Grabowski, *J. Atmos. Sci.*, *58*, 978 (2001).
[22] M. F. Khairoutdinov and D. A. Randall, *Geophys. Res. Lett.*, *28*, 3617 (2001).
[23] M. Satoh, T. Matsuno, H. Tomita, H. Miura, T. Nasuno and S. Iga, *J. Comput. Phys.*, 227, 3486 (2008).
[24] H. Tomita and M. Satoh, *Fluid Dyn. Res.*, 34, 357 (2004).
[25] M. Yamasaki, *J. Meteor. Soc. Japan*, 85B, 145 (2007).
[26] M. W. Moncrieff and E. Klinker, *Quart. J. Roy. Meteor. Soc.*,123, 805 (1997).
[27] W.C. Chao and S.-J. Lin, *J. Atmos. Sci.,* 51, 1282 (1994).
[28] B. E. Mapes, *J. Meteor. Soc. Japan,* 76, 29 (1998).
[29] K. Oouchi, *J. Meteor. Soc. Japan,* 77, 907 (1999).
[30] K. Oouchi, A. T. Noda, M. Satoh, B. Wang, S.-P. Xie, H.G. Takahashi and T. Yasunari, *Geophys. Res. Lett.*, 36, L11815 doi:10.1029/2009GL038271 (2009b).
[31] K. Yoneyama, Y. Masumoto, Y. Kuroda, M. Katsumata, K. Mizuno, Y. N. Takayabu, M. Yoshizaki, A. Shareef, Y. Fujiyoshi, M. J. McPhaden, V. S. N. Murty, R. Shirooka, K. Yasunaga, H. Yamada, N. Sato, T. Ushiyama, Q. Moteki, A. Seiki, M. Fujita, K. Ando, H. Hase, I. Ueki, T. Horii, C. Yokoyama and T. Miyakawa, *Bull. Amer. Meteor. Soc.*, 89, 1889 doi:10.1175/2008BAMS2519.1. (2008).
[32] H. Miura, M. Satoh, T. Nasuno, A.T. Noda and K. Oouchi, *Science*, 318, 1763 (2007).
[33] T. Nasuno, H. Miura, M. Satoh, A. T. Noda and K. Oouchi, *J. Meteor. Soc. Japan*, 87, doi:10.2151/jmsj.87.335 (2009).
[34] H. Fudeyasu, Y. Wang, M. Satoh, T. Nasuno, H. Miura and W. Yanase, *Geophys. Research Lett.* 35, L22808 doi:10.1029/2008GL0360033 (2008).
[35] K. Oouchi, A. T. Noda, M. Satoh, H. Miura, H. Tomita, T. Nasuno and S. Iga, *Sci. Online Lett. Atmos.*, 5, doi:10.2151/sola.2009- 017 (2009a).
[36] H. Taniguchi, W. Yanase and M. Satoh, *J. Meteor. Soc. Japan*, 88, 571 (2010).
[37] W. Yanase, H. Taniguchi and M. Satoh, *J. Meteor. Soc. Japan*, 88, 497 (2010).
[38] Q. Moteki, R. Shirooka, H. Kubota, T. Ushiyama, K. K. Reddy, K. Yoneyama, M. Katsumata, N. Sato, K. Yasunaga, H. Yamada, B. Geng, M. Fujita, M. Yoshizaki, H. Uyeda and T. Chuda, *J. Geophy. Res.*, 113, D14126, doi:10.1029/2008JD009793 (2008).
[39] H. Tomita, *J. Meteor. Soc. Japan*, 86A, 107 (2008a).
[40] K. Takata, S. Emori and T. Watanabe, *Global and Planetary Change*, 38 (1-2), 209 (2003).
[41] H. Tomita, *J. Meteor. Soc. Japan*, 86A, 121 (2008b).
[42] P. Liu, M. Satoh, B. Wang, H. Fudeyasu, T. Nasuno, T. Li, H. Miura, H. Taniguchi, H. Masunaga, X. Fu and H. Annamalai, *Mon. Wea. Rev.*,137, doi: 10.1175/2009MWR2965.1.

(2009).
[43] H. Yamada, Q. Moteki and M. Yoshizaki, *J. Meteor. Soc. Japan*, 88, 437 (2010).
[44] M. C. Wheeler and H. H. Hendon, *Mon. Wea. Rev.*, 132, 1917 (2004).
[45] B. Fu, T. Li, M. Peng, F. Weng, *Weather Forecasting*, 22, 76 (2007).
[46] J. Y. Gao, T. Li, *Mon.Wea.Rev.*, 139, 885 (2011).
[47] P. C. Hsu, P.C., T. Li, C.-H. Tsou, *J. Climate,* 24, 927 (2011).
[48] R. Krishnan, R., K. V. Ramesh, B. K. Samala, G. Meyers, J. M. Slingo and M. J. Fennessy, *Geophys.Res. Lett.*, 33, L08711, doi:10.1029/2006GL025811 (2006).
[49] X.Fu, B. Yang, Q. Bao and B. Wang, *Mon. Wea. Rev.*, 136, 577 (2008).
[50] J. Matthews, P. Singhruch and K. J. Heywood, *Clim. Dyn.* doi 10.1007/s00382-009-0631-7 (2009).
[51] M. Katsumata and K. Yoneyama, *J. Meteor. Soc. Japan*, 82, 1035 (2004).

In: Cyclones: Formation, Triggers and Control
Editors: K. Oouchi and H. Fudeyasu
ISBN: 978-1-61942-976-5
© 2012 Nova Science Publishers, Inc.

Chapter 12

PROJECTION OF FUTURE TROPICAL CYCLONE CHARACTERISTICS BASED ON STATISTICAL MODEL

Nobuhito Mori*
Kyoto University, Uji, Kyoto, Japan

Abstract

As a consequence of global warming, several climate models suggest that the frequency and intensity of tropical cyclones will change by the end of the 21st century. A brief review of future tropical cyclone characteristics and stochastic tropical cyclone models is summarized in the first half of this chapter. Future tropical cyclone projections are based on stochastic tropical cyclone models with global climate change effects incorporated, and the analysis of the averaged and extreme conditions of tropical cyclones is discussed in the second half of this chapter. The future projection of TC tracks and intensities were computed by stochastic tropical cyclone models as discussed in the second half of this chapter. Based on the analysis of general circulation models and stochastic tropical cyclone models, both a decrease in cyclogenesis frequency and changes in track have regional impacts to annual tropical cyclone frequency. Using mean statistical characteristics from a general circulation model to run a stochastic tropical cyclone model, it is possible to estimate the uncertainty associated with these future tropical cyclone changes.

1. Introduction

Research on global climate change due to enhanced greenhouse effects on the earth's environment is changing our understanding of these phenomena to allow for assessment, mitigation and adaptation strategies for the future development of human society. From 1870 to 2004 [1], the sea-level has risen at a rate of 1.7 mm $\pm$ 0.3 mm/yr, with a significant increase in the rate during the last ten years IPCC-AR4[2]. The major source of sea-level rise is thermal expansion of sea water due to an increase in temperature, thus sea-level rise

*E-mail address: mori@oceanwave.jp

is a static, or averaged, side issue of climate change influencing coastal and upper ocean regions. Sea-level rise greatly impacts human activity near the coastal zone [2] and simply exacerbates the vulnerability of coastal regions to other physical processes, such as storm surges and storm waves. Another direct consequence of sea-level rise is inundation of low-lying coastal areas, which is a long-term concern.

On the other hand, future changes in storm surge and ocean wave climates are a dynamic side issue of how climate change influences coastal regions [e.g. 3]. If extreme weather events become stronger in the future, it is necessary to seriously consider the effects of these dynamic phenomena to prevent and reduce the impact of coastal disasters. The long-term changes to storm surge and storm waves are important considerations for structures near the coastal zone. For example, a coastal breakwater is designed with a maximum storm surge level and pressure from a maximum wave condition, a so-called design wave, for a predetermined design lifetime. Generally, coastal areas that will be most affected by increases in storm surge are the heavily populated mega-delta regions of East Asia, South-East Asia, South Asia and the Gulf of Mexico. For example, the total land area of Japan is approximately 378,000 square kilometers; coastal areas with elevations 10 meters or less below sea-level account for about 19,000 square kilometers and are populated with more than 25 million people, or approximately 20% of the total population of Japan. The United States has similar areas, such as Louisiana and Texas. Tropical cyclone (TC) activity has major impacts to the delta regions located in the subtropical to middle latitudes. In these regions, a 1 hPa decrease in atmospheric pressure corresponds to a 1 cm change in the barotropic effect to storm surge, and wind-induced surge is proportional to the square of wind speed [e.g. 4]. Also, wind induced surge is related to the duration of a storm. Therefore, storm surge is not only sensitive to TC intensity, but also to the track of the cyclone. And, a certain number of TC tracks is necessary to make an assessment of storm surge risk at a particular location.

Future storm surge projections have been conducted for Europe. For example, Lowe and Gregory [5] and Woth [6] have analyzed future storm surges around the UK and in the North Sea regions, and they showed significant increases in storm surge elevations for the UK coast and the continental North Sea coast resulting from increases in the intensities of mid-latitude storms. For the subtropical to mid-latitude zones, however, future storm surge projections are limited due to the lack of knowledge of how TCs in these regions will change in the future. In these latitudes, the majority of coastal regions measuring a few hundred kilometers in area have limited historical records of TC landfall. Therefore, assessing the TC risk in a particular region is difficult even considering the present climate alone. One way to overcome the limited amount of actual events is to use statistical modeling of TC events. Statistical modeling of TCs from tropical cyclone development (denoted cyclogenesis hereafter) to tropical cyclone weakening (denoted cyclolysis hereafter) can enhance the number of events by one or two orders of magnitude from the original amount of data. Combining climate models with statistical TC models is a useful tool to assess the risks of future TCs.

The beginning of this chapter describes a brief review of future projections of TC characteristics due to global climate change. Then, a review of stochastic tropical cyclone models including our latest results is presented. Finally, how TC characteristics change in the future climate is discussed based on the stochastic tropical cyclone model results along with the analysis of general circulation models (GCMs).

2. Review of Future Tropical Cyclone Climatology

Beside the pioneer studies of understanding TC formation and structure [e.g. 7, 8], the study of TC climatology has progressed over the last decade. Many studies have been concerned with the global warming issue of future changes in TC characteristics such as frequency and intensity. In this section, a brief review of present and future changes in tropical cyclone characteristics will focus on the application of a stochastic tropical cyclone model. Therefore, location changes of future tropical cyclones will be summarized, along with a summary of the dynamics of future tropical cyclone changes as in Kunutson et al. [9].

The basic conceptual understanding of TCs suggests there could be relationship between TC activity and sea surface temperature (SST) (e.g. Webster et al. [10] for historical change). It is well known that SST $>$ 26°C is a requirement for TC formation, and there is a theoretical relationship between SST and maximum potential TC intensity [11]. Emanuel [12] discussed how global warming affects the TC intensity of historical record. Rising SST and a possible increase in TC activity in the Atlantic basin since 1950 have been influenced by global warming [e.g. 13]. On the other hand, Webster et al.[10] summarized that no global trends in the number of tropical cyclones have yet emerged against a background of increasing SST, although there was an increase in category 4-5 TCs in the past record. The interannual variability associated with the El Ninõ-Southern Oscillation (ENSO) and the North Atlantic Oscillation (NAO) could complicate the difficulty of the analysis.

Generally, assessing changes in TC climatology requires an understanding of tropical cyclone physics and model resolutions. While previous studies have conducted climate projections using several models, one issue that has yet to be clearly addressed is the quantitative detection of long-term trends of frequency and intensity of tropical cyclone under the warm-climate condition. The model resolution gives major impact to the climatic features of projected TCs. This lack of information reflects the fact that high-resolution global model projections remain extremely expensive; consequently, most climate runs are performed at a relatively coarse resolution (e.g. 100-200 km mesh). For example, Yokoi et al.[14] evaluated TC climatology using eight different World Climate Research Programme's Couples Model Intercomparison Project Phase 3 (CMIP3) models by analyzing the daily-mean atmospheric output for the Western North Pacific (WNP); these models have a horizontal resolution of T63 or T106. Five of the eight models reproduced realistic horizontal distributions of cyclogenesis with a large proportion occurring over the 10N-20N and 120E-150E areas. Based on 18 GCMs of CMIP3 and results of regional downscaling, Bender et al.[13] summarized a significant anthropogenic increase in the frequency of intense TCs, and a decrease in the total number of TCs in the Atlantic basin may emerge from background climate variability by the end of 21st century. The recent important review of this issue based on state-of-the-art models is summarized by Kunutson et al.[9]. The future projections based on theory and high-resolution dynamical models indicate that the global warming will cause the globally averaged intensity of tropical cyclones and it will be increased about 2-11% by 2100 [9]. They concluded that global-warming-induced increases of TC intensity are larger in higher resolution models than in coarse-grid models, it is plausible that existing models may systematically underestimate future intensity trends.

There is a clear tendency that a higher resolution model (60 km mesh or less) projects an increase in the frequency of the stronger tropical cyclones [e.g. 9]. Bengtsson et al.[15] investigated the differences in outputs between T63 (180 km mesh) and T213 (60 km mesh) resolutions, reporting that while the two different resolutions are similar in terms of future changes in the spatial patterns of tropical cyclone genesis frequency (TGF), they are critically different in terms of future changes in the frequency of intense TCs. As reported by Oouchi et al.[16], advances in computational resources have enabled climate projections using high-resolution atmospheric general circulation models (AGCMs) such as TL959, a triangular truncation 959 model with a linear Gaussian grid, which is equivalent to a 20 km mesh horizontally. Oouchi et al.[16] summarized possible changes in future TC climates using TL959 and warm-climate conditions; the tropical cyclone frequency was globally reduced by about 30 % compared to present-day-climate conditions. They also found that the warm-climate conditions increased the number of intense tropical cyclones on a global scale. Murakami and Sugi [17] also discussed the effects of model resolution on projected TC climatology in the WNP; the basin-scale annual mean cyclogenesis number and the spatial distributions of TGF in the WNP are not critically dependent on resolution when the domain resolution is finer than TL319 (60 km mesh) for intense TCs and TL159 (110 km mesh) for TGF. Regarding understanding TC related environmental factors, Murakami and Wang [18] analyzed future TC tracks in the North Atlantic Ocean (NAT) and noted a signal shift in TC locations is well predicted by how the projected genesis potential index (GPI; [19]) changes. They summarized that the main environmental factors that influence TC changes are captured by GPI, and that the changes in maximum potential intensity and vertical motion increases cyclogenesis in the eastern NAT. These factors are related to the enhanced convective activity in the eastern Atlantic intertropical convergence zone. For the northwestern region of the NAT, the factors that increase cyclogenesis include maximum potential intensity and reduced vertical shear. Using the higher resolution models, quantitative projection of future TCs and related analysis have been improved.

Regarding the influence of future TC change on a regional scale, Wang and Chan [20] suggested that the number of TCs that approach the Japanese archipelago may increase if the mean location of cyclogenesis shifts eastward by 10 degrees. On the other hand, compared to the studies regarding TC numbers, the studies on TC intensity change and other TC climatic aspects are limited. Yokoi and Takayabu [21] discussed the future change in the number of TCs due to warmer SST in the WNP based on an analysis using five CMIP3 models. Their results noted a slight shift eastward of the area in the WNP where cyclogenesis frequently occurs. Using high resolution GCM data, Murakami and Sugi [17] reported the cyclogenesis number in each ocean basin using the same TC detection method. According to their results, there are significant reductions of cyclogenesis number in the global (GL) domain, Northern Hemisphere (NH), Southern Hemisphere (SH), Eastern North Pacific (ENP), and South Indian Ocean (SIO) areas, and a general increase in the North Atlantic Ocean (NAT). Although the models with relatively fine resolution show similar future changes on hemispheric scales (e.g., GL, NH, and SH), they are inconsistent in several subbasins in terms of both the sign and magnitude of change.

Numerous studies have addressed the issue of changes in frequency and intensity of TCs on the global or basin-wide scale. Our interests are projection of future TC characteristics on the regional or country-wide scales. However, assessing TC risk on a regional scale is difficult due to the limited resolution of GCMs and the limited number of TC events over the computational period. Generally, a return period of 50-100 years is necessary for minimum TC pressure and maximum wind speed to assess TCs on regional scale. Therefore, statistical modeling is required to assess the risk of future TCs.

3. Review of Stochastic Tropical Cyclone Models

Statistical projections of TCs have been examined in engineering and insurance industries. There are two different approaches to design structures and assess the risks of TCs. The primitive engineering approach estimates landfall rates along the coastline based on historical records. This is called a site specific approach or a local model that provides TC statistics for a particular location of interest [22]. The statistics or probability density functions (pdfs) of TC parameters include central pressure, radius of maximum winds, direction, moving speed, and distance of closest approach; these are analyzed in the site specific approach. Applying a site specific approach to climate change is difficult because all TC parameters at all locations need to consider the influences of climate change.

The second approach to statistical projection of TCs involves a stochastic tropical cyclone model/storm track method [23] which is a type of Monte Carlo simulation. The stochastic tropical cyclone model (denoted STM hereafter) simulates a TC track by random draw from a temporal-spatial pdf of cyclogenesis locations based on a historical data set of TCs; it is possible to expand this approach to a basin-wide scale. The STM simulates the full track of a tropical cyclone, from cyclogenesis to cyclolysis, where cyclolysis includes the transition from a TC to an extra-tropical system The cyclogenesis and cyclolysis indicate generation and weakening of low pressure circulation physically. However, these are simply used for generation and dissipation of TC in the study of statistical projection of TC). Figure 12.1 shows one such historical TC data set, the International Best Track Archive for Climate Stewardship (IBTrACS), maintained by The National Oceanic and Atmospheric Administration (NOAA). The STM models TC translation as a function of TC location and moving speed at each time step. The changing ratios of TC direction, speed and central pressure are provided in pdf format based on the historical record of TCs as shown in Fig. 12.1. Using this approach, TC characteristics are updated at each time step as shown Fig. 12.2. Both the local model and the STM have been verified with the historical data set [24]. The STM has biases due to an inadequate or missing depiction of physical processes, but it shows good agreement with the local model if the target area has relatively few landfalls.

There are several approaches to simulate TC translation with STMs. One major approach is an autoregressive model and another approach is a non-parametric method as a time-series model. Vickery et al.[23] used an autoregressive model for increments in TC speed and direction with random disturbances. For example, changes in TC speed c and direction θ between time step i and $i+1$ are denoted by

$$\Delta \ln c = a_1 + a_2 x + a_3 y + a_4 \ln c_i + a_5 \theta_i + \varepsilon, \quad (1)$$

$$\Delta\theta = b_1 + b_2 x + b_3 y + b_4 c_i \ln c_i + a_5 \theta_i + b_6 \theta_{i-1} + \varepsilon, \quad (2)$$

where a_i is an empirical coefficient, x and y are the longitude and latitude, c_i is the TC moving speed, θ_i is the TC direction, and ε is a random number. The coefficient a_i is given by the basin-wide historical data set. This formulation is the standard description of an STM. James and Mason [25] used a first order autoregressive model (Markov process) for TCs near northeastern Australia, but modeled the latitudinal and longitudinal increments with spatially independent coefficients. The following model expresses the rates of change in TC location for one time step.

$$\Delta x_i = a_0 + a_1 \Delta x_{i-1} \tag{3}$$

$$\Delta y_i = b_0 + b_1 \Delta y_{i-1} \tag{4}$$

It is possible to assume a_0 as a local dependent variable to consider local TC characteristics. The tracks conform to the observed statistics of the historical dataset while maintaining their previous motions, but x and y are assumed to be independent of each other.

Emanuel et al.[26] also used the Markov model by sampling a transition matrix populated from historical data. Alternatively, Hall and Jewson [27] developed a non-parametric STM using the means and variances of latitudinal and longitudinal displacements remaining from anomalies of track displacements. Their model determines the coefficients by averaging near-neighbor historical data with a weighted function. Therefore, the result depends on the spatial scale of averaging. The central pressures of TCs can be modeled independently of the tracks, although most studies use the same procedure to model TC tracks and central pressures. A higher order autoregressive process is required for central pressure due to a nonlinear tendency for a TC to decrease intensity rapidly at higher latitudes. Vickery et al.[23] used the Markov model for the relative difference of central pressure P:

$$\Delta \ln P_{i+1} = c_0 + c_1 P_i + c_2 P_{i-1} + c_2 P_{i-2} + \text{SST effects} \tag{5}$$

where c_i is an empirical constant that depends on latitude, intensity and direction. Since the central pressure is influenced by sea surface temperature (SST) along with other physical properties, this additional contribution to P has been included. James and Mason [25] used maximum potential intensity (MPI) for the deterministic part of the central pressure change:

$$\Delta P_i = c_0 + c_1 \Delta P_{i-1} + c_2 e^{-c_3 [P_i - MPI(y_i)]} \tag{6}$$

This formulation incorporates the latitudinal dependence of mean TC intensity as a function of MPI. The latitudinal dependence of the mean MPI allows for the TC to decrease intensity rapidly at higher latitudes. To account for central pressure change in the latitudinal direction, we developed a different approach, which will be described later in this chapter.

The autoregressive STM requires joint pdfs for each variable in the computational domain. For example, joint pdfs for direction, speed and pressure are required at each grid point in the autoregressive STM. This is difficult to calibrate and is also hard to handle considering global warming effects. Therefore, a non-parametric STM is useful to make maximal use of historical data to determine pdfs without over fitting the model and thereby becoming too complex [27]. We developed a non-parametric type of STM based on principal component analysis (PCA) [28]. The advantage of using PCA for the STM is that two or more variables are normalized in a unified way to simple relationships to minimize the restriction function and cost function without employing a preassigned distribution. The PCA is applied to obtain variation of a value x_1 considering the first and second principal components. The mathematical procedure to compute variation of a value can be summarized as follows. Target variables of PCA in the present study include a value x_1 to translate the TC into an area (cell) and its variation x_2 to translate out. Principal component axes Z_1 and Z_2, which cross average the value of variables, (x_1, x_2), can be expressed as

$$\begin{aligned} Z_1 &= a_{11}(x_1 - \overline{x_1}) + a_{12}(x_2 - \overline{x_2}) \\ Z_2 &= a_{21}(x_1 - \overline{x_1}) + a_{22}(x_2 - \overline{x_2}) \end{aligned} \tag{7}$$

where Z_1 and Z_2 are the first and second principal components and a_{ij} is a variance or covariance coefficient of x_i and x_j, and $\bar{x}_i$ denotes the mean of x, respectively. In principal component analysis, information loss may take place since several variables are summarized to a linear expression in Eq.(7). The values of a_{11} and a_{12} will be selected to maximize the variance of Z_1. For the Z_2 axis, a_{21} and a_{22} need to be chosen so that the variance of Z_1 is at a minimum while Z_1 and Z_2 are orthogonal. The variance and covariance of explanatory variable (x_1, x_2) can be expressed as (s_{11}, s_{22}) and (s_{12}) respectively. The variance of Z_1, or $Q(a_{11}, a_{12})$, for the new axis becomes as follows

$$Q(a_{11}, a_{12}) = s_{11}a_{11}^2 + 2s_{12}a_{11}a_{12} + s_{22}a_{12}^2. \tag{8}$$

The next step is an estimation of a_{ij} from given information. The cost function F minimizes the estimation error and is introduced using eigenvalues. Under the control condition of $a_{11}^2 + a_{12}^2 = 1$, the functions $F(a_{11}, a_{22})$ and $F(a_{21}, a_{22})$ can be written using eigenvalues of the variance-covariance matrix λ_1 and λ_2.

$$\begin{aligned} F(a_{11}, a_{12}) &= Q(a_{11}, a_{12}) - \lambda(a_{11}^2 + a_{12}^2 - 1) \\ F(a_{21}, a_{22}) &= Q(a_{21}, a_{22}) - \lambda(a_{21}^2 + a_{22}^2 - 1) \end{aligned} \tag{9}$$

The partial derivatives of $F(a_{11}, a_{22})$ and $F(a_{21}, a_{22})$ minimize information loss and give the following

$$\begin{aligned} \frac{\partial F(a_{11}, a_{12})}{\partial a_{11}} &= 2(s_{11}a_{11} + s_{12}a_{12} - \lambda_1 a_{11}) = 0, \\ \frac{\partial F(a_{11}, a_{12})}{\partial a_{12}} &= 2(s_{12}a_{11} + s_{12}a_{12} - \lambda_1 a_{12}) = 0, \\ \frac{\partial F(a_{21}, a_{22})}{\partial a_{21}} &= 2(s_{11}a_{21} + s_{12}a_{22} - \lambda_2 a_{21}) = 0, \\ \frac{\partial F(a_{21}, a_{22})}{\partial a_{22}} &= 2(s_{12}a_{21} + s_{12}a_{22} - \lambda_2 a_{22}) = 0. \end{aligned} \tag{10}$$

The eigenvalues, λ_i, can be estimated by the variance and covariance value, s_{ij}. The coefficient a_{ij} and the principal component Z_i can be estimated using the above procedure. Finally, the normal distributions of new information for two axes Z_i, along with the joint pdfs can be obtained by Eq.(7). Using the joint probability distribution, the input information of the approaching TC at a particular cell provides the variation of the TC parameter for the Monte Carlo simulation method. The non-parametric STM based on the PCA algorithm only requires an input variable and its variation, although the autoregressive STM requires joint pdfs for all related variables. Therefore, the joint pdf can be estimated by a simple algorithm. Specially the non-parametric STM can be simplified mathematically, if the Gaussian function is assumed for the pdf. Once the relationship between the TC location (x_i, y_i) and the increments $(\Delta x_i, \Delta y_i)$ is known, the Euler explicit scheme is used for the time integration.

Simulated cyclogenesis in an STM is conducted in several different ways. Vickery et al.[23] sampled historical cyclogenesis locations directly. Emanuel et al.[26] sampled a smoothed temporal-spatial dependent pdf of the historical record. Rumpf et al.[29] used a near-neighbor approach to develop and sample a cyclogenesis kernel pdf (Gaussian or other function). Instead, for both cyclogenesis and cyclolysis, we used a smoothed pdf of the historical data set to enhance the accuracy when compared to the historical data set.

Figure 12.3 shows an example of simulated TC tracks by STM [30]. The pdfs for the STM simulation were given by IBTrACS as shown in Fig. 12.1 and the simulation was conducted for a 50 year interval. Qualitatively, the overview of the TC tracks in each basin is reasonably distributed. Figure 12.4 shows the grid-averaged annual mean frequency of TC translation (passed TC) for IBTrACS as simulated by STM for a 10,000 year interval. The simulated peak in the WNP shows good agreement with observed data, and simulated distributions in other basins also show reasonable reproducibility. The global grid averaged error against the historical data is 0.06 /yr and indicates reasonable accuracy. The direction, moving speed and intensity were verified against both spatial patterns and local characteristics, respectively [30].

Using an STM for future climate projections is one option to assess tropical cyclone risks. If the STM satisfies the TC characteristics in the present climate condition, then a sufficient number of TC events can be obtained at arbitrary locations. If appropriate future TC conditions are provided, the future projection of TCs using an STM provides a tool to quantitatively evaluate risk along a given coast line. Potential future statistical conditions of TCs will be discussed in the next section.

4. Projection of Future Tropical Cyclone Activity

4.1. Future Tropical Cyclone Statistics

To use an STM for future climate projections, certain statistical characteristics such as the shift in cyclogenesis locations and other climate change parameters are required as given conditions. The future changes of cyclogenesis and cyclolysis locations and other TC statistics of each basin are discussed here. And, the ensemble model technique is applied to reduce statistical uncertainty.

To simulate the A1B Scenario from 2000 to 2100, eight different GCMs are used in the analysis as shown in Table 12.1. Two Meteorological Research Institute of Japan (MRI) models, MRI-AGCM-3.1S and MRI-AGCM-3.2S [31], were developed for CMIP5 and are the finest resolution models among those selected. The other AOGCMs are included in CMIP3. The MRI models are the atmospheric general circulation models (AGCM), and the remaining are atmospheric-ocean coupling general circulation models (AOGCM).

Table 12.1. List of GCMs for analysis

Model name	resolution	ω [s^{-1}] (vorticity threshold to extract)
MRI-AGCM-3.1S	TL959	3.13×10^{-4}
MRI-AGCM-3.2S	TL959	3.54×10^{-4}
CCCMA CGCM3	T63	8.84×10^{-5}
CSIRO 3.0	T63 L18	6.76×10^{-5}
CSIRO 3.5	T63 L18	7.66×10^{-5}
GFDL CM2.0	2.5 x 2.0 degrees	1.00×10^{-6}
MIROC3.2 (HIRES)	T106 L56	1.58×10^{-5}
MPI ECHAM5	T63 L31	5.71×10^{-5}

Table 12.2. Ensemble averaged future cyclogenesis centroid shift (unit:degree)

Method	Dir.	NIO	**WNP**	**ENP**	**NAT**	**SIO**	SPO	<WNP-SIO>
E. Mean	Lon.	-1.04	2.84	-4.36	2.73	-2.78	0.01	3.18
W.E. Mean	Lon.	-0.76	2.30	-4.25	3.82	-2.43	0.16	3.20
E. Mean	Lat.	0.54	0.45	-0.12	1.36	-0.65	0.42	0.65
W.E. Mean	Lat.	0.76	0.26	0.03	1.15	-0.58	0.37	0.50

The high-resolution AGCMs yield better TC characteristics and less uncertainty given the prescribed SST information. These models comprise the ensemble to analyze future change, i.e., at the end of 21st century, to TC characteristics.

The TC-detection method adopted in the present study is a modified version of that used by Oouchi et al.[16]. Oouchi et al.[16] employed six criteria to identify TCs and compare the simulated annual global-mean cyclogenesis number with observed data (see [17]). The resolution-dependent relative vorticity at 850 hPa ($\zeta 850$) ensures that the annual observed global-mean TC number (e.g., 84.8/yr for the period 1979-2003) is attained for the present-day simulation for each given model resolution. The cyclogenesis locations were counted for each 5 degree $\times$ 5 degree grid area over the global domain. This detection method identifies observed TCs reasonably well [17]. The TC positions were counted over 2.0 degree $\times$ 2.0 degree grid points every 6 hours. The total count is defined as the frequency of TC occurrence, which indicates the probability of a TC occurring in a designated basin. When the count is applied to the genesis location only (e.g., when initially identified according to the detection method), the total number is defined as the frequency of cyclogenesis.

The statistical variables used in this study are the following: $E(x)$ denotes the average of variable x; $E(x_{ij})$ and $\sigma(x_{ij})$ denote the grid averaged and standard deviation of variable x_{ij}, respectively; $\langle x \rangle$ denotes the longitudinal average of x; $\sigma_{lat}(x)$ denotes the longitudinal standard deviation; $|x|^m$ and $\sigma_m(x)$ denote the model ensemble mean and standard deviation, respectively. In the following section, the ensemble of x uses three different definitions: $E(x_{ij})$, $\langle x \rangle$ or $|x|^m$. The IBTrACS dataset was used for validation and weighted ensemble averaging.

Table 12.3. Ensemble averaged future cyclolysis centroid changes

Method	Dir.	NIO	**WNP**	**ENP**	**NAT**	**SIO**	SPO	<WNP-SIO>
E. Mean	Lon.	-1.36	2.86	-5.14	3.58	-1.98	0.07	3.39
W.E. Mean	Lon.	-1.40	2.50	-4.41	3.17	-1.99	-0.60	3.02
E. Mean	Lat.	0.90	0.74	0.41	1.63	-1.04	0.67	0.96
W.E. Mean	Lat.	1.08	1.06	1.33	1.52	-1.97	1.62	1.47

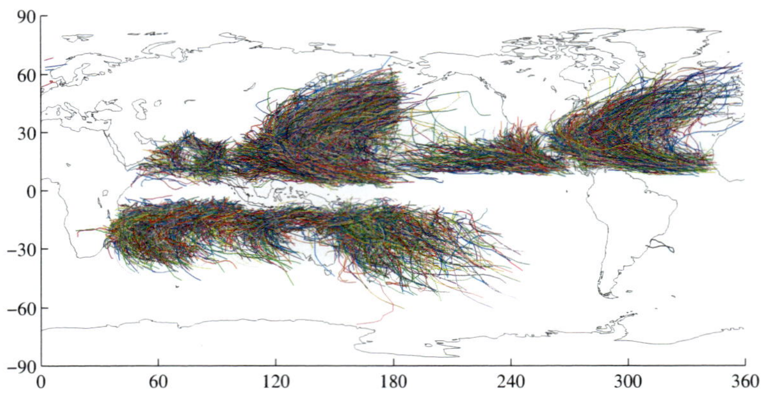

Figure 12.1. Historical TC tracks from 1950 to 2009 by IBTrACS (different color indicates different track)

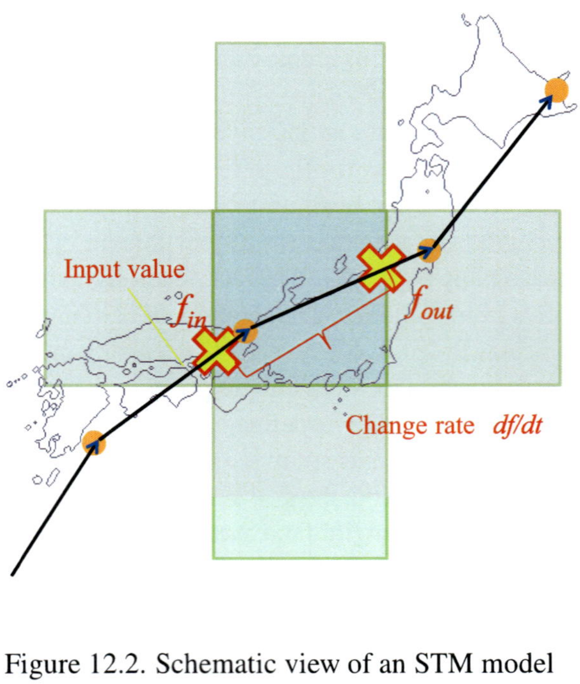

Figure 12.2. Schematic view of an STM model

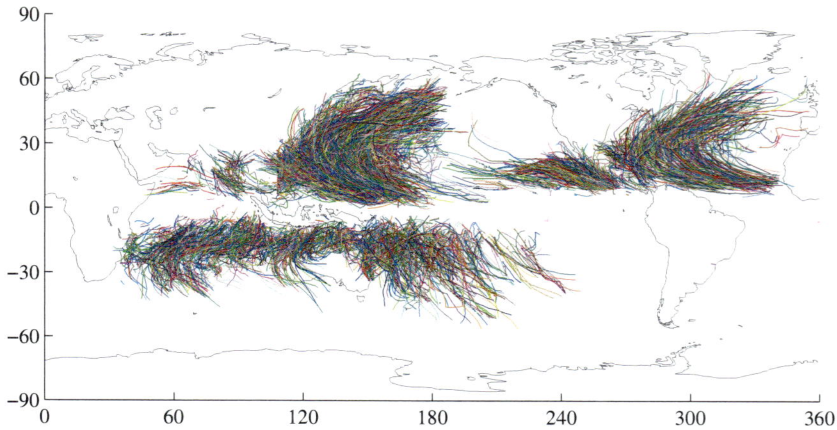

Figure 12.3. Simulated TC tracks by Global STM (different color indicates a different track)

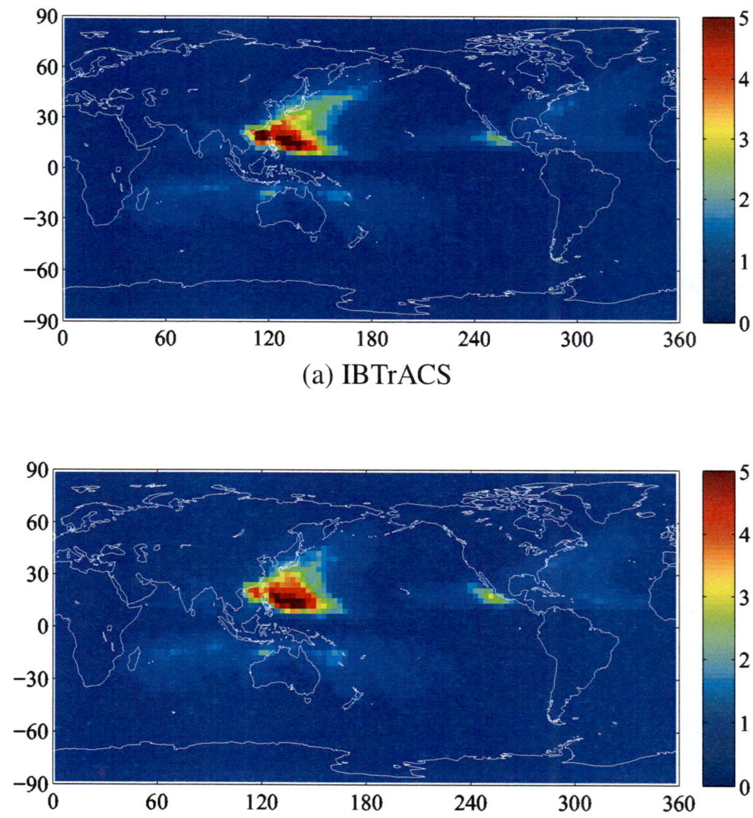

(a) IBTrACS

(b) STM

Figure 12.4. Grid averaged annual TC translation number (/yr)

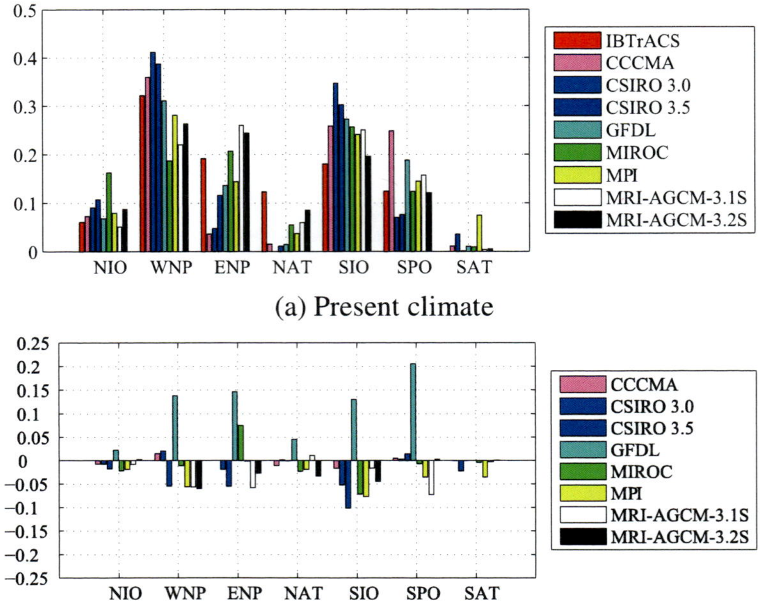

(a) Present climate

(b) Future change from present climate condition

Figure 12.5. Normalized cyclogenesis frequency and its future change for each basin

The TC processes for the STM include cyclogenesis, cyclolysis, moving speeds, directions, and central pressure changes. To identify how these processes change from present-day climates to future projected climates, the mean and standard deviation were analyzed to identify signatures in TC changes. Significant changes were observed in the cyclogenesis number, cyclogenesis location, cyclolysis location and average central pressure. A summary of how these TC characteristics change is provided in the following section.

Figure 12.5 shows cyclogenesis numbers for each ocean basin. The upper panel indicates present climate cyclogenesis numbers and the lower panel indicates its future change from the present climate condition, respectively. The values are normalized total cyclogenesis numbers in the present climate. Therefore, the sum of the present climate values equals one for each model. Besides the GFDL model, most of the models indicate a negative change in cyclogenesis number in the future climate. There are significant differences between the models for each basin. The model differences are relatively large in the WNP, ENP and SIO. The 5-10% decrease in the number of TCs in the future climate is consistent with previous studies [e.g. 21].

Figure 12.6 illustrates the ensemble-averaged centroid shifts for cyclogenesis. The upper and lower panels indicate the shifts in the cyclogenesis centroids in the longitudinal and latitudinal directions, respectively. The left-side bars are the simple ensemble-averaged centroid shift estimated as the mean of the centroid of each model, and the right-side bars are the weighted ensemble-averaged centroid shift using the relative distance of the centroid between the model and the IBTrACS dataset, $|\vec{x}^{\text{present}}|^m - \vec{x}^{\text{IBTrACS}}$, using a Gaussian

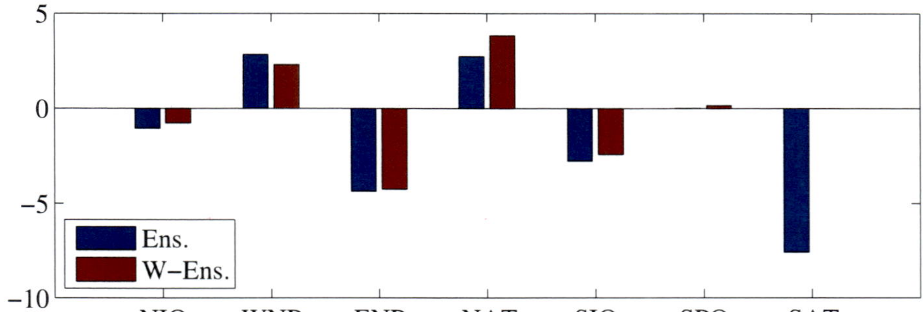

(a) Longitude direction (unit: degree)

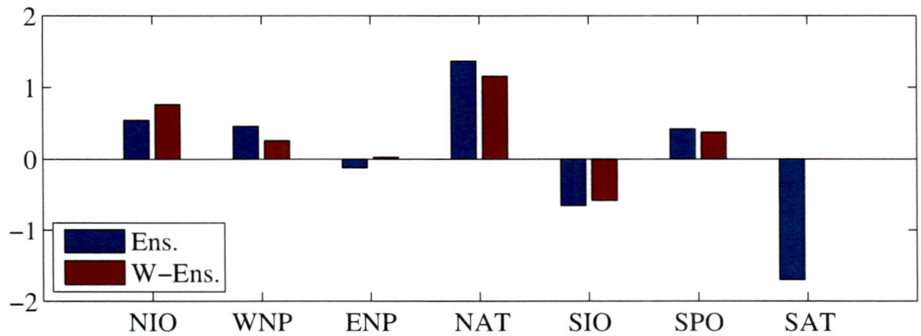

(b) Latitude direction (unit: degree)

Figure 12.6. Ensemble averaged future cyclogenesis centroid shifts (left-side bar: simple ensemble mean, right-side bar: weighted ensemble mean)

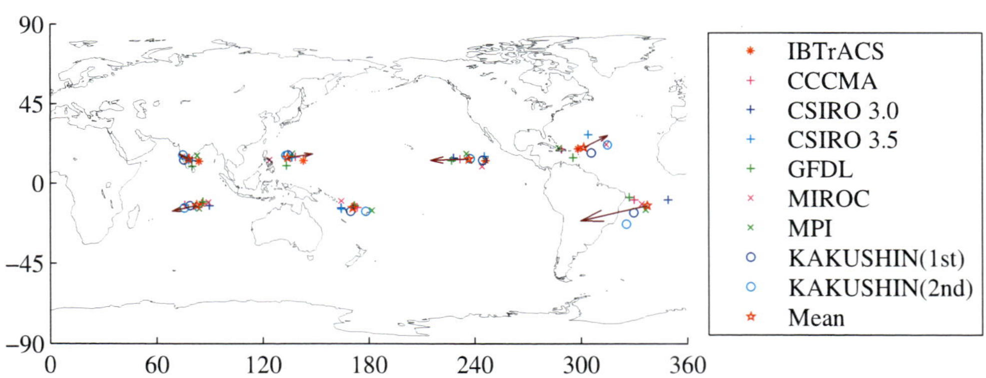

Figure 12.7. Future change of cyclogenesis centroid shift for each basin

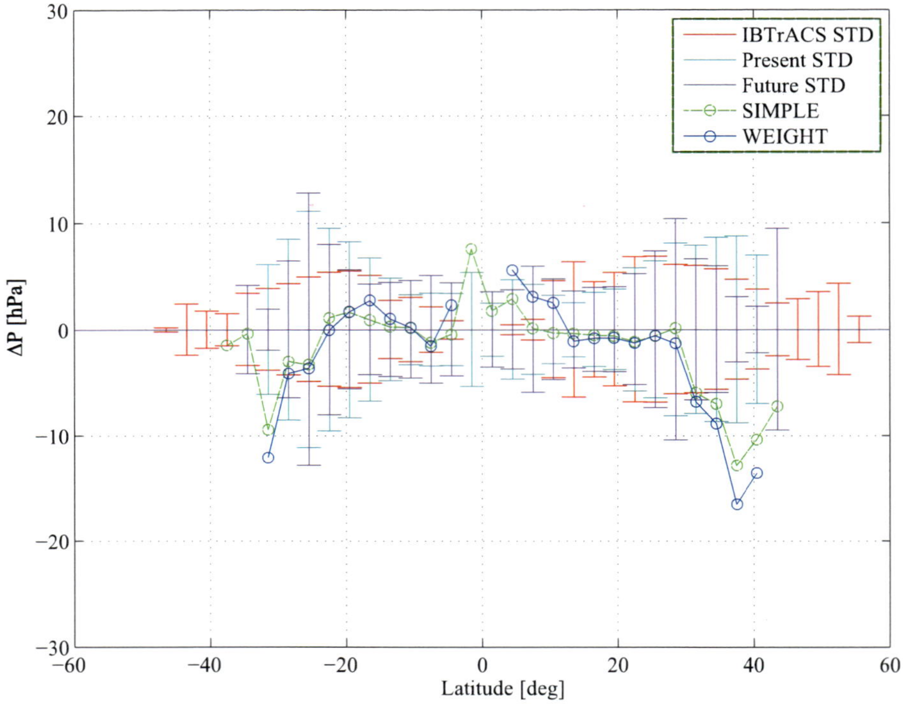

Figure 12.8. Future change of longitudinally averaged central pressure for TCs at various latitudes (dashed line: average, solid line: weighted ensemble average)

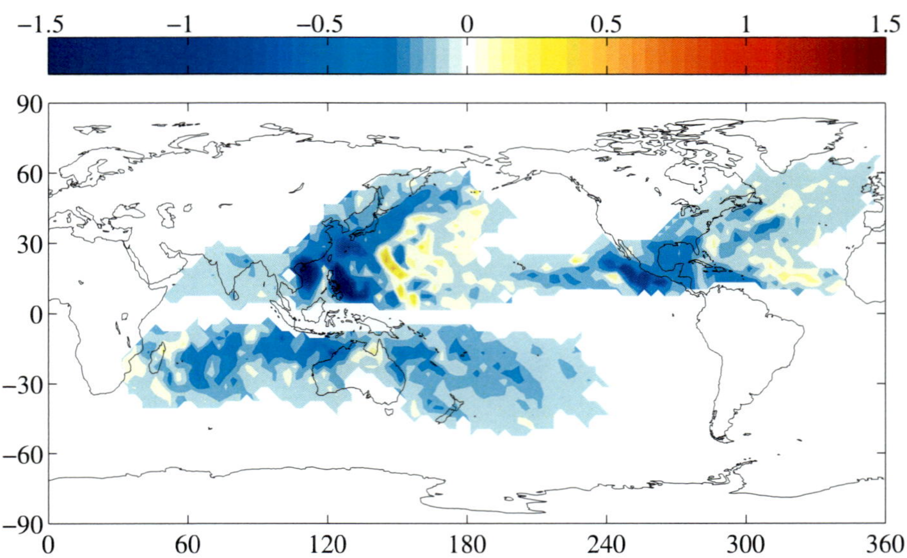

Figure 12.9. Future change of grid averaged annual TC frequency (/yr)

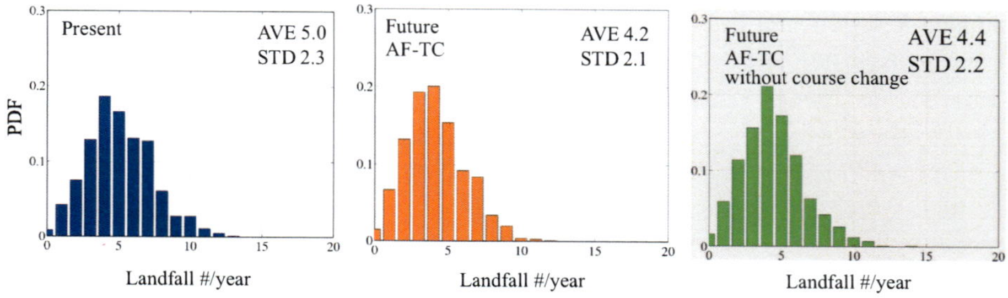

Figure 12.10. Projection of TC landfall number for Japan (left: present climate, center: future climate, right: future climate without TC course change)

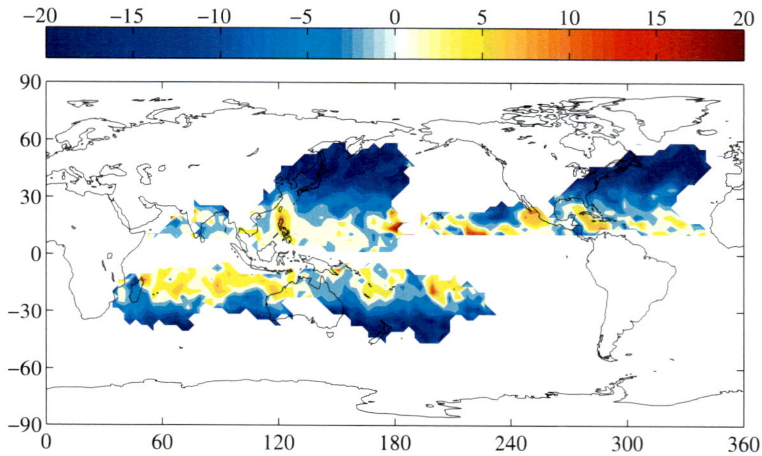

Figure 12.11. Future change of grid averaged central pressures (hPa)

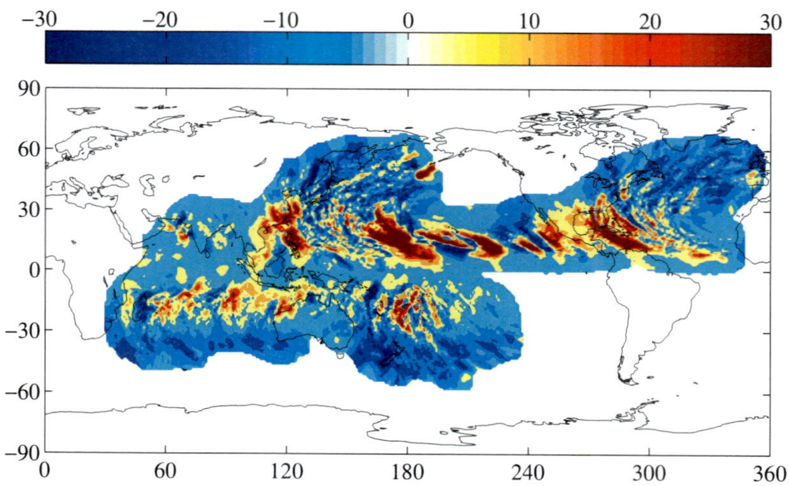

Figure 12.12. Future change of minimum central pressures for 100 year return period (hPa)

function as a kernel function. The weighted ensemble average of the future centroid shift, $|\Delta\vec{x}|^m$, is given by

$$|\Delta\vec{x}|^m = \int \frac{\Delta\vec{x}}{\sqrt{2\pi}\,\sigma_m(\vec{x}^{\text{present}})} \exp\left[-\frac{1}{2}\left(\frac{\vec{x}-\vec{x}^{\text{IBTrACS}}}{\sigma_m(\vec{x}^{\text{present}})}\right)^2\right] dx \tag{11}$$

where the standard deviation of a Gaussian function is given by the model standard deviation of the present climate, $\sigma_m(\vec{x}^{\text{present}})$, and $\Delta\vec{x}$ is the centroid difference between the present and future climates of each model. Therefore, the model with minimal reproducibility error in the present climate has maximal influence on the weighted ensemble averaged centroid shifts in the future climate.

A clear trend is observed in how the cyclogenesis centroid shifts in future projections. The future cyclogenesis centroids move toward the center of the ocean basin in the longitudinal direction and toward the poles in the latitudinal direction, with one exception being the SPO. These results are consistent with previous studies. The future centroid shift in the longitudinal direction is several times greater than the centroid shift in the latitudinal direction (see Fig. 12.7). For example, the cyclogenesis centroid in the WNP moves toward the east-northeast direction, and the centroid in the ENP moves toward the west-northwest direction. Second, the future cyclogenesis centroid shift has the same order of relative error which is defined as the difference between the observed (IBTrACS) centroid and the model ensemble averaged centroid for the present climate condition (not shown). The uncertainty associated with these cyclogenesis characteristics in these basins is significant and not trivial. Therefore, the changes in the expected future cyclogenesis locations has sufficient uncertainty to warrant careful projections of future tropical cyclones, and discussions should be conducted quantitatively basin by basin.

Figure 12.7 shows the ensemble averaged future cyclogenesis centroid shift for the present climate for each ocean basin. The star $\star$ indicates the weighted ensemble averaged cyclogenesis centroid, and the other symbols illustrate the cyclogenesis centroids, all for the present climate. The arrow indicates the ensemble averaged future cyclogenesis centroid shift from the present climate ($\star$), and the length of the arrow is five times the actual value. Most of the major ocean basins (WNP, ENP, NAT and SIO) indicate a move of the future cyclogenesis centroid toward the center of the ocean basin, although there are no significant shifts in the NIO and SPO basins. The SAT basin shows different characteristics but the number of tropical cyclones that occur in this area is minimal.

Table 12.2 and Table 12.3 provide a summary of the future cyclogenesis and cyclolysis centroid changes for the present and future climates. The notations of E. and W.E. in the tables indicate the simple ensemble averaged value and the weighted ensemble averaged value. The right-most column, $\langle \text{WNP} - \text{SIO} \rangle$, indicates the average value of the major basins WNP, ENP, NAT and SIO. There are few differences between the ensemble averaged and the weighted ensemble averaged values for these cases, but for cyclolysis the changes differ by 50 % in the latitudinal direction. This difference is a result of the coarse model resolution mentioned previously. For both cyclogenesis and cyclolysis, the centroid shifts in the longitudinal direction by approximately 3 degrees. However, in the latitudinal direction, the centroid changes vary between the cyclogenesis and cyclolysis scenarios. Future cyclolysis centroid shifts are greater than cyclogenesis shifts on average; cyclolysis centroid shifts are 48 % larger by the ensemble averaged method and 194 % larger by the weight ensemble averaged method. These statistics indicate that the tracks of future TCs will migrate toward the poles by about 0.5 - 1.5 degrees relative to those of the present climate, and will translate toward the center of the basin by about 3 degrees. Although we discussed only mean values, these values in future TC shifts are basically the same or smaller for models with resolutions less than the MRI models. Therefore, higher resolution models such as MRI models are necessary for detailed discussions, and they will be improved in CMIP5

Many studies have discussed how the intensity of TCs change. The central pressure, speed and direction of a cyclone primarily depend on movement in the latitudinal direction. Although global warming causes a decrease in the cyclogenesis number and a longitudinal shift of TCs, it is useful to examine how the basic statistical characteristics of TCs change with respect to latitudinal shifts. The statistical parameters for central pressure, moving speed and direction were analyzed by averaging across the longitudinal direction; both the mean and standard deviation of longitudinally averaged values are discussed here.

There were no significant differences between the present and future climates for the longitudinally averaged statistics except for central pressure. Figure 12.8 illustrates the future change of the model ensemble averaged values for longitudinally averaged central pressure as a function of latitude. The ensemble mean values in the figure indicate that the mean central pressure changes from the present to future climates: the solid line indicates the weighted ensemble average, and the dashed line indicates the simple ensemble average, $|\langle E(\Delta P_{ij})\rangle|^m$, respectively. The bars indicate the standard deviation of IBTrACS and the model standard deviations of the present climate and future climate, respectively. The ensemble averaged central pressure change for the future climate, $|\langle E(\Delta P_{ij})\rangle|^m$, is monotonically decreased in the polar direction. A change in future mean pressure can be observed at latitudes greater than 20-30 degrees, with a decrease of about 15 hPa at 40 degrees latitude. There is no difference between the weighted ensemble averaged and ensemble averaged results; therefore, the value of $|\langle E(\Delta P_{ij})\rangle|^m$ is neither model nor weighted function sensitive. The model standard deviation of central pressure, $\sigma_m(\langle E(\Delta P_{ij})\rangle)$, is smaller in the lower latitudes and increases toward the poles. The values of the model standard deviation of mean central pressure, $\sigma_m(\langle E(\Delta P_{ij})\rangle)$, in both the present and future climates are greater than the longitudinal standard deviation of IBTrACS, $\langle E(\Delta P_{ij}^{\mathrm{IBTrACS}})\rangle$. Therefore, the uncertainty associated with changes in central pressure is not negligible, and the estimated future pressure change, ΔP, of a tropical cyclone in the middle latitudes can include large bias. Although there is no physical interpretation of this yet, SST increases in the future climate may decrease TC intensity and increase the duration of TCs as cyclolysis locations shift toward the poles. Therefore, the pressure change, ΔP, and the increased shift in cyclolysis locations are intuitively predicted. This discussion requires further analysis combining how the TC characteristics change with the dynamic change in environmental conditions and remains a topic of future study.

Future changes in both the dynamic and thermodynamic conditions in tropical oceans influence cyclogenesis frequency and the locational shift toward the center of the basin. In general, the SST in the center of a tropical sea is warmer than in the outer regions [32], and a TC developing in the center of a warm tropical sea would endure longer [21, 33]. Therefore, changes in the intensity of future TCs are related to the centroid shifting toward the center of a basin. The dynamic relationship of the TC centroid shift is not discussed here; that issue will be explored in future studies.

Global warming impacts on TC characteristics have been analyzed using the MRI-AGCM-3.XS and CMIP3 models. Future changes in cyclogenesis and cyclolysis in each ocean basin are discussed along with other TC climatic changes. Statistically, the most salient change in TC climates include shifts of the cyclogenesis and cyclolysis centroids toward the center of the ocean basin. Using the statistical signatures of future climate change analyzed here, the future projection of TCs will be applied to the STM in the next section.

4.2. Future Tropical Cyclone Projection by Stochastic Tropical Cyclone Model

The future projection of TC tracks and intensity were computed by the STM. The statistical properties of TCs in the future climate are summarized as follows.

1. Cyclogenesis frequency decreases (Figure 4).
2. Cyclogenesis centroids shift toward the center of a basin (Figure 5).
3. Cyclolysis centroids shift toward the center of a basin (not shown).
4. Central pressures of TCs decrease quadratically in the polar direction (Figure 7).
5. Remaining TC characteristics are unchanged in the future climate.

All of these statistical changes are assumed to be the results of future TCs based on the model ensemble averaged values analyzed in Section 4.1; this itemized list of the future TC condition is referred to as the assumed future TC condition (AF-TC) hereafter.

The changes listed in the AF-TC were computed by the STM over a time-frame of 10,000 years on a global scale. The difference between the future condition and the present condition amounts to significant changes in both TC frequency and intensity. For example, the future change of annual averaged TC frequency at each grid is shown in Fig. 12.9 (the present climate results are shown in Fig. 12.4). As a result of the given condition of Item 1 in the AF-TC, the total number of TCs decreases globally in the future climate. Additionally, the TC frequency in Fig. 12.9 decreases in most regions except for some areas in the WNP and NAT. The averaged annual TC frequency change in the future climate is ± 0.2/yr which corresponds to 5% -10% of the annual TC frequency depending on the region. These areas of cyclogenesis frequency increase are caused by the TC track shifting toward the center of the ocean basin. The WNP and NAT have peaks in each basin as shown in Fig. 12.4, and the shifts of those peaks produce increases in TC frequency in front of the distribution in the WNP and NAT. Therefore, both cyclogenesis frequency decrease and track change are important conditions for TC risk assessment locally. For example, Fig.12.10 shows the sensitivity of the TC landfall number for Japan. The STM simulation was conducted over 10,000 years and more than 50,000 TCs traversed Japan. The left and center panels illustrate the STM results in the present climate and future climate (AF-TC) conditions, respectively. The right panel illustrates the STM results in the future climate without TC course change, which means only the cyclogenesis number change is considered. The averaged and standard deviations of landfall for Japan in the present climate are 5.0 and 2.3, respectively, and these values agree well with the observed numbers in our definition of landfall. On the other hand, the TC landfall number decreases by 16% in the STM projection of the future climate. However, the TC landfall number would be decreased by only 12% if the TC course change is neglected in the future climate. Although cyclogenesis change is a major factor behind the change in future TC numbers, the TC course change produces a significant impact, 25% in this case, to TC landfall numbers depending on the location.

Figure 12.11 illustrates averaged central pressure of TC change from the present to future climate. According to the AF-TC Item 4, the mean central pressure is assumed to decrease in the polar direction. The averaged central pressure directly reflects this assumption. There are minor effects due to central pressure change in the longitudinal direction. However, the extreme change in central pressure in the future climate shows different characteristics. The extreme minimum TC central pressure with a 100 year return period is calculated by extreme value analysis. The wind field is estimated by the super gradient wind scheme [34]. The future change of the minimum central pressure with a 100 year return value shows both positive and negative changes in the future climate condition. The spatial distribution of extreme minimum central pressure is different from the averaged central pressure as shown in Fig. 12.11. Overall, the extreme minimum central pressure will be smaller (i.e., stronger TCs) at latitudes above 30N and 30S. However, the spatial variation is significant for the extreme value and has a striped pattern in the longitudinal direction. The extreme value of minimum central pressure shows positive change from subtropical to mid-latitudes around 30 degrees, and it shows negative change in the polar direction. These complex changes in the extreme values result from a combination of AF-TC Items 1-4. Although the cyclogenesis and cyclolysis shifts and the mean central pressure changes in the latitudinal direction reduce the central pressure of TCs in the middle latitudes, the cyclogenesis and cyclolysis shifts in the longitudinal direction either enhance or reduce the extreme central pressure change of TCs in the future climate.

The statistical projection presented here is limited to the numerical values of the mean and extreme central pressures of the TC. The wind speed was given by the gradient of the pressure distribution and is slightly different from the central pressure, and is therefore sensitive to synthesis typhoon models. Additionally, only Item 1 of the AF-TC was considered here, but the other itemized conditions are also possible and produce uncertainty with the future TC projection. Reducing the uncertainty of future TC projections is necessary and needs to be discussed.

Summary

The first half of this chapter focused on the future change of tropical cyclone (TC) climatology and stochastic tropical cyclone models (STMs). With GCMs, large degrees of uncertainty exist in basin-scale TC climatology in the future climate, such as with the cyclogenesis, cyclolysis, and central pressure changes for each basin. On the other hand, the STM based on the present climate is a reasonable tool to assess risk associated with TCs, such as landfall number on a regional scale.

Validation of the STM on a basin-wide scale is still under development; this is useful to assess TC risks that result from climate change and to incorporate future climate effects in the STM. On the other hand, global warming impacts on TC characteristics were discussed and analyzed using several CMIP3 and CMIP5 models. Our interests include changes to TC frequency, intensity, and other TC climatology changes. The major statistical changes to TC climatology are shifts in the cyclogenesis and cyclolysis centroids toward the center of the ocean basin.

As concerns the regional impact of future TC change (e.g. for Japan), statistical downscaling is necessary to reduce the uncertainty of the TC tracks. The second half of this chapter discussed the future projection of TC tracks and intensity, as computed by the STM. Both the decrease in cyclogenesis frequency and the change in TC tracks yield regional impacts to the annual TC frequency. The extreme TC intensities and wind speeds illustrate complex spatial patterns, and they become significantly more pronounced above 30 degrees latitude. Although the future projection with the assumed ensemble averaged future TC condition was discussed here, the direct use of a single GCM in lieu of the STM is an alternative choice to assess the risk of of future TCs [35, 36]. GCM results are sensitive and show some extreme behavior of future TC characteristics as opposed to the results of this study. To reduce the uncertainty associated with a future projection, both ensemble averaged future TC conditions and direct use of GCM results are important. A quantitative comparison of several methods will be discussed in the near future.

Finally, in the future climate, changes in the dynamic and thermodynamic conditions in tropical oceans have influences on cyclogenesis frequency and locational shifts toward the center of the ocean basin. The dynamic relationship behind TC centroid shift is not discussed here. However, Chan [37] and others have addressed that tropical cyclones form more eastward during El Ninõ in the Western North Pacific, and these shifts are associated with the SST based El Ninõ index [20]. Therefore, the projection of ENSO in the future climate has an indirect impact on the projection of TC activity. These dynamic issues are intrinsic to the discussion of the risk assessment of TCs and remain a topic of future study.

Acknowledgment
The author appreciates Drs. Hajime Mase, Tomohiro Yasuda, Sota Nakajo and Hiroyuki Murakami for providing suggestions and materials. The author acknowledges the developers of the CGMs used in this study and The World Climate Research Programme's CMIP3 project, and especially the Meteorological Research Institute, Japan.

References

[1] J. Church and N. White, *Geophy. Res. Lett.*, **33**, 1, L01602 (2006).

[2] IPCC, "Chapter 11, IPCC fourth assessment report (AR4)", http://www.ipcc.ch/ (2007).

[3] N. Mori, T. Yasuda, H. Mase, T. Tom, and Y. Oku, *Hydrological Research Letters*, **4**, 0, 15 (2010).

[4] P. Nielsen, *Coastal and Estuarine Processes*, Advanced Series on Ocean Engineering, World Scientific Publishing Company (2009).

[5] J. Lowe and J. Gregory, *Philosophical Transactions of the Royal Society A: Mathematical, Physical and Engineering Sciences*, **363**, 1831, 1313 (2005).

[6] K. Woth, *Geophy. Res. Lett.*, **32**, 22, L22708 (2005).

[7] S. Manabe, J. Holloway, and H. Stone, *J. Atmos. Sci*, **27**, 4, 580 (1970).

[8] W. Gray, *Meteorology over the tropical oceans*, 155–218 (1979).

[9] T. Knutson, J. McBride, J. Chan, K. Emanuel, G. Holland, C. Landsea, I. Held, J. Kossin, A. Srivastava, and M. Sugi, *Nature Geoscience*, **3**, 3, 157 (2010).

[10] P. Webster, G. Holland, J. Curry, and H. Chang, *Science*, **309**, 5742, 1844 (2005).

[11] K. Emanuel, *Nature*, **326**, 6112, 483 (1987).

[12] K. Emanuel, *Nature*, **436**, 7051, 686 (2005).

[13] M. Bender, T. Knutson, R. Tuleya, J. Sirutis, G. Vecchi, S. Garner, and I. Held, *Science*, **327**, 5964, 454 (2010).

[14] S. Yokoi, Y. Takayabu, and J. Chan, *Climate dynamics*, **33**, 5, 665 (2009).

[15] L. Bengtsson, K. Hodges, M. Esch, N. Keenlyside, L. Kornblueh, J. LUO, and T. Yamagata, *Tellus A*, **59**, 4, 539 (2007).

[16] K. Oouchi, J. Yoshimura, H. Yoshimura, R. Mizuta, S. Kusunoki, and A. Noda, *Journal of the Meteorological Society of Japan*, **84**, 2, 259 (2006).

[17] H. Murakami and M. Sugi, *SOLA*, **6**, 0, 73 (2010).

[18] H. Murakami and B. Wang, *Journal of Climate*, 2699–2721 (2010).

[19] K. Emanuel and D. Nolan, in "Proc. 26th Conf. on Hurricanes and Tropical Meteorology", 240–241 (2004).

[20] B. Wang and J. Chan, *Journal of Climate*, **15**, 13, 1643 (2002).

[21] S. Yokoi and Y. Takayabu, *Journal of the Meteorological Society of Japan*, **87**, 3, 525 (2009).

[22] L. Russell, *Journal of the Waterways, Harbors and Coastal Engineering Division*, **97**, 1, 139 (1971).

[23] P. Vickery, P. Skerlj, and L. Twisdale, *Journal of Structural Engineering*, **126**, 10, 1222 (2000).

[24] T. Hall and S. Jewson, *Journal of Applied Meteorology and Climatology*, **47**, 361 (2009).

[25] M. James and L. Mason, *Journal of Waterway, Port, Coastal, and Ocean Engineering*, **131**, 181 (2005).

[26] K. Emanuel, S. Ravela, E. Vivant, and C. Risi, *Bulletin of the American Meteorological Society*, **87**, 3, 299 (2006).

[27] T. Hall and S. Jewson, *Tellus A*, **59**, 4, 486 (2007).

[28] T. Yasuda, H. Mase, S. Kunitomi, N. Mori, and Y. Hayashi, in "Proceedings of the 32nd International Conference on Coastal Engineering, ASCE", (2010).

[29] J. Rumpf, H. Weindl, P. Höppe, E. Rauch, and V. Schmidt, *Mathematical Methods of Operations Research*, **66**, 3, 475 (2007).

[30] S. Nakajo, N. Mori, T. Yasuda, and H. Mase, *Journal of Japan Society of Civil Engineers, Ser. B2 (Coastal Engineering)*, **67**, in press (in Japanese) (2011).

[31] A. Kitoh, T. Ose, K. Kurihara, S. Kusunoki, and M. Sugi, *Hydrological Research Letters*, **3**, 49 (2009).

[32] M. Halpert and C. Ropelewski, *Atlas of tropical sea surface temperature and surface winds* (1989).

[33] J. Chan, *Tellus A*, **59**, 4, 455 (2007).

[34] T. Fujii and Y. Mitsuta, *Annuals, Disas. Prev. Res. Inst., Kyoto Univ*, 229–239 (in Japanese) (1986).

[35] T. Yasuda, H. Mase, and N. Mori, *Hydrological Research Letters*, **4**, 0, 65 (2010).

[36] K. Nishijima, T. Maruyama, and M. Graf, in "Proceedings of 11th International Conference on Applications of Statistics and Probability in Civil Engineering", (2011).

[37] J. Chan, *Monthly Weather Review*, **113**, 4, 599 (1985).

INDEX

#

A

B

C

D

E

F

G

H

I

J

K

L

M

N

O

P

Q

R

S

T

U

V

W

Y

Z